Terrestrial orchids have a wide appeal, but unfortunately they rank among the most vulnerable of all plant species, and little is known about how they reproduce in nature. This book contains a detailed survey of the biology of terrestrial orchids, from seed dispersal to establishment and life of the adult plant, based on comparisons of field and culture experiments. The unusual mode of obtaining energy by means of mycorrhiza is examined and evaluated in terms of plant structure and functions and the impact of this mycotrophic nutrition on orchid evolution. The book makes it clear that an understanding of germination, life histories and seasonal pheno-logy in natural habitats is essential for the success of culture methods, propagation and conservation. The final chapter is a systematic presentation of the life history, endophytes and propa-gation of 36 genera of terrestrial orchids.

Environmental plant physiologists will find this a stimulating book; for all those who are involved in orchid horticulture the book is indispensable.

Terrestrial orchids

from seed to mycotrophic plant

Terrestrial orchids
from seed to mycotrophic plant

Hanne N. Rasmussen

Senior Scientist, Forest & Landscape, Faculty of Life Sciences,
University of Copenhagen, Denmark

CAMBRIDGE UNIVERSITY PRESS
Cambridge, New York, Melbourne, Madrid, Cape Town, Singapore, São Paulo

Cambridge University Press
The Edinburgh Building, Cambridge CB2 8RU, UK

Published in the United States of America by Cambridge University Press, New York

www.cambridge.org
Information on this title: www.cambridge.org/9780521451659

First published 1995
This digitally printed version (with corrections) 2008

A catalogue record for this publication is available from the British Library

Library of Congress Cataloguing in Publication data

Rasmussen, Hanne N.
Terrestrial orchids from seed to mycotrophic plant / Hanne N. Rasmussen.
    p.   cm.
Includes bibliographical references (p.   ) and index.
ISBN 0 521 45165 5 (hardback)
1. Orchids. 2. Orchids – Propagation. 3. Orchids – Physiology. I. Title.
QK495.065R38 1995
584′15–dc20 94-35241 CIP

ISBN 978-0-521-45165-9 hardback
ISBN 978-0-521-04881-1 paperback

Contents

Acknowledgements

My interest in orchid mycorrhiza was first aroused when the techniques for symbiotic germination were introduced to me by Mark Clements, then at the Royal Botanic Gardens, Kew. I owe him many thanks for this demonstration and for further help in setting up an orchid propagation group at the University of Copenhagen. During the following years my students, Bo Johansen and Torben Andersen, and I gradually left the notion of symbiosis as a cure-all for propagation problems and began to focus on how environmental factors affect germination. For the progress made in those years and for a fine collaboration I am grateful to all the members of the informal 'orchid research group' at the Botanical Laboratory, including students and technicians as well as scientists, and to other colleagues who were helpful in many ways. My stay at the Smithsonian Environmental Research Center for 17 months enabled me to carry out field studies that would hardly have been possible elsewhere and the numerous results of which are only now being fully analysed. I am grateful to Dennis Whigham and the rest of the staff at SERC for this outstanding opportunity. Access to field localities on Öland, Sweden, was kindly granted to me by the Ecological Research Station of Uppsala University, made possible by Bertil Kullenberg.

During the work on this book I have enjoyed stimulating discussions on orchid matters with a number of colleagues and students of whom I can only mention a few, including Finn N. Rasmussen, Gunnar Seidenfaden, Bo Johansen, Torben Andersen, Bo Jørgensen, Dennis Whigham, John Freudenstein, Douglas Gill, Warren Stoutamire and Larry Zettler. Many of these have also supported the work in a number of practical ways. I also wish to thank librarians in three institutions, the Botanical Central Library in Copenhagen, The Smithsonian Institution Library and the library at Aarslev, who showed admirable zeal in finding the rare books I wanted.

Work with relevance to this book was mainly carried out at the Botanical Laboratory, University of Copenhagen, and at the Smithsonian Environmental Research Center, Maryland, USA; the text was finally prepared while I have been in my present position at the Danish Institute for Plant and Soil Science. I owe special thanks to my new colleagues for patience, encouragement and practical assistance during the last phase of the work and to Margaret Greenwood Petterson for her valuable counselling in matters of language and style. The experimental work was supported by funding from the University of Copenhagen, Smithsonian Institution, Danish Agricultural and Veterinary Research Council, the Smithsonian Environmental Sciences Program, Danish Fulbright Commission and Foundation of 26th of May.

Aarslev, Denmark
May 1994   H.N.R.

Introduction

Background

Surely no plant group surpasses the orchid family in its appeal to scientific curiosity and imagination. The untiring professional and public interest in the Orchidaceae is reflected in a considerable number of volumes published each year on orchids; in particular, the great structural variety in the orchid flower is well represented in the literature.

Compared with the floristic study of this large and taxonomically complex group, other types of investigation lag behind. Of the many unusual features that can be noted in the orchid family their mode of life in association with fungi has probably attracted the least attention, which is surprising since it is without doubt a fundamental aspect of orchid biology.

The orchid family is the only large group of higher plants that makes consistent use of an alternative nutritional system. Orchid mycorrhiza differs from other major types of mycorrhiza in that the fungus supplies the plant with energy. During some of its life stages the orchid can rely entirely on mycotrophy for nutrition, while during other stages the plant makes use of both mycotrophy and phototrophic nutrition, either alternately or the one supplementing the other.

The great diversity of orchid habitats, which include places where paucity of soil or light precludes most plant life, is best understood in the context of orchid mycotrophy. In fact, many of the outstanding features of the orchids, such as the complexity of floral structures, the diversity of plant–pollinator interactions and the unusual characteristics of the seeds, can all be viewed as functional adaptations in relation to or dependent on the mycotrophic lifeform (Chapter 12). This in itself justifies a treatise on orchids centred on their mycorrhiza.

If we are to understand the terrestrial, mainly temperate lifeforms it is of particular importance to consider their biology in a mycotrophic context. Terrestrial orchids are often represented as being more dependent on their endotrophic fungi than epiphytic species are, since terrestrial seedlings remain underground and are mycotrophic for months or even years, whereas epiphytic seedlings, which have immediate access to light, can begin to photosynthesize at an early stage.

Another reason for dealing exclusively with the extratropical orchids of the Northern Hemisphere is that these species occur in the densely populated industrialized regions of the world and for this reason are of special interest both to the general public and to conservationists. Considerable economic interests are associated with the commercial production of many of these species, but with current techniques propagation is difficult and often slow. Many temperate-zone species have decreased dramatically in numbers in their natural habitats, mainly as a result of changes in agricultural practice and urbanization, and their conservation is now focused on as a symbol of our efforts towards the preservation of biological diversity in general. A number of projects have been initiated in different parts of the world for preserving the local orchid flora by means of propagation and the formation of seed banks, and by research on the reproductive biology of orchids (Clements, 1982a; Fay & Muir, 1990; Dixon & Sivasithamparam, 1991; Johansen & Rasmussen, 1992, amongst others). Although the priorities of the conservationist are not identical with those of the propagator a knowledge of seed and seedling biology is essential to both groups and there is clearly a need for information to be compiled and extended. A rational approach to species conservation must be based on this kind of information.

The early life stages of terrestrial orchids can only be effectively studied if field observations are combined with laboratory studies, since the minute seeds and underground seedlings are extremely difficult to observe in nature. If interpreted with due caution, the information obtained from symbiotic and asymbiotic *in vitro* cultures can be generalized to apply to the processes that occur in nature. New insight can be reached when the available information on the structure and phenology of seeds, germination, seedlings and adult mycorrhizal plants as observed in the field is compared with physiological data gained chiefly in the course of laboratory propagation and cultivation.

The studies by Bernard (1909) and Burgeff (1909, 1936) greatly increased our knowledge of orchid seeds and mycorrhiza and included reviews of all data that were available at the time. These works remain basic to our studies today. More recently the physiology of orchid seeds has been summarized (Arditti, 1967, 1979), and in a handbook on the cultivation of orchids from seed many physiological aspects of orchid biology were discussed (Fast, 1980). In these publications the emphasis, however, has been on asymbiotic propagation techniques and on the epiphytic tropical species. A recent horticultural survey of the terrestrial extratropical species dealt only briefly with seeds and seedling stages (Cribb & Bailes, 1989).

Objectives

A survey of the existing literature in a field of research always helps in outlining the subjects that are in particular need of investigation. In my opinion considerable progress in the study of orchid biology can be achieved if field studies and culture experiments are carried out with the purpose of supplementing each other, and if statistically tested evidence is viewed in an ecophysiological and evolutionary context. Some key areas in which I would personally like to encourage more research are: (*a*) the biology of the orchid endophytes, (*b*) the polymorphism of germination requirements in orchid seeds and (*c*) the effect of environmental factors on organogenesis.

(*a*) Although living pure cultures of orchid endophytes are comparatively easy to obtain, and the synthesis of orchid mycorrhiza is possible by allowing seeds to germinate in co-culture with a mycelium, this convenient experimental system has not been used as much as it deserves. We need to know far more about the identity and specificity of the fungi, as well as their physiological role in relation to the plant and the environment (Hadley, 1982a). Moreover, we know very little about the competitive advantages of orchid mycorrhiza or about the significance of fungi in speciation and geographical distribution of orchids. As regards conservation, there is little point in preserving seed banks of terrestrial orchid species unless living cultures of compatible fungi are also preserved (Johansen & Rasmussen, 1992); the chance of establishing viable populations by re-introducing propagated plants depends on the ability of the introduced specimens to give rise to spontaneous seedlings, and this requires the presence of suitable fungi. Orchid endophytes are difficult to trace in the soil and difficult to identify as potential symbionts unless they are actually extracted from orchid tissue, and this may become increasingly difficult as orchid populations decline in size. Although most orchid fungi are probably capable of a life outside of the orchids, it would be tedious to search for them among the vast number of soil fungi. By presenting the available data on the location and extent of the infection as well as its seasonal variation in various orchid species I hope to facilitate the investigation of the endophytes.

(b) Seed and seedling populations of wild species are often polymorphic and do not respond to experimental manipulation with all-or-none reactions. Statistical analysis of the physiological responses has been surprisingly little used, although the minute orchid seeds are remarkably well suited for setting up large-scale experiments that yield sound statistical evidence. Terrestrial species have the reputation of being difficult to raise from seed. While the problems that some species present have been solved by either asymbiotic or symbiotic methods, it is becoming increasingly clear that terrestrial extratropical species do not form a homogeneous group with respect to germination and seedling requirements. An understanding of the variation in requirements of the seeds and the varying degrees of fungal compatibility and dependency is essential not only for improving the methods of propagation but also for tracing the evolutionary origin of these traits.

(c) In the holarctic zone the year is divided into seasons, which may be assumed to be of major importance to plant development. Nevertheless, few available studies deal with the responses of orchid seeds and seedlings to changes in temperature, humidity, day length and other climatic factors, as compared with the numerous studies in which the chemical factors of the culture substrate have been investigated. The subterranean microclimate and biosphere should also be taken into account in the search for environmental signals governing plant development. This would be a promising approach to some of the technical problems connected with seed dormancy, loss of viability and failing organogenesis *in vitro*.

Literature and taxa

In spite of many issues that still need more investigation, the literature on the terrestrial orchids, in particular as regards propagation, is already extensive. The first chapters of this book comprise a general treatment of the seed biology of holarctic orchids and development of mycotrophic plants based mainly on studies of these species; data from work on tropical epiphytes or southern extratropical taxa are included where relevant for the discussion but without attempting a comprehensive treatment.

I have generally avoided using secondary sources except for tracing primary sources. The references that I cite range from basic scientific studies and monographs to brief notes on casual observations on plants in culture or in the field. As can be expected, many studies, particularly older ones, do not meet the requirements for modern scientific investigation, and a large number of my references are from journals that do not submit their articles to peer review. I have tried to extract information impartially from each source while discussing the conflicting evidence. Apart from presenting a framework of well-corroborated information, an important purpose of surveying literature is to reveal the gaps and ambiguities in our current information, and point out areas that invite further scientific investigation.

I have placed special emphasis on including works that may be otherwise inaccessible to many readers. Much of the literature is not in English. A number of unpublished studies and doctoral theses have been made available to me through the Botanical Library at the University of Copenhagen; some recent ones have been sent to me by colleagues and are cited here with their permission. Also included are numerous unpublished data produced in laboratories where I have worked, first at the Botanical Laboratory at the University of Copenhagen, Denmark, and later at the Smithsonian Environmental Research Center in Maryland, USA.

Chapter 13 is subdivided into sections on individual genera which comprise specific data on life history, mycorrhizal status and techniques of propagation from seed. The taxa included here are restricted to the holarctic region occupying all extratropical parts of the Northern Hemisphere, as in Diels & Mattick (1958). There are no entries for genera for which I have found few or no data of the required type, even if this policy has meant excluding some otherwise well-known genera such as *Hexalectris* and *Isotria*. The notes on the geographical distribution and habitats of each genus serve only to provide a background for the phenological and ecophysiological data. Exhaustive floristic information is readily available in regional floras. The nomenclature has as far as possible been adapted to modern usage, but I have favoured the use of broad taxa, thus reducing the status of small or monotypic genera, simply as a practical arrangement to restrict the number of entries. This does not necessarily reflect any taxonomic standpoint on my part. Names used here and synonyms frequently encountered in the literature cited are listed in Appendix B.

1

Properties of 'dust' seeds

The germination biology of 'dust' seeds such as those of orchids is not very well known because the minute size makes it difficult to follow their movements from the time they are released from the fruit. Since the first stages of seedling development in terrestrial orchids are underground the young plants are usually not observed until some time after germination. This leaves a whole phase in the life cycle of orchids virtually uncharted. A quantitative approach to seed ecology has only recently become possible with the introduction of a technique for field sowing and the retrieval of seeds (Rasmussen & Whigham, 1993).

In spite of these difficulties it is possible to make some reasonable assumptions about seed dispersal and predation on the basis of the geographical distribution of plants and the physical properties of the seeds, and a great deal is known about the physiology of germination from observations *in vitro*, dealt with in greater detail in Chapter 4 and 6.

Orchid seeds are produced by the thousand in fruits that usually develop into dry capsules. The seeds of terrestrial orchids measure from 0.07 to 0.40 mm across and from 0.11 to 1.97 mm in length, including the testa (Harvais, 1974; Arditti *et al.*, 1979, 1980, Healey *et al.*, 1980) and those of epiphytic species tend to be even smaller (Fig. 1.1; Stoutamire, 1983). This places orchid seeds among the smallest known in the plant kingdom.

Seeds of tropical species often weigh less than 1 μg. Those of the holarctic species are in comparison somewhat heavier: seeds of

Length, mm

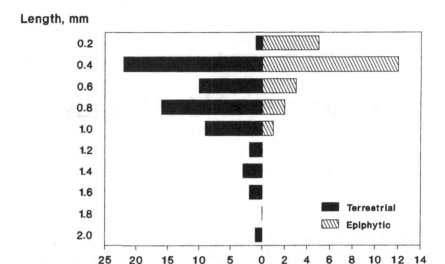

Fig. 1.1. Seed length of 67 terrestrial, primarily temperate North American taxa, compared with 23 epiphytic, mainly tropical taxa. Data from Stoutamire (1983).

Goodyera repens and *Cephalanthera damasonium* weigh about $2\,\mu g$, those of *Limodorum abortivum* $5.7\,\mu g$ and those of *Gymnadenia conopsea* $8\,\mu g$ (Neger, 1913; Burgeff, 1936; Koch & Schulz, 1975). Since about 526 seeds of *Cypripedium acaule* weigh 1 mg the mean weight is *c.* $1.9\,\mu g$ (Stoutamire, 1964). While the reported seed weights give us some idea of the order of magnitude, it is difficult to compare the individual values as some may be based on fresh weight immediately after harvest, others on weight after storage with varying degrees of dehydration.

There is a large volume to weight ratio in orchid seeds on account of the inflated air-filled testa that often has long tapering ends, i.e. the seeds are scobiform (Fig. 1.2; Hirt, 1906). The same seed physiognomy is found in several other plant families and is characteristic of wind dispersal, although the seeds are usually much heavier and larger than those of orchids.

A few orchids, among these *Galeola septentrionalis*, have fleshy fruits, that are dispersed by animals. The seed weight, $22\,\mu g$

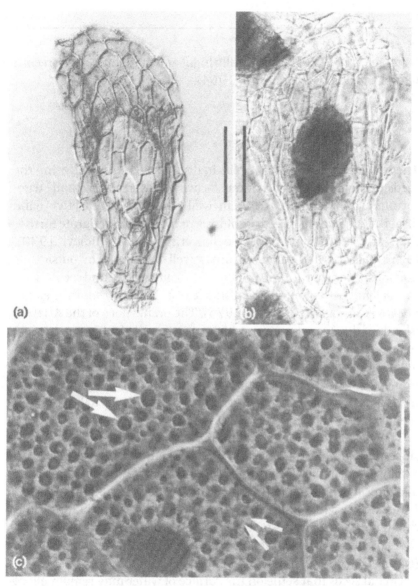

Fig. 1.2. Orchid seeds. (*a*) Fresh seed of *Liparis loeselii*, stained in Sudan IV. (*b*) Seed of *Liparis loeselii* stained in Sudan IV after the surface had been treated in calcium hypochlorite (Ca(OCl)$_2$) for 1 hour and seeds incubated on oat medium with symbiotic fungus D159–2 for several weeks. The lipids stored in the embryo now stain. (*c*) Nutrient storage in the embryo of *Dactylorhiza majalis* consisting of protein bodies (large arrows) and lipids (small arrows). Scale bars represent (*a*) and (*b*) 0.2 mm, (*c*) 20 µm (*c*) reproduced from Rasmussen (1990) with permission of the *New Phytologist*.

(Hamada, 1939), is also unusually high, and may constitute a record within the family (Nakamura, 1964).

1.1 Dispersal

The mature capsule usually splits in dry weather, thus releasing the seeds under optimum conditions for wind dispersal. Size and structure of orchid seeds make them well equipped for effective wind dispersal when dry. Indeed orchids seem to be able to migrate further than any other wind-dispersed flowering plants (Ridley, 1930). *Orchis militaris* has colonized areas well over 100 km outside its previously known range in England (Willems, 1982) and seedlings of *O. simia* have been found at a distance of *c.* 250 km from the nearest known seed source (Crackles, 1975). The orchid flora of the Atlantic islands of the Azores comprises three species, all of which are closely related to species occurring in continental Portugal about 1500 km away (Ridley, 1930). The long-distance dispersal that is inferred from this geographical distribution is so far the most extreme known in orchids. Dispersal over distances of 5–10 km seems to be common (Lehaie, 1910).

The minute size not only allows the orchid seed to travel far, but since they are also produced in such large numbers the seeds can locally achieve a fairly dense coverage around the mother plant, thus allowing a scanning of the surroundings for suitable micro-sites. Large seed production seems to be a common trait of plant groups that have a saprophytic–mycotrophic or parasitic seedling stage and therefore are very exacting in requirements for the germination site (Rauh *et al.*, 1975, and references therein).

Orchid seeds that settle on the surface of water may remain afloat for weeks (Ziegenspeck, 1936). The testa is covered with a water-repellent lipoid layer and buoyancy is further improved by the rough sculpturing of the testa which traps air bubbles, and by the air spaces between the capacious testa and the embryo (Ziegenspeck, 1935; Carlson, 1940; Rauh *et al.*, 1975). Imbibition takes place slowly in water, the rate increasing somewhat when detergent is added. Only when the lipoid surface is chemically attacked by rinsing in ethanol, ether or other lipophilic solvents do the seeds take up water fairly rapidly (Fig. 1.2a, b; Ziegenspeck, 1935).

This resistance to the uptake of water is more pronounced in terrestrial species than in epiphytes (Stoutamire, 1983) and is probably one of the greatest obstacles to success when germinating terrestrial species *in vitro*. Slow and differential imbibition could be instrumental in spreading out the time it takes for a population of seeds to germinate, thereby contributing to a polymorphic germination strategy (Ziegenspeck, 1935).

The floating capacity of orchid seeds may be important for dispersal in bog species. The South African species, *Disa uniflora*, and three related species are dispersed by water (Kurtzweil, 1994), but there are no reported observations of water dispersal in holarctic species, with the possible exception of *Cypripedium calceolus* whose fruits have been observed to close up when dry (Böckel, 1972). According to this report the valves of the capsule separate in rainy weather and release the seeds in raindrops.

The water-repellent testa will tend to keep seeds dry in the soil for considerable periods of time, perhaps enabling them to resist microbial attack better. When seeds have been treated with hypochlorite ($NaOCl$ or $Ca(OCl)_2$), which affects the lipoid surface and in part destroys the water-repelling properties, the testa decays completely in about 6 months in the soil whereas the testa of untreated seeds remains more or less intact (unpublished data). When untreated seeds are allowed to germinate in the soil the testa is ruptured but it does not entirely disintegrate (Fig. 6.1a, p. 114; Rasmussen & Whigham, 1993).

In general, extremely small seeds germinate only under the influence of light on or near the surface of the soil, which has the adaptive significance of allowing the seedling leaves to attain the functional stage above ground before the meagre resources of seed-reserve nutrients have been exhausted (e.g. Grime, 1979). In contrast, the heterotrophic orchid seedlings are independent of light and there is thus no need for them to produce above-ground parts. It is perhaps significant that unlike other seedlings those of orchids do not lengthen in their first phase of growth but increase in breadth ('Croissance par épaississement': Bernard, 1909).

In many terrestrial orchids germination is impeded or prevented by illumination (Chapter 4). Darkness requirements and the very slow hydration of the seeds could both be seen as mechanisms for ensuring that the seeds will be completely buried before they germinate (Lucke, 1981; Rasmussen & Rasmussen, 1991), but field observations of young seedlings of terrestrial orchids are few and rarely state the exact depth at which they were found. Stojanow (1916) observed seedlings of *Orchis morio* in a depth of 3–5 cm and of *O. ustulata* 5–10 cm below the surface in deep humus. Curtis (1943) observed *Cypripedium* seedlings at a depth of 2–5 cm below the surface, noting that it would have taken a considerable time for them to have reached this depth, whether the seeds had been washed down by rainwater or whether they had become covered by accumulating peat and humus. Seedlings of *Platanthera blephariglottis* and *P. ciliaris* that were observed germinating in bogs by Case (1964) were situated in the dead brown *Sphagnum* beneath the living moss. That the seeds have germinated immediately after dispersal seems precluded, and this is consistent with a considerable body of evidence from the field and from experiments with seeds (Chapters 3, 4 and 13). Many species are confined to habitats with shallow soil, particularly in the Mediterranean area. Protocorms in the upper layers of the soil may die from drying out during summer (*Orchis mascula, Dactylorhiza maculata*, Möller, 1987a; unpublished data) and this could be of significance in limiting population size in habitats with shallow soil, but the same species seem to germinate in great numbers (densities of 400–500 m^{-2} have been recorded: Möller, 1987a). Leeson and co-workers (1991) found numerous protocorms of *Dactylorhiza fuchsii* in moss and grass tufts that could give some protection against drying out. Severe seasonal drought in the natural habitat would increase the adaptive value of a life history strategy with a relatively short protocorm stage and the early transfer of biomass to more drought-resistant organs such as tubers (Chapter 10).

The terrestrial species investigated by Stoutamire (1974) appeared to have two different germination strategies. One group included bog species characterized by synchronous and immediate germination, with seedlings that turned green rapidly, as would be appropriate in a seasonally predictable habitat with favourable conditions for germination at dispersal time. This seedling behaviour suggests that germination is usually superficial, is not inhibited by light, and takes place in an environment where there is little risk of desiccation. The other group comprised species that germinated over a long period, as would be adaptive in variable and somewhat unpredictable environmental conditions where the most suitable germination strategy is the variety of reactions that is achieved with a polymorphic seed population. Many examples of seed polymorphism are given in the following.

1.2 **Reserve nutrients and seed mycotrophy**

It is usually claimed that orchid seeds do not contain sufficient reserve nutrients to enable them to germinate and produce an autotrophic seedling, symbiosis with fungi being thus a necessity. However, the available data show that orchid species vary in the degree to which they depend on fungi (Chapter 4) and, as suggested by Harrison (1977), it is possible that their reliance on fungi in germination is due to their inability to metabolize nutrient reserves rapidly rather than an insufficient supply.

The amount of nutrients stored in the mature seed, though of necessity small, is concentrated, the embryo cells containing lipid and protein bodies, and rarely starch grains, to the almost entire exclusion of other organelles (Fig. 1.2c; Manning & Van Staden, 1987; Yeung & Law, 1992). In some species the embryo cells also contain glucoprotein bodies, and there may be globoid inclusions of phytin in the protein bodies (Chapter 2; Manning & Van Staden, 1987; Rasmussen, 1990; Richardson *et al.*, 1992).

Many species, for instance several species of *Dactylorhiza* and *Orchis*, are reported to germinate in water and to remain alive for some days or weeks without receiving any external nutrients (Table 4.3, p. 62–63, and Chapter 13). Vermeulen (1947) used the expression 'waiting time' for the period after germination during which the seedlings can subsist on their own reserves while waiting for a compatible infection. Species that are able to germinate in water can apparently mobilize at least part of their food reserve immediately after the hydration of the embryo cells. The actual process of germination therefore takes place without infection.

Other species, such as *Cypripedium reginae* (Harvais, 1973, 1982) and *Goodyera repens* (Downie, 1940, 1941), require the addition of soluble sugars if they are to germinate *in vitro* without fungal infection. *Galeola septentrionalis* needs mineral salts as well, particularly nitrogen compounds, before it can germinate (Nakamura, 1962), and other species require cytokinins or certain as yet unidentified organic nitrogen compounds (Table 4.3, p. 62). Many less exacting species, although capable of germinating without special additions to the substrate, show an increase in germination percentage in their presence, which means that individual seeds within a seed batch differ somewhat in their nutritional requirements.

It is assumed that in nature any compounds necessary for germination are supplied by the fungi to the embryos, although it has also been suggested that direct absorption from the soil could be a significant factor (Chapters 6 and 8). Both soluble carbohydrates and nitrogen compounds are usually in short supply in the soil and are readily absorbed by the microorganisms in the soil and recycled (e.g. Campbell, 1977). Although embryos slowly take up nutrients from the substrate *in vitro* it is scarcely conceivable that the orchid embryo, with its small and mainly non-absorbent surface (Chapter 6), could effectively compete with soil microorganisms. On the other hand, the large external network of hyphae to which the infected seedling is connected provides an extensive interface with the soil.

1.3 Seed loss and predation

In typical populations the amount of seed produced is so great that this would give rise to massive increases in the number of individuals if it were not for limitations acting on germination and seedling establishment. Dispersal by wind entails losses because many seeds either fasten in the vegetation and never reach the ground or land on unsuitable substrates. In a stable or increasing population the abundance of seed has to compensate for these losses so that enough seedlings are available for the selection of the most suitable genotypes at the seedling stage (Grime, 1979). My own preliminary observations in field sowing experiments indicate a considerable loss of young seedlings before or immediately after infection (Rasmussen & Whigham, 1993). When cultivated *in vitro* adjacent seedlings compete before any shortage of carbohydrates in the substrate arises, but the competition is mutual, reducing the growth rate of all seedlings in denser sowings (Rasmussen *et al.*, 1989; Tsutsui & Tomita, 1989). If these observations are applicable to natural conditions it means that seedlings that are well spaced will tend to grow faster and will thus have a better chance of survival.

Compatibility with the available fungi would seem to be one of the most important selective factors in seedling establishment, but little is known from nature since germination in soil has seldom been observed. Sibling seeds germinated *in vitro* often vary considerably in their response to infection. Extremely broad or even bimodal seedling size distribution can be observed *in vitro*, and this seems to be connected with the differential ability of the seedlings to maintain a beneficial relationship with the fungus in question (Alexander & Hadley, 1983). Furthermore, under *in vitro* conditions the various endophyte strains differ strikingly in the extent to which germination is stimulated and seedling growth rate increased (Chapter 5).

Seeds of species that germinate in spring must remain quiescent in the ground for more than 6 months after dispersal, and during this time interaction with the soil fauna is possible. The activity of burrowing animals may displace the seeds to soil strata that are more or less suitable for germination, or the seeds could be consumed. It seems probable that seeds of this size either incidentally pass through animals that process the soil or, since the reserves of lipids and proteins form a concentrated meal, are actively sought after by foraging members of the soil fauna. Earthworms are thought to act as internal seed vectors in the dispersal of a number of plant species with small seeds (McRill & Sagar, 1973). Millipedes that I have kept in captivity consumed seeds of both *Tipularia discolor* and *Goodyera pubescens*. About 50% of the seeds of *Goodyera* became unviable but the remainder passed through the gut apparently intact and were able to germinate (unpublished data). This survival rate is perhaps not surprising, since orchid seeds can withstand drastic chemical surface treatment in the laboratory. In species whose seeds require severe surface treatments in order to germinate in the laboratory, predation may in fact have a beneficial effect on germination. Partial decomposition of the testa by bacteria could also in some species promote germination, although in others it might constitute a loss to the seed population (Veyret, 1969).

2

Seed development

2.1 Post-pollination events

When pollen is placed on the stigma of an orchid flower the pollen tubes begin slowly to grow down through the style, eventually reaching the ovules several days later (Table 2.1). The placentas are only slightly differentiated at the time of pollination, the ovules not being beyond megasporogenesis (Fredrikson, 1991), and ovule development proceeds only in those flowers that have received pollen. Thus no further biomass is invested in unpollinated flowers, which is obviously advantageous where, as is often seen in orchids, pollination occurs infrequently (e.g. Gill, 1989) and each developing fruit represents an enormous investment of resources because of the vast numbers of seeds.

In most holarctic species fertilization takes place from 1 to 2 weeks after pollination, but in some species, particularly those belonging to *Cypripedium*, up to 13 weeks may elapse (Table 2.1). There is some variation in different reports on the same species and Hildebrand (1863) noted a difference between plants growing in the field and specimens of the same species cultivated in pots, indicating that environmental factors affect the duration of the process.

In most species embryogenesis lasts about 2 weeks (Veyret, 1974). The species of *Cypripedium* differ in this respect as well; for instance, embryos of *Cypripedium calceolus* var. *pubescens* consist of no more than 12 cells 49 days after pollination, i.e. about 21 days after fertilization (Table 2.1, Light & MacConaill, 1990). In most orchids

Table 2.1. *Optimum time for excising immature seeds for culture in relation to time of fertilization and seed maturity (indicated in number of days after pollination)*

	Ferti-lization	Optimum excision	Maturity	References
Cypripedium acaule		60		Withner (1953)
C. calceolus	31			Hildebrand (1863)
C. calceolus		42		Lucke (1982)
C. calceolus		49–53		Malmgren (1989a)
var. *pubescens*	28			Duncan & Curtis (1942)
var. *pubescens*		60		Withner (1959)
var. *pubescens*		49	110	Light (1989)
var. *parviflorum*	26–33			Carlson (1940)
var. *parviflorum*	35			Hildebrand (1863)
var. *parviflorum*		56		Pauw & Remphrey (1993)
C. candidum		42		Pauw & Remphrey (1993)
C. reginae		42		Lucke (1978a)
C. reginae		42		Frosch (1985)
C. reginae		56		Pauw & Remphrey (1993)
C. reginae		60		Withner (1959)
C. reginae			85–95	Ballard (1987)
Cypripedium sp. (spp.?)			105	Lucke (1971b)
Epipactis atrorubens	90			Guignard (1886)
Cephalanthera damasonium	43			Hildebrand (1863)
Limodorum abortivum	24			Guignard (1886)

Species				Reference
Neottia nidus-avis	<9			Hildebrand (1863)
Listera ovata	10			Guignard (1886)
L. ovata	8–9		30	Hildebrand (1863)
L. ovata			30	Lucke (1971b)
Platanthera bifolia	21			Fredrikson (1991)
P. chlorantha	24			Hildebrand (1863)
P. hyperborea		35		Rasmussen (unpublished)
Dactylorhiza maculata	14			Guignard (1886)
D. maculata	17–18			Hildebrand (1863)
D. majalis	17			Hildebrand (1863)
D. majalis	20			Guignard (1886)
Gymnadenia conopsea	15			Hildebrand (1863)
G. conopsea	15			Guignard (1886)
G. conopsea		28–35		Malmgren (1989a)
Nigritella nigra		21–28		Malmgren (1989a)
Orchis coriophora	9			Hildebrand (1863)
O. mascula	21–31			Hildebrand (1863)
O. militaris	30			Hildebrand (1863)
O. militaris		35–42		Rasmussen (unpublished)
O. militaris			77	Farrell (1985)
O. morio	13			Hildebrand (1863)
O. morio	15			Guignard (1886)
O. morio		28		Malmgren (1989b)
O. morio		42		Frosch (1980)
O. simia	13			Guignard (1886)
O. ustulata	8–10			Guignard (1886)

Table 2.1. (cont.)

	Ferti-lization	Optimum excision	Maturity	References
Anacamptis pyramidalis	8–9			Hildebrand (1863)
A. pyramidalis	8–10			Guignard (1886)
Barlia robertiana		49–56	77	Rasmussen (unpublished)
Himantoglossum hircinum	19			Hildebrand (1863)
H. hircinum	<24			Guignard (1886)
Ophrys argolica		35–49	49	Rasmussen (unpublished)
O. bombuliflora		42–56	70	Rasmussen (unpublished)
O. fusca		58–63	63	Rasmussen (unpublished)
O. insectifera	20			Hildebrand (1863)
O. insectifera		35–49		Borris & Albrecht (1969)
O. insectifera		42–49		Malmgren (1989a)
O. holoserica	90			Guignard (1886)
O. sphegodes			<60	Lucke (1971b)
Ponerorchis graminifolia	12–13	35–40		Nagashima (1989)
Arethusa bulbosa		>54		Yanetti (1990)

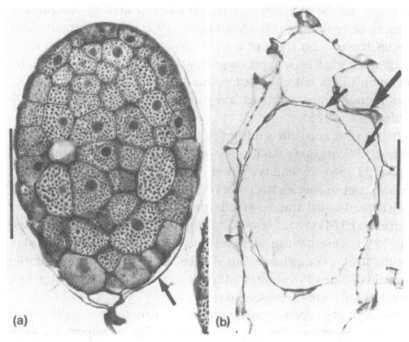

Fig. 2.1. (*a*) Cellular organization of the embryo of *Dactylorhiza majalis*
surrounded by carapace (arrow); ungerminated seed, semi-thin section
stained in PAS–aniline blue black. (*b*) Section through the testa (large
arrow) and carapace (small arrows), embryo removed. From seed of
Dactylorhiza majalis, stained in toluidine blue. Scale bars represent
0.1 mm. (*a*) Reproduced from Rasmussen (1990) with permission of the
New Phytologist.

endosperm development is suppressed (Afzelius, 1916; Swamy,
1949; Cocucci & Jensen, 1969) and the zygote divides to form a
globolar-ovoid embryo with little structural polarization. When
present, the suspensor is unicellular or forms a uniseriate string of
cells in the orchid groups dealt with here (Wirth & Withner, 1959).
By the time the seed has matured the suspensor has degenerated
(Veyret, 1974).

The orchid embryo is an ovoid body, usually with little or no
structural polarity, and there is no external nutrient tissue such as
endosperm (Fig. 2.1*a*). In *Bletilla striata* (and fewer than 10 other
orchid species) a small collar of tissue representing a cotyledon is

produced at the chalazal end of the embryo and shows varying degrees of rudimentation (Nishimura, 1991). However, in most species there is no trace of a leafy structure in the mature seed. Although a small leafy appendage may develop after germination, by definition this is not a cotyledon since it is not present in the seed, even though it may be structurally homologous to one (Batygina & Andronova, 1988).

In some species the embryo is extremely simple. The mature embryo of *Epipogium aphyllum*, for instance, always comprises 8 cells (Geitler, 1956), i.e. embryogenesis is completed after three mitotic cycles. In most species the embryo is somewhat more complicated, with the estimated number of cells ranging from 29 in *Calypso bulbosa* (Harvais, 1974) to *c.* 200 in *Dactylorhiza majalis* (Fig. 2.1*a*; Rasmussen, 1990). One of the most highly developed orchid embryos, that of *Bletilla striata*, comprises about 734 cells, assessed by applying the formula of ellipsoid volume to the number of cells cut in a median longitudinal section (illustrated in Veyret, 1974).

Seed quality depends on a number of factors during pollination and seed ripening. The source of the pollen is important. Self-pollination may result in a reduction in the seed quality although obligate out-crossing species are rare among European species. In nature the barriers to self-pollination are usually ecological or structural rather than physiological (Dafni & Bernhardt, 1990; B. Johansen, unpublished data). In some species, selfing done by hand yields roughly the same quality of seeds as does cross-pollination (Harvais, 1980; Ballard, 1987), but compatibility patterns can vary between populations within a species (Whigham & O'Neill, 1991).

It is hardly surprising that the results of germination tests carried out under standard conditions vary with parent genotypes and compatibility, but the result from successive harvests with the same parentage can also differ (e.g. Downie, 1959*a*; Fast, 1974; Harvais, 1980; Ballard, 1987). The number of viable seeds depends on the effective pollination, for instance whether adequate pollen was deposited on the stigma, which could be problematic in species with loose pollinia, such as *Cypripedium* (Ballard, 1987). The effectiveness of pollination could also be influenced by the age of the flower at pollination time, though no such effect has been demonstrated in *Cypripedium* (Ballard, 1987).

As regards qualitative differences in germination requirements from one year to another, there is little experimental evidence. However, Harvais (1982), who regarded the orchid seed as a precociously released proembryo, suggested that the longer the seeds were allowed to develop in the capsules the fewer would be the chemical requirements for germination. Hence, the year-to-year variation could reflect differences in external conditions during seed maturation which either extend or curtail the time spent in the capsule.

This interesting theory seems to be contradicted by all the evidence gained from *in vitro* culture of seeds taken from immature capsules, which generally germinate more readily than ripe seeds (see below). However, cultures of immature seeds are usually grown on chemically complex media that may provide for almost any chemical requirement, and the comparative success of immature seeds could be the result of other factors such as an increase in the rate of imbibition. If Harvais' theory could be applied to orchid seeds in general it would certainly help to explain some of the inconsistencies in germination results, but it would require painstaking experimental work to demonstrate such an effect convincingly.

2.2 Maturation

As the orchid embryo matures a change takes place in the seed reserves: immediately after fertilization starch-filled amyloplasts are observed in the embryo cells, but the starch has usually disappeared by the time the seeds ripen (Manning & Van Staden, 1987; see, however, *Calypso bulbosa*, Yeung & Law, 1992); lipids begin to accumulate after fertilization and persist in the mature seeds together with protein bodies (Poddubnaya-Arnoldi & Zinger, 1961). As with other seeds ultimate maturation is mainly a process of dehydration. Vacuolation, which is observed in the young embryo cells, disappears with approaching maturity (Manning & Van Staden, 1987).

The moisture content of mature orchid seeds is about 11% (Pritchard, 1984).

The outer integument enlarges considerably as the seeds develop, and separates almost completely from the inner integument which remains as a closely fitting envelope around the embryo (Carlson, 1940; Veyret, 1969). Only a few cell strands connect the two integuments at maturity. The mature testa, that forms the well-known reticular-foveate pattern, consists of dead cell walls from the outer integument. The cells of the inner integument are almost completely resorbed, but a thin layer persists, the so-called carapace (Fig. 2.1*b*; Veyret, 1969), which is hydrophobic and sheathes the embryo, being interrupted only at the micropyle. Veyret (1969) noted that seeds with a particularly well developed carapace, for instance species of *Cephalanthera* and *Epipactis*, germinated with difficulty. In some readily germinating seeds, viz. those of *Orchis morio*, *Neottinea maculata* and *Serapias lingua*, the carapace was incomplete. These observations suggested that the carapace is an important barrier to the uptake of water. The thickness of the carapace varied with the provenance of the seeds (Veyret, 1969).

The cell walls that make up the testa consist of cellulose and other plant cell wall components and are lignified (Carlson, 1940; Nakamura, 1964). Since the carapace is formed from cuticular remains it presumably contains cutin and various wall polysaccharides. A thin-layer chromatographic analysis of an ethanolic extract has also indicated the presence of suberin in the seeds (Eiberg, 1970). The testa varies from sand-coloured over shades of brown to almost black, deepening in colour as the seed matures.

2.3 Cultivation of immature seed

Immature orchid seeds often germinate more readily *in vitro* than do ripe seeds. 'Embryo culture' has become something of a routine method for a number of tropical orchid species (Withner, 1953) and it has also been highly recommended for European species (Borris, 1969; Fast, 1982, Malmgren 1989*a*). Not only did Borris (1969) improve germination results for species such as *Dactylorhiza majalis*,

D. maculata, *D. incarnata* and *Gymnadenia conopsea* whose mature seeds germinate rather readily, but he also achieved success with some species whose mature seeds had failed to germinate under the same conditions, for instance *Platanthera* spp. and *Cypripedium calceolus*. However, the technique apparently did not work with seeds from the Mediterranean area (Borris, 1969) or with seeds from species such as *Cephalanthera longifolia* or species of *Epipactis* that are considered to be really difficult to germinate (Lindén, 1980, 1992). These differences in response to immature culture are probably connected with the kind of dormancy that develops in the growing seeds.

The desiccation that takes place in the intact maturing fruit is interrupted when ovules are excised and transferred to a culture medium with low osmolarity. Since declining water potential is probably a regulating factor for protein accumulation in ripening seeds, rehydration of the immature seed *in vitro* generates a developmental change from protein storage to protein mobilization, i.e. germination (Kermode, 1990). If orchid seeds are excised while the suspensor is still functioning, and the seed envelope has not yet acquired its moisture-repellent qualities, there are few obstacles to rapid rehydration. The stage of metabolic quiescence that follows desiccation can thus be bypassed.

When immature seeds are to be cultivated they are often excised while the embryo cells are still fully hydrated and mitotically active. Light & MacConaill (1990) considered that for *Cypripedium calceolus* var. *pubescens*, the optimum time for excision coincides with the stage when the embryos consist of only 9–12 cells, i.e. considerably smaller than mature size. Borris (1969) recommended excising the seeds while the testa still consists of living cells and the embryos have attained about two-thirds of their final size. The same probably applies to species for which excision within 7 days after fertilization is recommended (*Orchis militaris*, *Ophrys insectifera*), but in many studies the size of the embryo or the number of cells at the time of excision is not specified. The only way to judge how far the embryos have developed is a comparison with other studies describing the time that elapses from pollination to fertilization and on to seed maturity (Table 2.1).

Studies on several species of *Cypripedium* indicate that optimum germination can be obtained when seeds are excised between 42 and 60 days after pollination; after this time the result decreases until the seeds mature about 85–110 days after pollination. The optimum period corresponds to the interval of between 14 and 32 days after fertilization, assuming that fertilization occurs roughly 4 weeks after pollination in the genus *Cypripedium* taken as a whole (Table 2.1). During this period the embryo cells are still proliferating, as indicated by the study of *C. calceolus* var. *parviflorum*, starch has disappeared from the embryo cells, but the suspensor cells are still living. The testa is lignified but has not yet become suberized (Carlson, 1940).

However, it is apparently not always essential to extract the embryos of *Cypripedium* spp. during a particular interval. In embryos of *Cypripedium reginae* Ballard (1987) found no relation between time of excision and germination in samples taken after 46 days (Fig. 2.2). Some of the best results were obtained with seeds that had been excised from 45 to 60 days after pollination, but the data showed considerable unpredictable variation between different pollinations and, in contrast to other studies, there was no clear decrease in germination in samples taken from near-mature capsules.

These conflicting results could be explained by assuming an interrelation with a cold requirement. Since Ballard's (1987) material suggested that the cold stratification requirement increases as the embryo matures, lack of chilling may be responsible for a low germination result when seeds are excised at a late stage in their development. In the investigation in which no optimum excision time could be found the seeds were all subjected to a cold stratification.

Some investigators prefer to harvest orchid seeds as late as 1 week before dehiscence rather than to use seeds from ripe capsules (e.g. Fast, 1982; Mitchell, 1989). Nagashima (1989) found that in a number of species the optimum excision time was immediately before maturity, which suggests that physiological changes, probably processes that increase the depth of dormancy or complete the formation of the carapace, continue to take place until the fruit ripens.

It has been found possible to inoculate immature seeds of *Listera*

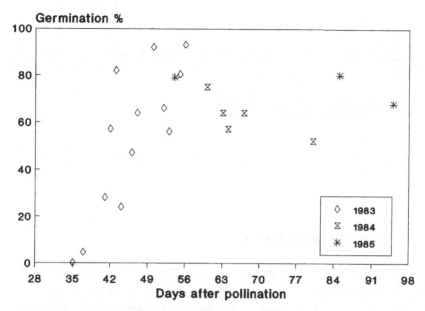

Fig. 2.2. Germination of immature seeds of *Cypripedium reginae*. The abscissa shows excision time measured as number of days after pollination. Sowings were made on Knudson C medium and seeds cultured in darkness, for the first 2 months at 5 °C, then for 3–4 months at 17–22 °C, germination was recorded when past its peak. Data from Ballard (1987).

ovata while they were still on the placenta and allow them to develop into symbiotic seedlings, and finally into adult plants (Rasmussen *et al.*, 1991 and unpublished data). We have put this technique forward as a possible method for testing fungal compatibility in species that tend to be difficult to germinate, but later experiments with other species did not meet with equal success and we have not yet been able to determine whether orchids in general possess this capacity for forming mycorrhiza from immature embryos.

3

Seed survival

3.1 Time of germination

In field sowing experiments performed in Maryland, USA, I observed the first signs of germination in *Goodyera pubescens* at the beginning of May, i.e. in spring, and in *Galearis spectabilis* and *Corallorhiza odontorhiza* at the beginning of June (Rasmussen & Whigham, 1993). The seeds had been harvested during the previous autumn and sown in November.

All other evidence on germination *in situ* is based on observations made at various times of the year of seedlings that were assumed to be young, although their exact age could not be determined. Casual observations of single protocorms are almost without value, but more reliable data are available from year-round studies of a given population which allow the size distribution of seedlings at different seasons to be compared. The general impression is that most of the holarctic species germinate in spring (Table 3.1). Ames (1922) found young seedlings of the American species *Goodyera pubescens* in spring, which agrees with my own observations from field sowings, and seedlings of many European taxa have also been found in spring, exceptions being some species of *Dactylorhiza*, *Orchis*, *Himantoglossum*, *Spiranthes*, *Epipogium* and *Corallorhiza*, the young seedlings of which were dug up in late summer or autumn (Table 3.1; see also Chapter 13).

A very large proportion of the holarctic species, except for those mentioned in Table 3.1, have not been recorded as protocorms in nature. A wider application of the field sowing technique that has recently been described (Rasmussen & Whigham, 1993) should yield more information on when seeds germinate and how rapidly the seedlings develop under near-natural conditions.

It is perhaps surprising that the seeds germinate in spring. In the first place, in most temperate vegetation types there is a considerable input of dead biomass during the autumn and early winter which stimulates intensive fungal activity in the soil and should provide ideal conditions for mycotrophic seedlings to begin their development. Secondly, spring germination implies that the seeds are equipped with dormancy or after-ripening mechanisms that prevent them from germinating immediately after their release in summer or autumn and also that they can survive for a fairly long period in the soil. On the other hand quite a number of the species germinate readily *in vitro* with no indication of seed dormancy, and moreover a rapid loss of germinability has been observed in some orchid seeds stored under laboratory conditions, suggesting that these seeds would be short-lived in the soil and would need to germinate immediately after dispersal (Van Waes, 1984).

However, there is growing evidence of dormancy mechanisms in temperate orchid seeds. Dormancy is broken by certain temperature regimes, lengthy imbibition, chemical breakdown of the testa and signals from fungi (Chapters 4 and 5). When setting up *in vitro* cultures an investigator may unwittingly break some of these dormancy mechanisms through the surface sterilization of the seeds or the addition of certain compounds to the culture medium.

Furthermore, there is no clear correlation between reported retention of viability during laboratory storage and field observations of germination. In some species of *Dactylorhiza* whose protocorms have been observed in autumn (Vermeulen, 1947; Möller, 1990*b*), the seeds survive longer in dry storage than those of most other orchids – up to a year or more (Van Waes, 1984; Johansen & Rasmussen, 1992). In contrast, some of the species that have been observed as protocorms in spring (e.g. *Epipactis helleborine, Listera ovata*) are reported to have very short storage survival – about 4–8 weeks (Van Waes, 1984).

Imbibed seeds kept on agar can remain viable for many months, as

Table 3.1. *Time of germination, mainly as estimated from the time of year when small seedlings have been found in the wild*

	Spring	Summer	Autumn	Winter	References[a]
Cypripedium acaule	+	+			Curtis (1943)
C. calceolus	+			Dec	Irmisch (1853)
C. calceolus	+	+			F & Z (1926a)
var. parviflorum	+	+			Curtis (1943)
C. reginae	+	+			Curtis (1943)
Epipactis helleborine	+				Ziegenspeck (1936)
Neottia nidus-avis	+				Bernard (1899)
Listera ovata	+				F & Z (1926b)
Spiranthes cernua			+		Ames (1921a)
Goodyera pubescens	May				R & W (1993)
Goodyera pubescens	+				Ames (1922)
Platanthera bifolia	+				F & Z (1927c)
P. chlorantha	+				F & Z (1927c)
Dactylorhiza fuchsii			Oct		Leeson et al. (1991)
D. maculata			Oct		Möller (1990b)
D. sambucina	+				F & Z (1927c)
Dactylorhiza spp.		+			Vermeulen (1947)
Coeloglossum viride	+				F & Z (1927c)
Orchis mascula		Aug			Möller (1987b)

Species				Reference
O. mascula	+			Ziegenspeck (1936)
O. militaris	+			Irmisch (1853)
O. militaris	+			Ziegenspeck (1936)
O. militaris	+			Möller (1989)
O. pallens		+		Ziegenspeck (1936)
Orchis spp.		+		Vermeulen (1947)
Himantoglossum hircinum	+			Ziegenspeck (1936)
Ophrys apifera			+	Fabre (1856)
Ophrys spp.			+	F & Z (1927c)
Liparis loeselii			+	Mrkvicka (1990a)
Corallorhiza odontorhiza			Jun	R & W (1993)
C. trifida	+			F & Z (1927a)
C. trifida	+			Weber (1981)
Epipogium aphyllum	+			F & Z (1927a)

Note: [a]F & Z, Fuchs & Ziegenspeck; R & W, Rasmussen & Whigham.

testified by lengthy *in vitro* experiments (Lucke, 1971*a*; Ballard, 1987; Rasmussen, 1992*b*; Stoutamire, 1992). Since seeds lying in the soil must be assumed to be fully imbibed, storage on agar corresponds more closely to the conditions of the natural seed bank than does the conventional cold, dry seed storage (Villiers, 1975). Evidence from long-term *in vitro* studies thus lends support to the possibility that some orchid seeds remain viable in the soil for several months and germinate in the spring, which is in fact what I have observed in *Goodyera pubescens*, *Corallorhiza odontorhiza* and *Galearis spectabilis*. All three species survived in the soil as seeds for more than 6 months, and evidence of widespread seed decay was first observed after almost 12 months (Rasmussen & Whigham, 1993).

3.2 Survival in storage

The decrease in the germination capacity of stored seeds has often been interpreted as loss of viability which is usually caused by genetic damage in the form of random deterioration of macromolecules, primarily in the chromosomes. In imbibed tissue such damage is continuously repaired, but in dried seeds the damage tends to accumulate (Villiers, 1975) unless the moisture content is so low as to minimize all physiological activity. The least satisfactory storage conditions are those where the moisture content is about 20%, which is high enough to cause cellular damage but too low to allow effective biochemical repair (Villiers & Edgcumbe, 1975). By the time a suboptimal set of storage conditions has caused a marked loss of seed viability there is, in addition, a considerable amount of genetic damage in the surviving seeds, so that among the seedlings that do actually develop there is a large proportion with low vitality and genetic defects (e.g. Roberts, 1973).

Recommended storage conditions for orchid seeds combine a moisture content of *c.* 5% and low or sub-zero temperatures (Pritchard, 1985*b*; Seaton & Hailes, 1989). An alternative worth investigating, at least for short-term storage, would be to store imbibed seeds under physical conditions that prevent them from germinating, for instance in light (Chapter 4).

Moisture content

Seed moisture content is technically difficult to determine accurately since orchid seeds are so light, but a few data exist. While the capsule is still closed the mature seeds may contain as much as 23% moisture, but when released they retain about 11% if not artificially dried (Pritchard, 1984, 1985b).

When a seed sample is dried, usually in a desiccator, an equilibrium is established between seed, air and desiccant, the final amount of seed moisture being dependent on the nature of the desiccant and the temperature at which equilibration takes place. In *Cattleya* seeds that were equilibrated at room temperature the moisture content obtained was 5.6% over a saturated solution of $CaCl_2$ and 2.2% over silica gel, these being the two most widely used desiccants (Seaton & Hailes, 1989). In a number of other species Pritchard (1984) reduced the seed moisture to c. 5% by equilibrating at 2 °C over silica gel.

Seaton & Hailes (1989) obtained the highest survival rate for *Cattleya* seeds after drying them over $CaCl_2$, achieving about 5% moisture content in the seeds; the higher degree of dehydration (c. 2%) that was achieved with silica gel proved to be inferior for germination in combination with all storage regimes tested.

Optimum seed moisture of c. 5% can thus be achieved by equilibrating the seeds either over a saturated aqueous solution of $CaCl_2$ at room temperature or over silica gel at 2 °C. Immediately after equilibration the containers with the seeds in should be sealed to make them airtight, and there should be a minimum of air enclosed. Unripe seeds from green capsules also need to be dried before they are stored (Davis, 1946; Fast, 1982; Mitchell, 1989). The fruit tissue should be discarded as it may house spores and viruses (Davis, 1946; Mitchell, 1989).

However, when mature seeds of a range of European species were desiccated over silica gel in a refrigerator and should accordingly have contained about 5% moisture, viability still declined dramatically within a year (Van Waes, 1984). It is possible that their storage life was shortened by excessive drying, but since the investigation of dormancy patterns in orchid seeds is far from complete it is difficult to tell whether loss of germinability is perhaps sometimes actually due to secondary dormancy brought on by storage conditions. Secondary dormancy could be part of a flexible germination strategy governed by the access of water that would enable the seeds to enter a state of dormancy with more specific requirements for germination in the event of severe drought. This theory fits in with observations on many orchids, freshly harvested seeds often yielding much higher germination percentages than seeds that have been stored for a short time (e.g. Van Waes, 1984).

Experiments with stored seeds of *Calanthe discolor* point to the induction of deeper dormancy, rather than to loss of viability, during dry storage. Miyoshi & Mii (1987) found that the germination percentage was reduced to 5%, from 30% in fresh seeds, when they dried the seeds for 1 month. They could partially reverse this effect by protracted soaking before sowing the seeds, or by using a liquid substrate. If the seeds were stored without making any attempt to dry them the germination percentage remained at the same level as for fresh seeds.

Furthermore, the drop in germination percentage observed in dry seeds could in part be due to the fact that dry seeds germinate more slowly. In *Eulophia alta*, Pritchard (1985*b*) found that when seeds with an initial moisture content of 20% were sown, twice as many germinated as when the seeds had an initial moisture content of 5%. This result was recorded 4 weeks after sowing, but when germination was recorded again later the drier seeds had reached the same level as that of the moister seeds. If the imbibition process is considerably protracted in seeds from dry storage, standard germination tests could easily yield inferior results without this being a true indication of lower viability.

Storage temperature

When seeds of a selection of temperate and tropical species were dried until the moisture content dropped to 5% it was found that they could be exposed to low temperatures ($-20\,°C$ or $-196\,°C$) without the germination responses being adversely affected (Pritchard, 1984). Lucke (1985) stored seeds of tropical species at $-20\,°C$ for up to 8 years, but he gave no details of the final germination percentages. He found, however, that the old seeds took 2–4 weeks longer to germinate than fresh seeds.

However, in long-term testing of *Cattleya* seeds, Seaton & Hailes (1989) found that the germination percentage decreased more rapidly at $-18\,°C$ than at room temperature, and at both temperatures approached zero within 200–300 days. The seeds retained their germinability longest when stored at $5\,°C$, 30% of them still germinating in the seventh year of storage.

In other words there are no clear guidelines for the optimum temperature regime during the storage of dried orchid seeds. Differences in temperature tolerance between orchid groups could in part be related to their adaptation to different climates. Lucke (1985) noted that there was a decline in germination when seeds of tropical species were subjected to repeated freezing and thawing. Pritchard (1984), on the other hand, raised the level of germination of *Orchis morio* to 80% with four freeze/thaw cycles, compared with about 60% in the controls.

3.3 Viability testing

As long as our knowledge of dormancy patterns in orchid seeds is still far from complete, it is difficult to distinguish non-viable seeds from those that are dormant. It would be ideal if the quality of a seed batch could be determined not in conjunction with a germination test, but by an independent test for viable embryos.

There is a standardized use of certain aqueous stain solutions for viability testing of plant seeds, but these stains are only effective if they are brought into direct contact with the embryo cells, which is usually achieved by dissecting the seeds. Orchid seeds, however, are too small to be manipulated or cut individually. If the staining test is performed on intact seeds the water-repellent testa prevents uniform absorption of the stain, and the colour of the testa can also interfere with the interpretation of the test (Eiberg, 1970).

The testa can be removed by placing soaked orchid seeds in a drop of water between microscope slides and then gently rotating the slides so that the embryos are propelled out through the tapering ends of the testa (Pritchard, 1985a). The testa can also be removed by ultrasonic treatment (Miyoshi & Mii, 1988). Instead of these mechanical treatments, Van Waes and Debergh (1986a) used long sterilization times in hypochlorite to soften and partially rupture the testa and to extract the water-repellent lipid coating. The effect was increased by adding detergent to the hypochlorite solution and subsequently rinsing and soaking the seeds in water for up to 24 hours (Van Waes, 1984). Sterilization with $Ca(OCl)_2$ generally yields better staining results than those obtained with NaOCl. Other sterilizing agents such as hydrogen peroxide (H_2O_2) and alcohol are not suitable (Table 3.2).

Triphenyltetrazolium chloride (TTC) and fluorescein diacetate (FDA) have both been used for testing viability in orchid seeds, TTC being the stain officially recognized by the International Seed Testing Association (Hartmann & Kester, 1975). In living cells the colourless TTC solution is reduced to a stable red insoluble compound. Van Waes (1984) pretreated orchid seeds in $Ca(OCl)_2$ with added detergent, rinsed them for 24 hours in water, and stained them in 1% TTC with a pH of between 6.5 and 7.0 for 18–24 hours in darkness at 30 °C (the solution is extremely sensitive to light, Mackay, 1972). Since the strongly oxidizing hypochlorites could interfere with the reduction of TTC, the seeds should be rinsed thoroughly after surface sterilization. The need for rinsing decreases with increasing length of pretreatment, presumably because of the progressive decay of the surface structure of the testa, but no more than 24 hours is required (Van Waes, 1984).

FDA enters the living cells where it is hydrolysed by esterase to yield the fluorescent stain fluorescein. If the cell membranes are intact this stain accumulates in the cells. After the seeds have been surface sterilized and rinsed in water, a drop of the suspension of seeds is mixed with an equal quantity of a 0.5% (weight/volume) solution of FDA in absolute acetone (Pritchard, 1985a). The reaction takes place within 10 minutes, and the result should be recorded within the following 10 minutes before the fluorescence fades.

The pretreatment in hypochlorite affects the ability of the seeds to

Table 3.2. *Staining with triphenyltetrazolium chloride (TTC) after various types of pretreatment of the seeds*

	Dactylorhiza majalis	Epipactis helleborine
Absolute ethanol (30 min)	0	0
H_2O_2 10%	0	0
H_2O_2 20%	0	0
H_2O_2 30%	0	0
NaOCl 2%	0	0
NaOCl 5%	7.8 ($\pm$1.6)	0
$Ca(OCl)_2$ 5%	10.3 ($\pm$1.8)	1.3 ($\pm$0.9)
$Ca(OCl)_2$ 10%	9.8 ($\pm$0.9)	1.9 ($\pm$1.1)
Abs. ethanol (5 s) + 1% Tween 80	0	0
H_2O_2 20% + 1% Tween 80	0	0
NaOCl 5% + 1% Tween 80	11.4 ($\pm$2.5)	1.5 ($\pm$0.6)
$Ca(OCl)_2$ 5% + 1% Tween 80	18.5 ($\pm$3.2)	4.6 ($\pm$1.3)

Notes:
Selected data from Van Waes (1984, table XIII).
H_2O_2, hydrogen peroxide; NaOCl, sodium hypochlorite; $Ca(OCl)_2$, calcium hypochlorite.
Values are percentage of stained seeds, mean of 4 replicates $\pm$ SE of mean. Seeds were pretreated for 2 hours and subsequently rinsed in sterile water for 24 hours; the staining procedure was standardized.

take up water thus facilitating germination as well as staining (Fig. 3.1). The number of seeds that stain roughly corresponds to the number that could germinate after a similar pretreatment. The seed viability measured should, of course, reflect the highest level of germination that can be achieved in a seed batch; it is thus essential to determine the optimum pretreatment that will give the highest possible reaction to the stain (Pritchard, 1985a; Van Waes & Debergh, 1986a). A pretreatment that is longer than optimum, however, does not give a reliable staining reaction because apparently the chemical destruction brought about is sufficient to reduce the staining reaction and thus no longer indicates the initial viability of the seeds that are treated (Fig. 3.1).

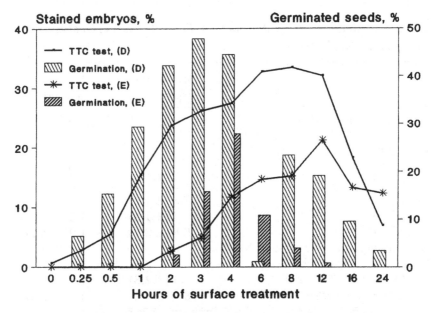

Fig. 3.1. Stainability and germinability as a function of surface sterilization treatment of seeds in *Dactylorhiza maculata* (D) and *Epipactis helleborine* (E). Outside the optimum range of treatments both tests reveal only part of the seed viability (germination potential) of a seed batch. TTC staining was done after surface sterilization in 5% NaOCl + Tween 80 (1% v/v). Asymbiotic germination was on BM1 (D) and BM2 (E) (see Appendix A). Data from Van Waes (1984, tables XXI and XXII).

4

Requirements for germination

Under suitable aeration and temperature conditions non-dormant plant seeds need only access to water to be able to germinate, but since orchid seedlings are heterotrophic to varying degrees their seeds often have several requirements from outside besides water. These are undoubtedly met by the endophyte when germination takes place in nature, or indirectly by the substrate that the fungus lives on.

Some other requirements for germination of orchid seeds are associated with dormancy. The currently available evidence suggests that several types of dormancy, both exogenous and endogenous, occur in the seeds of holarctic orchids. In spite of the apparently delicate nature of the translucent testa its impermeability to water represents a form of physical dormancy that can be overcome by prolonged soaking or by scarification. There are different types of physiological dormancy that can be broken by darkness, temperature regime, atmospheric conditions, or certain external chemical signals. Finally, morphological dormancy, expressed as a requirement for the embryos to after-ripen, probably also occurs in orchid seeds.

The treatments and substrate additives that are necessary to obtain germination *in vitro* can thus be divided into those that are needed for seeds to germinate in symbiotic culture, i.e. such that break dormancy, and the more extensive set of conditions that must be met if the seeds are to germinate asymbiotically. The difference between these sets would seem to directly represent the effects of the fungus, but this conclusion may not be valid, since several external stimuli may elicit the same physiological response, for instance when different metabolites influence different steps in the same group of reactions.

4.1 Scarification

Rinsing in hypochlorite not only serves the purpose of sterilizing the seeds but also increases the percentage of seeds that germinate (Harvais & Hadley, 1967b; Eilhardt, 1980; Harvais, 1980; Lindén, 1980). Harvais & Hadley (1967b) raised the germination percentage in *Dactylorhiza purpurella* by sterilizing the seeds until they lost their dark colour and noted that seedling growth was not adversely affected by this treatment. Eilhardt (1980) observed that mature seeds that had been surface sterilized sometimes germinated better than near-mature seeds excised from a sterilized capsule. By varying the time the seeds were immersed in 2% sodium hypochlorite (NaOCl) between 10 and 45 minutes, Lindén (1980) found that germination increased with increasing duration of sterilization in *Dactylorhiza maculata*, *Gymnadenia conopsea* and *Orchis morio*, up to a maximum above which the germination percentage fell. This obser-vation has been amply supported by later work (Van Waes, 1984; Van Waes & Debergh, 1986b; Rasmussen, 1992b and unpublished data), i.e. the germination percentage is positively correlated with duration of sterilization until the optimum is reached above which germination declines, sometimes dramatically (Fig. 4.1). The same effect can be observed in symbiotic culture (Fig. 4.2; Rasmussen, unpublished data).

The optimum sterilization time for germination is usually longer than the time required for decontaminating the seeds (Table 4.1), and is much more narrowly defined than the optimum for viability staining (Fig. 3.1, p. 38). For reasons not yet determined, both viability staining and germination results are usually better after sterilization in calcium hypochlorite ($Ca(OCl)_2$ than in NaOCl.

Both hypochlorites are oxidizing agents, NaOCl being stronger than $Ca(OCl)_2$. As salts of a weak acid, hypochlorites are strongly

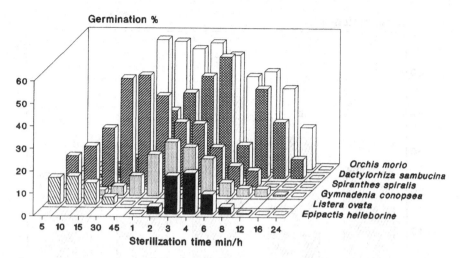

Germination %

Orchis morio
Dactylorhiza sambucina
Spiranthes spiralis
Gymnadenia conopsea
Listera ovata
Epipactis helleborine

Sterilization time min/h

Fig. 4.1. Optimum sterilization for germination of a range of species. Surface sterilization was in 5% NaOCl + Tween 80 (1% v/v). Asymbiotic germination was on BM1 and BM2 respectively (see Appendix A). Data from Van Waes (1984, table XX).

alkaline in solution. Since $Ca(OCl)_2$ has almost exactly twice the molecular weight of NaOCl, a 5% solution (weight/volume) of either salt in deionized water gives a *c.* 0.7 N solution with a pH of *c.* 12. $Ca(OCl)_2$ is generally used in a saturated (*c.* 7.5%) or slightly weaker (5%) solution (Stoutamire, 1963; Van Waes, 1984) whereas NaOCl is used in concentrations ranging from 0.2% to 5% (Purves & Hadley, 1976; Lucke, 1971*a*; 1978*b*; Eilhardt, 1980; Frosch, 1982). Commercial bleaches with trade names such as Chlorox and Domestos which contain 4.25–5.25% NaOCl (Arditti, 1982*b*) are often used as they are or in diluted form.

Chlorine evolves from the volatile solutions, which may reduce the sterilizing effect if the solution is allowed to stand. This is a factor that could be of significance when a weak solution is used in conjunction with lengthy sterilization. Several workers recommend preparing the diluent immediately before use (e.g. Arditti, 1982). Harvais (1980) has drawn attention to the relation between number of seeds, duration of sterilization and concentration of the solution. Alkalinity can in fact measurably decrease during surface sterilization if a large number of seeds are treated in a small volume of sterilant. Since the

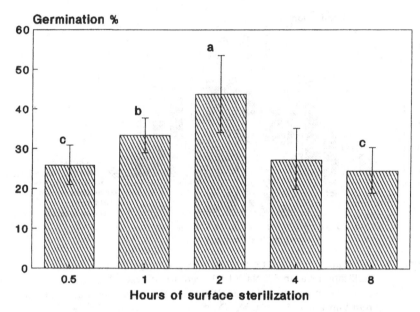

Fig. 4.2. Symbiotic germination as function of duration of surface sterilization, in 5% $Ca(OCl)_2$ with 0.05% Tween 80. Seeds of *Liparis lilifolia* were sown on W2 medium (a modification of oat medium omitting soluble carbohydrates and replacing oats with chopped wood), inoculated with a symbiotic fungus M31 isolated from *L. lilifolia*, and incubated at 20 °C in darkness. Mean and 95% significance intervals of 10 replicates are shown; *a*, b and c indicate differences with at least 95% significance. Results recorded after 9 weeks of incubation at 20 °C (arcsin transformed original data).

volume of solution and amount of seeds are rarely indicated by investigators it is difficult to compare different sterilization procedures, although a surplus of hypochlorite may usually be assumed.

Hydrogen peroxide (H_2O_2) is sometimes used as an alternative sterilizing and oxidizing agent (Masuhara & Katsuya, 1989). It is a weaker sterilant than NaOCl and at effective concentrations is more harmful to the seeds (Lucke, 1978b; Van Waes, 1984; Rasmussen unpublished data). McAlpine (1947) added 0.03% H_2O_2 direct to the substrate so as to avoid the need for surface sterilization, but this method does not seem to have been used to any great extent. Ethanol is also sometimes used for sterilization of seeds, but from what is now known about the stimulating effect on germination, there is no good

Table 4.1. *Sterilizing and scarifying effects of surface treatment in sodium hypochlorite (NaOCl), indicated as the number of contaminated vessels out of 6 replicates after varying durations of treatment*

	Hours								
	$\frac{1}{12}$	$\frac{1}{6}$	$\frac{1}{4}$	$\frac{1}{2}$	$\frac{3}{4}$	1	2	3	4
Listera ovata	3	1+	1	0	−	−	−	−	−
Spiranthes spiralis	3	1	1	0	+	−	−	−	−
Dactylorhiza maculata	4	1	0	−	−	−	−	+	−
Orchis morio	4	1	2	0	+	−	−	−	−
Dactylorhiza incarnata	6	3	2	0	−	−	+	−	−
Platanthera chlorantha	5	3	3	1	0	−	−	−	+
Dactylorhiza sambucina	6	3	3	0	−	−	−	−	+
Orchis coriophora	6	3	3	0	−	−	−	−	+
Anacamptis pyramidalis	5	4	1	0	−	−	−	+	−
Gymnadenia odoratissima	5	4	2	0	−	−	−	−	+
Gymnadenia conopsea	6	4	1	0	−	−	+	−	−
Dactylorhiza praetermissa	6	4	2	0	−	−	+	−	−
Epipactis atrorubens	6	6	6	6	5	4	1	0	+
Epipactis helleborine	6	6	6	6	6	3	2	1	+

Notes:
Data from Van Waes (1984, tables XIX and XX).
Treatment was with 5% (v/w) NaOCl + Tween 80 (1% v/v). '+' indicates the treatment yielding the highest germination.

reason to hesitate using hypochlorites for surface sterilization except perhaps in a few cases (see Section 13.9).

Like the hypochlorites, H_2O_2 and alcohol are oxidants, but they are not basic and unlike the hypochlorites are not reported to promote germination. Other strong bases, such as sodium hydroxide (NaOH), can have the same effect on germination as hypochlorite; Eiberg (1970) found that when the seeds had been immersed in 4 N NaOH for 15 minutes germination was higher than in the controls. This suggests that it is the alkalinity of the solution used for pretreatment that is of importance for improving the germination, not its oxidative effect.

For any effective chemical treatment of the testa in aqueous solutions the addition of a detergent is necessary since by containing both hydrophilous and lipophilous groups the detergent improves the contact between water and a lipoid surface such as the seed coat. The most widely used detergent, with the trade name Tween 80, is a sorbitolanhydride. Eiberg (1970) observed that the germination percentage decreased when the concentration of Tween exceeded 0.5%, but other investigators have used up to 1% without noting any ill effects. Harvais (1980) found that the detergent Teepol is injurious to orchid seeds at a concentration of 2 ppm. Tween 80 is so far the only detergent that is widely used for sterilizing orchid seeds, but additional comparative information on the effects of various detergents would be useful.

A further improvement of the scarification is obtained by treating the seeds in 0.5–2% sulphuric acid (H_2SO_4) before applying the hypochlorite solution (Malmgren, 1993), but the strength of the acid and the duration of the pretreatment must be modified according to the species. This surface treatment improved germination in *Orchis militaris*, *O. purpurea* and *Anacamptis pyramidalis* (see Sections 13.17 and 13.20). Another recent development is to treat lightly sterilized seeds in a solution of wall degrading enzymes for up to 15 days before sowing (Lindén, 1992), a procedure which has been tested with some success in species of *Cypripedium*.

Butcher & Marlow (1989) recommended removing the testa by mechanical means for better germination results (mainly with tropical species), but this is a tedious process since it must be done with sterile instruments. The entire outer testa can be removed by ultrasound, which also increases the germination (Miyoshi & Mii, 1987). When the testa was ruptured by grinding the seeds in sand, however, there was no perceptible effect on germination (Stoutamire, 1974). The effectiveness of mechanical scarification may depend on the extent to which the inner integument, the carapace, is damaged. Miyoshi & Mii (1987) suggested that ultrasonic treatment also rendered the carapace more permeable. The increase in germination percentage observed by Pritchard (1984) in *Orchis morio* after repeated freezing and thawing could also be the result of seed coat rupture.

What natural processes in the soil could be comparable to surface treatment for hours in concentrated bleach, or actual removal of the testa, is a puzzling question. Access of water to the embryo itself seems to be the basic requirement that has to be met. In seeds with chemically and mechanically intact testa the entry of water through the micropyle and suspensor by abiotic means would be a very slow process, if possible at all. In symbiotic cultures, however, it can be seen that the hyphae of the endophyte grow through the micropyle or pierce the testa and surround the embryo inside the testa before they actually begin to penetrate the embryo. Although there is no known enzymatic activity in the fungi that dissolves suberin, they do produce other wall-degrading substances that may attack the carapace (Table 5.3, p. 94), so that the fungi may assist in germination by making water accessible to the embryo and by overcoming the water-repellent effect of the testa. However, since lengthy sterilization also increases symbiotic germination the chemical scarification apparently brings about more effects than can be achieved solely by activities of the symbiont. The possibility of bacteria, as suggested by Veyret (1969), or animals in the soil being involved in the chemical breakdown of the orchid seed testa has already been mentioned (Chapter 1), but as yet very little is known about the changes taking place in the seeds while they lie ungerminated in the soil. Mechanical decomposition by weathering processes, such as fluctuations in temperature causing freezing and thawing, obviously also occurs.

4.2 Soaking and leaching

Long-term soaking may produce an effect similar to that of removing or partially decomposing the testa. Burgeff (1936, 1954) was the first to prepare seeds for sowing by soaking them for several months, but it is not clear what was meant when he stated that 'regelmässiger Keimung' (literally: regular germination) could be achieved in this way only with seeds of *Cypripedium, Gymnadenia* and *Platanthera*. Seemann (1953) devised a double flask in which the seeds could be aseptically rinsed, for several months if necessary, but this technique does not seem to have been widely used. Lengthy soaking alone has not been found to solve all problems connected with breaking dormancy, such as encountered in *Epipactis palustris* (Stoutamire, 1974; Lindén, 1992).

Other investigators have noted the advantage of using a liquid medium either for pretreatment or as the sowing substrate (Liddell, 1944; Kano, 1968; Miyoshi & Mii, 1987). Curtis (1936) obtained a germination of about 50% in *Cypripedium acaule* using liquid media after several attempts to germinate seeds on solid substrates had failed. I have found that liquid media are inferior for germinating seeds of *Dactylorhiza majalis*, however, and practically no seeds germinated at all if the medium was stirred.

Apart from increasing the permeability of the testa, it is also conceivable that soaking at the same time leaches out inhibiting substances. Mixed sowings of seeds that germinate readily and seeds of problematic species have not resulted in any inhibition in the former (Eiberg, 1970; Stoutamire, 1974), but other observations suggest that inhibiting substances in fact do exist in orchid seeds. The endogenous level of abscisic acid (ABA) is about 5 times higher in seeds of the problematic *Epipactis helleborine* than in those of the readily germinating *Dactylorhiza maculata* (van der Kinderen, 1987), and Van Waes (1984) found that after seeds of *Dactylorhiza maculata* had been surface sterilized for 2 hours there was practically no ABA to be detected; however, exogenously applied ABA inhibited germination in these seeds, suggesting that high endogenous ABA levels are associated with dormancy.

4.3 Temperature

Both rate and percentage of germination are dependent on temperature. When seeds of *Dactylorhiza incarnata* were incubated in darkness in deionized water to which 0.05% detergent had been added 50% of them germinated within 11 days at 23–24 °C, while about 20 days at 20–21 °C or 24 days at 27–28 °C were required for the same result to be achieved (Eiberg, 1970). For most species the temperature that yields the highest germination percentage lies between 22 and 25 °C, but some germinate best below 20 °C, such as the species of *Cypripedium* (Stoutamire, 1974). At temperatures both above and

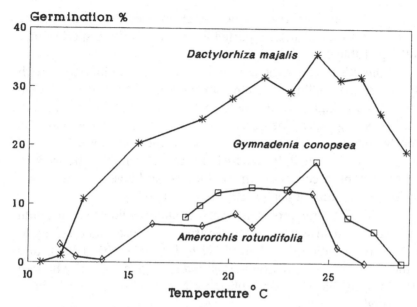

Fig. 4.3. Temperature optimum for germination of three species in symbiosis with *Tulasnella calospora* (D47–7) cultured in darkness on oat medium. Means of 10 replicates are shown, recorded 6 weeks after sowing. Original data.

below the optimum some of the viable seeds obviously remain resting (Harvais & Hadley, 1967*b*; Rasmussen *et al.*, 1990*a*). With the seasonal fluctuations of temperature in nature this reaction pattern ensures that germination in a seed population can be spread out over a period of time.

The optimum temperature was about the same for germinating the arctic species *Amerorchis rotundifolia* as for cold-temperate species such as *Dactylorhiza majalis* and *Gymnadenia conopsea* (Fig. 4.3). The only difference seems to be that the temperate species show a greater tolerance of high temperatures.

Cold stratification

The limited amount of practical information available on the use of cold stratification for breaking dormancy in orchid seeds mainly concerns species that are otherwise notoriously difficult to germinate *in vitro*. Success has been reported for several of these species after chilling (Table 4.2).

In *Epipactis palustris* the germination percentage increased with the duration of chilling, from almost zero after only 4 weeks at 5 °C, to a significantly higher level after 8 weeks, and a still higher level after 12 weeks (Fig. 4.4). The effect of chilling in this species was also clearly dependent on the presence of a symbiotic fungus, and on previous incubation at 20 °C which presumably enables the seeds to imbibe, establish contact with the fungus, and possibly after-ripen before cold treatment begins (Rasmussen, 1992*b*).

Roughly the same procedure, i.e. incubation followed by gradual cooling to cold stratification temperatures was used successfully in asymbiotic germination of *Arethusa bulbosa* (Yanetti, 1990), and some of the best germination in *Cypripedium reginae* noted by Ballard (1987) occurred in cold-stratified seeds that had previously been incubated.

Most reports of failure of seeds to respond to chilling seem to be associated with the application of the low temperatures immediately after sowing, and with treatments that lasted 8 weeks or less. However, some deviant results have also been reported and the germination percentage even after long chilling periods can still be fairly low (Table 4.2), indicating either that seed viability is low or that other dormancy mechanisms are involved.

Evidence obtained by Ballard (1987) suggests that cold requirement develops during the process of seed maturation, which might mean that external conditions while the seeds ripen could determine the degree to which the seeds become cold requiring, hence accounting for some of the observed provenance and year-to-year variation in germination responses. If a seed sample is polymorphic with respect to chilling, a portion of the seeds would germinate at room temperature, and the requirements of the remaining seeds could pass unnoticed, the poor result perhaps being ascribed to low seed quality.

Apart from its effect on the germination process, chilling has also been found to improve growth in *Epipactis palustris* immediately after germination. Cold treatment apparently activated the metabolic systems that are responsible for organogenesis in the seedlings and

Table 4.2. *Cold stratification of orchid seeds*

	Procedure	Result	References
Cypripedium acaule	3–5 months at 5°C, then 25°C	70%	Coke (1990)
C. calceolus	1–8 weeks at −5 to 5°C, dry or moist	0	Fast (1974)
C. calceolus	Control, no chilling	1%	
	3 months at *c.* 5°C	8%	Ballard (1990)
	4 months at *c.* 5°C	16%	
C. calceolus	3–5 months at 5°C	50%	Coke (1990)
C. candidum	8 weeks at 5°C	0	Stoutamire (1990)
C. candidum	2 months at 4°C	Erratic	Pauw & Remphrey (1993)
C. reginae	>2 months at 5°C	→90%	Ballard (1987)
Epipactis helleborine	8 weeks of cooling	0	Van Waes (1984)
E. palustris	Several months at 2–4°C	+	Borris & Albrecht (1969)
E. palustris	3 months at −10 or 3°C	0	Linden (1980)
E. palustris	8 weeks at −1 to 10°C	0	Lucke (1981)
Cephalanthera rubra	3 months at 5°C	(+)	Weinert (1990)
Spiranthes spiralis	8 weeks at −1 to 10°C	0	Lucke (1981)
Goodyera pubescens	Several months at 5–10°C	+	Stoutamire (1974)
Dactylorhiza majalis	3–4 months	+++	Borris (1970)
Dactylorhiza sp.	3 months at 3°C	0	Riether (1990)
Arethusa bulbosa	Gradually cooled, then 3–4 months at 4–6°C, then slowly warmed	+	Yanetti (1990)
Aplectrum hyemale	3–5 months at 5°C, then 3–5 months at 25°C	60%	Coke (1990)
Tipularia discolor	3–5 months at 5°C, then 3–5 months at 25°C	95%	Coke (1990)

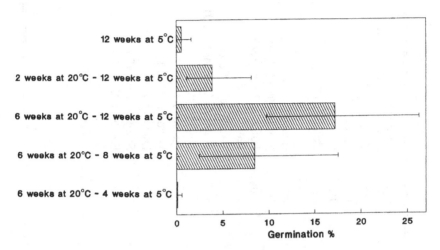

Fig. 4.4. Germination of *Epipactis palustris* after varying duration of incubation at 20 °C and cold stratification at 5 °C. Culture was on oat medium with fungus D412–1. Germination percentages were recorded after 3 additional weeks at 20 °C. Means of 15 replicates with 95% confidence intervals are shown (arcsin transformed data). Reproduced from Rasmussen (1992b), with permission of *Physiologia Plantarum*.

the formation of a stable symbiotic association with the fungus (Rasmussen, 1992b; see Chapter 9).

The physiological effect of cold stratification seems to be either to decrease the level of endogenous ABA or to increase the content of endogenous cytokinin, both of which can stimulate germination. Such changes have been shown to occur in seeds of a number of other plant families (Rudnicki, 1969; van Staden *et al.*, 1972). Exogenously applied ABA inhibits seed germination generally, whereas cytokinin has a stimulating effect that can wholly or partially replace the effect of cold stratification (Rudnicki, 1969; Pinfield & Stobart, 1972; Webb & Wareing, 1972). When a species of *Fraxinus* with a requirement for cold stratification was compared with a related species without dormancy, it was found to have higher levels of endogenous ABA before cold stratification than the latter (Sondheimer *et al.*, 1968). The study of the orchid species *Dactylorhiza maculata* and *Epipactis helleborine* referred to above (Section 4.2) may reflect similar differences between orchid species.

Considerable amounts of ABA can be present in the testa of the seeds of other plants, the content sometimes being so great that removal of the testa or leaching will abolish or greatly reduce the need for chilling (Pinfield & Stobart, 1972; Webb & Wareing, 1972). The marked responses of orchid seeds to the various surface treatments (see above) suggest that ABA could be present in the testa.

The ecological effect of cold requirement in orchid seeds is to prevent them from germinating immediately after dispersal. Chilling, freezing and thawing cycles and varying periods of soaking during winter may have a cumulative effect that brings about a softening of the testa, leaches out inhibitors to germination, and initiates physiological changes in the embryos.

Since a number of the holarctic orchid species are reported to germinate in spring, further investigation of cold stratification would be of interest, both in connection with germinating problematic species and in the search for seed polymorphism in species that usually germinate poorly under standard conditions in the laboratory. In pilot tests on species with unknown germination requirements, it would be easy to cool down the cultures for a period of at least 12 weeks if the usual incubation at about 20 °C does not induce germination, or if germination is poor.

Warm stratification

Since orchid embryos are poorly developed at the time of dispersal it is conceivable that a period of after-ripening is necessary, and observations of protracted germination responses *in vitro* (Nakamura, 1964; Lucke, 1971a; Stoutamire, 1992; Rasmussen, 1992b) may provide indirect evidence of this. However, the developmental changes taking place in incubated seeds that germinate extremely slowly have not been investigated.

Seeds of *Galeola septentrionalis* need to be incubated for about 10 months before they germinate *in vitro*. During this period the seeds should experience at least 70 days at 30 °C for optimum germination and temperatures falling to 25 °C for more than 55 days are inhibitive. Temperatures slightly higher than optimum (32–34 °C) during incubation also inhibit germination and result in abnormal seedling development (Nakamura, 1962, 1964).

In the cold-demanding seeds of *Epipactis palustris*, incubation at 20 °C before chilling significantly heightened the effect of chilling, and incubation for 6 weeks gave significantly better results than incubation for 2 weeks (Fig. 4.4; Rasmussen, 1992*b*). This may reflect the need for after-ripening before the seeds are ready to respond to low temperatures.

Fluctuations in diurnal temperatures

The seeds of some plants respond to diurnal fluctuations in temperature. To my knowledge tests using fluctuating temperatures have only been attempted once with orchid seeds (*Dactylorhiza majalis*: Rasmussen & Rasmussen, 1991). Little or no fluctuation in temperature was found to yield the best results. Since temperatures tend to become more stable with increasing depth below the surface of the soil, this observation suggests that if the seeds are well buried in the soil they germinate better, which is consistent with the fact that the seeds are usually negatively photoblastic, as described below.

4.4 Light

The effects of light on germination in orchids have not been fully explored. Taxonomic differences and differences in habitat appear to be reflected in the light sensitivity of the seeds. Australian terrestrial species are germinated in 16 hour photoperiods as a regular procedure (Clements, 1982*a*), and tropical species can be germinated in daylight (Thomale, 1957; Lucke, 1975). Although Thomale (1957) noted no marked difference between the results obtained in darkness and under moderate illumination he advised against allowing direct sunlight to fall on the culture vessels.

Most investigators of holarctic terrestrial orchids prefer to incubate seeds in darkness and find that even low light intensities, either continuous or in photoperiods, inhibit germination (Harvais &

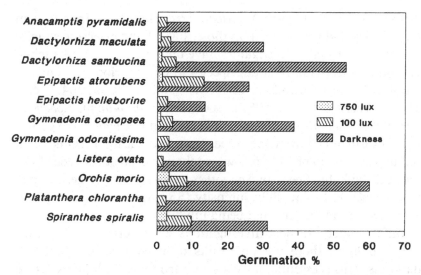

Fig. 4.5. Effect of light on asymbiotic germination in 11 species (BM medium and modifications: Appendix A). Based on data from Van Waes (1984, table XXVII).

Hadley, 1967*b*; Fast, 1978, 1980; Mead & Bulard, 1979; Lindén, 1980; Van Waes & Debergh, 1986*b*; Ballard, 1987; Rasmussen *et al.*, 1990*b*). Eleven European species tested by Van Waes (1984) all germinated poorly at a light intensity of 100 lux (*c.* 0.3 W m^{-2}) and virtually no seeds germinated with stronger irradiation in 14 hour photoperiods (Fig. 4.5). Incubated seeds of *Cypripedium calceolus* var. *parviflorum* that were intermittently subjected to light failed to germinate (Riley, 1983), and at least 14 days in continuous darkness were required to obtain the highest germination in *Dactylorhiza majalis* (Rasmussen *et al.*, 1990*b*). In this species light proved to be almost completely inhibiting, irrespective of whether it was a constant illumination or 16 hour or 8 hour photoperiods. While germination in one batch of seeds was inhibited in green light even at a low intensity (0.02 W m^{-2}: Rasmussen & Rasmussen, 1991) this did not apply to all seed samples of the species, the reaction apparently also being in part dependent on the provenance of the seeds.

However, the observations made by Stoutamire (1974) and Riley

(1983) suggest that North American terrestrial species that grow in bogs are less light sensitive than those that inhabit drier and shadier localities, and some investigators have obtained good results when germinating seeds of holarctic species in photoperiods or constant low light intensity (Knudson, 1941; Stoutamire, 1963, 1964; Lucke, 1971a; Wells & Kretz, 1987). Species of *Cypripedium, Goodyera, Spiranthes, Platanthera, Malaxis, Calopogon* and *Pogonia* were germinated by Curtis (1936) without specification of the light regime used, but apparently no precautions were taken to protect the seeds from natural daylight. Seeds of *Spiranthes magnicamporum* germinated fully (99%) under low light intensity (55 μmol m^{-2}s^{-1}: Anderson, 1991), germination thus not being inhibited.

Unfortunately in most investigations where orchids are germinated in light, there are no results from controls incubated in darkness. This is essential if we are to learn whether light exposure has a positive effect on germination in any of the terrestrial orchids, though so far there is little evidence that this is so. When experiments are set up to compare germination in light and darkness, the result in darkness is usually better, and in some experiments all the seeds sown in light failed to germinate (Eiberg, 1970; Fast, 1978; Van Waes & Debergh, 1986b).

One exception may be *Bletilla striata*, however, where germination was found to be significantly ($P = 0.05$) higher in low irradiation (0.2 W m^{-2}) of mixed white light than in darkness (Ichihashi, 1990). An attempt to identify the active component of the white light was not successful, however. Coloured light of different wavelengths failed to raise the germination percentage significantly above that of the control kept in darkness, which is difficult to explain.

We found that in *Dactylorhiza majalis* a pretreatment in light for up to 10 days subsequently raised the level of germination in darkness (Rasmussen *et al.*, 1990b), a reaction that we observed several times in this species but not in the other species we have tested (*Orchis mascula, Gymnadenia conopsea*: Rasmussen & Rasmussen, 1991). Van Waes (1984) studied a number of European species and observed no stimulating effects when the seeds were pretreated for 28 days in 14 hour photoperiods before they were incubated in darkness – in contrast the decrease in germination, compared with the control in darkness, was greatest when the light intensities were highest (up to 2500 lux, *c.* $7.5\,W\,m^{-2}$, with the same selection of species as in Fig. 4.5). Reaction to pretreatment in light is apparently not a widespread phenomenon, but recently it was noted in seeds of *Platanthera integrilabia*, whose germination was significantly increased (Zettler & McInnis, 1994).

When sowing in soil, Darnell (1952) claimed that seeds of terrestrial orchids should be sown on the surface, and attributed his good germination results to the light exposure. Since there is little indication that the species are light demanding other explanations must be looked for. Subsequent watering from above may in fact have washed many of the seeds into the soil at an early stage.

The reactions of orchid seeds to light appear to be fairly complicated, and subject to specific variation. Further investigation is clearly required, both to better understand the natural adaptations of the seeds, and to improve germination procedures for species of horticultural interest. Germination in *Dactylorhiza majalis* was not notably depressed by the amount of light that was needed for handling the seeds, but high sensitivity towards light could be one of the problems associated with the species that are now considered difficult to germinate. It would be of practical interest to determine which wavelengths the seeds respond to so that they can be handled in safe-light. It is not known whether phytochrome is active in the light perception of orchid seeds. Preliminary observations indicate sensitivity especially in the red area of the spectrum, which may have important consequences for the establishment of seedlings in relation to vegetation cover (Eiberg, 1970; Rasmussen & Rasmussen, 1991).

4.5 Atmosphere

Some investigators have obtained unusually high germination percentages when the culture vials were hermetically sealed (Nakamura, 1962; Kano, 1968; McIntyre *et al.*, 1974) or when the seeds were covered by the agar medium (Curtis, 1943; Weinert, 1990). In symbiotic germination tests of *Dactylorhiza majalis* carried out in our laboratory in Copenhagen we observed significant differences in

Germination %

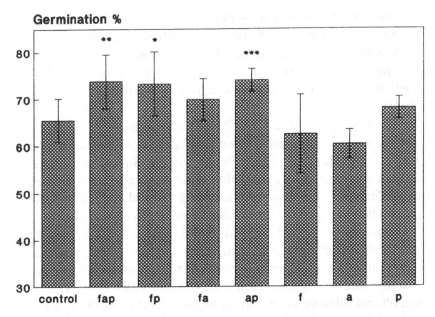

Fig. 4.6. Effect of a confined atmosphere on germination. The culture dishes were sealed with plastic film (f), aluminium foil (a) or zip-lock plastic bags (p) and combinations of these. In all cases when the film or foil were combined with zip-lock bags, germination percentage was significantly higher than the control (unwrapped Petri dishes). Data are means of 10–18 replicates with 95% confidence intervals (arcsin transformed data). Significant differences at 95% (*), 97.5% (**) and 99% (***) are shown. Original data.

germination percentages according to how the culture dishes were sealed and packed (Fig. 4.6). Both sealing of the dishes with household-type plastic film and packing them in aluminium foil increased the germination percentage in conjunction with stacking them in closed 'zip-lock' plastic bags.

These observations indicate that the ambient atmosphere affects the germination of orchid seeds. The germination process is aerobic, and the orchid fungi also require oxygen (O_2). Both organisms produce carbon dioxide (CO_2) by respiration. Whether the culture is symbiotic or asymbiotic the O_2 content of the air in the culture vessel tends to decrease and the CO_2 to increase, the levels being dependent on how much exchange is possible with the outside air. This also applies to air pockets in the soil.

The O_2 content in the soil atmosphere is usually lower than that above ground: as low as 16% in the upper layers of the soil as compared with $c.$ 21% in air (Stiles, 1960). Though low or high levels of O_2 induce germination in the seeds of some plants, the available data do not suggest that orchid germination is much affected, except when inhibited by hypoxia. Stoutamire (1964) obtained no improvements in the germination of terrestrial orchid seeds when he tested the effect of low concentrations of O_2, and Fast (1982) found that germination was considerably reduced by low O_2 pressure. In *Galeola septentrionalis* germination decreased markedly when O_2 concentrations below 5% were tested, but the atmospheric concentration (21%) proved to be too high for optimum germination (Nakamura *et al.*, 1975). Eiberg (1970) investigated the effects of pure O_2 on the seeds of *Dactylorhiza maculata* and found that germination did not differ from the result obtained in $c.$ 20% O_2; but lower levels of O_2 that would correspond to conditions within the soil were not tested.

The CO_2 content in the air of surface soil may amount to 3–4% as compared with 0.03% above ground (Stiles, 1960) and may locally be still higher in the vicinity of soil organisms such as orchid symbionts. It seems reasonable to expect high CO_2 levels to act as an environmental signal to the orchid seeds and to stimulate germination, but very little has been done to investigate this interesting possibility. In *Galeola septentrionalis* optimum germination was obtained with concentrations of $c.$ 8% CO_2 (Nakamura *et al.*, 1975). Unfortunately, there are no comparable data for other species.

Part of the increase in germination that occurs *in vitro* when the seeds are inoculated could be the result of the high level of CO_2 that inevitably develops in symbiotic cultures. Sometimes germination is also promoted in asymbiotic cultures that have inadvertently become infected with bacteria or fungi. In symbiotic sowings of *Dactylorhiza majalis* I tried to raise the CO_2 concentration further by adding the gas to a confined space in which the Petri dishes were incubated, and found that the germination percentage decreased whenever CO_2 was added. However, since the resulting CO_2 levels were not measured in this experiment they may have deviated considerably from the concentrations encountered by seeds in nature.

The ethylene level in the soil commonly ranges from 0.07 to $10 \, \mu l \, l^{-1}$ (Smith & Russell, 1969). It is produced by a number of soil fungi (Ilag & Curtis, 1968; Lynch & Harper, 1974; Abeles, 1985) including some that have been isolated from orchid mycorrhiza (*Tulasnella calospora, Ceratobasidium cornigerum*: Hanke & Dollwet, 1976). Ethylene accumulates at some depth when the level of O_2 is low (Lynch & Harper, 1974), and the concentration fluctuates with the season and tends to be highest in soils with a high content of organic matter (Dowdell *et al.*, 1972; Smith & Dowdell, 1974; Alexander, 1977). If the seeds could detect ambient ethylene concentration this could provide them with information as to their position in the soil, the time of year or the organic content of the soil. There are only a few data on effects of ethylene on orchid germination, but the results are interesting and should stimulate further study.

When seeds of *Dactylorhiza majalis* were incubated in dishes to which exogenous ethylene had been added there was a highly significant increase in the germination percentage, both in symbiotic and asymbiotic cultures, even at low concentrations, down to $1 \mu l l^{-1}$ (1 ppm, Fig. 4.7; Johansen & Rasmussen, 1992), which lies within the range normally encountered in the soil atmosphere. It seems possible that high ethylene concentrations may occur locally, for instance in the small air pockets inside the inflated testa which are invaded by the fungi. Apparently none of the high concentrations we used in our experiments had adverse effects on the seeds. In some plants the germinating seeds themselves produce ethylene which has an autostimulatory effect on germination (Taylorson & Hendricks, 1977), and ethylene production has also been observed in seedlings of tropical orchids germinating asymbiotically, where it reached concentrations of up to $125 \mu l l^{-1}$ (Hailes & Seaton, 1989). This suggests a considerable tolerance of high local concentrations of the gas.

4.6 Germination substrate

As pointed out by Withner (1959), an astounding array of culture media are capable of sustaining orchid seedlings *in vitro*. The reason could be that the seeds and seedlings utilize only a fraction of the

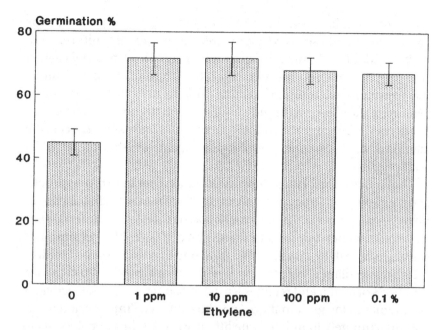

Fig. 4.7. Effects of ethylene on germination in *Dactylorhiza majalis*. Symbiotic culture on oat medium inoculated with fungus D47–7 in an atmosphere supplemented with amounts of ethylene varying from 1 ppm to 0.1%. The jars in which the dishes were kept (unwrapped) were treated with ethylene immediately after sowing and ventilated on day 21 with sterile atmospheric air. Germination percentages were recorded on day 42. All treatments differ significantly from the control ($P < 0.01$, arcsin transformed data). Mean germination percentages of 16–18 replicates with 95% confidence intervals are shown. Original data.

ingredients in the substrate. In an intact living system such as the germinating seed or the seed–fungus association absorption from the substrate is mainly an active process; such systems tend to have a high tolerance of varying concentrations of exogenous substances up to levels that give rise to problems of high osmolarity or toxicity.

Many of the substrate recipes have been developed through the empiric modification of Knop's or Pfeffer's solutions of mineral salts. The main trend as regards media for terrestrial orchids has been towards reducing the concentrations of mineral salts and increasing the amount of organic compounds (see Appendix A). Awareness of the association between seeds and fungi in nature has prompted the development and use of complex media for asymbiotic culture, often with the addition of undefined mixtures of organic compounds with a content of vitamins and growth regulators, the aim being to simulate the assumed contributions of the fungus. It is often difficult to determine the effects of the individual ingredients and, as pointed out by Hayes (1969), amongst others, there is no single factor that can be identified as a specific germination agent in the mycorrhizal relationship.

The composition of most substrates is such that they promote both germination and subsequent seedling development. This is convenient for the purpose of plant production but yields little information on requirements for germination, since a suitable germination substrate may differ considerably from one that satisfies the requirements for seedling development. It is in fact much easier to compare the effects of media and regimes if a clear distinction is made between requirements for germination and those for seedling development, but unfortunately in much of the literature, including some pioneering publications, this has not been done. For instance, Knudson's classic statement that orchid seeds would not germinate without the addition of soluble sugars refers to seedling development rather than to germination: 'When a nutrient medium is used containing only the essential salts and agar there is no germination [sic]. The seeds of *Cattleya* and those of certain other genera begin to grow and develop a green color. After about 2 months growth ceases and despite the apparent presence of chlorophyll no further growth is made' (Knudson, 1929). 'No germination' in this context clearly means that seedlings emerged from the testa and grew for about 2 months before they ceased to develop further. With this unorthodox definition of germination there is an obvious danger of misinterpretation which can be carried on to other works. For instance, that orchids require soluble sugars to be able to germinate is implicitly expressed in many texts, such as when the types of sugars that seedlings can use for their development are referred to as sugars 'that support germination' (Arditti, 1967, 1979).

In some terrestrial orchids the seeds do need carbohydrates from outside when they germinate, but in many others they can in fact germinate in pure water (Table 4.3), in the sense that the testa is ruptured and the seedlings begin to grow, although neither water nor a mineral salt solution will suffice to support seedling development for any length of time. Downie's pioneering work (1940–9) established the minimum requirements for germination in a number of species, her aim to identify the essential factors having been followed by some later workers. The general impression emerging from a compilation of their results is one of diversity, some species germinating readily, others having variously complex requirements for external nutrients (Table 4.3).

In studies where germination percentages are presented it becomes obvious that chemical requirements are rarely absolute in the sense that none of the seeds germinate in their absence, i.e. it is almost never a question of an all-or-none reaction. The germination percentage in a species usually rises with increasing complexity of the media, which reflects that individual seeds within a seed population vary somewhat in their germination requirements. This makes it all the more difficult to identify the active ingredients in the substrate that actually promote germination.

It is problematic to rely on the subjective assessment of germination percentages (i.e. when an actual count is not made). Regulated sowing is impracticable in orchids and ungerminated seeds can be hidden by large seedlings, which leads to an over-estimate of the germination. Since larger seedlings tend to conceal a relatively greater number of the ungerminated seeds, there is a risk that growth-stimulating components in the substrate will be given undue credit for promoting germination. Only quantitative data that can be treated statistically can help to give a reliable picture of substrate requirements in the germination process itself.

Carbohydrates

Experimental data show that some terrestrial orchid species require soluble sugars, usually sucrose or glucose, if their seeds are to germinate *in vitro* (Table 4.3). In other species the need for exogenous sugars arises after the seeds have germinated.

Table. 4.3. *Minimalizing the culture medium for asymbiotic germination in vitro*

	W	M	S	S+M	OrN	Other	References[a]
Cypripedium reginae	0	0		+	++	(Kinetin) +++	Harvais (1973, 1982)
Epipactis gigantea	(+)				++		Eiberg (1970)
E. atrorubens					+	(BA) +++	V & D (1986b)
E. helleborine					+	++	V & D (1986b)
Cephalanthera rubra						++	V & D (1986b)
Listera ovata	0	0		0		(Potato extract) +	Downie (1941, 1949b)
L. ovata	0						Eiberg (1969)
L. ovata				+	++		V & D (1986b)
Spiranthes spiralis				+	++		V & D (1986b)
Spiranthes spp.	+						Stoutamire (1964, 1974)
Goodyera repens	0	0		+			Downie (1940, 1941)
G. repens						++	Hadley (1982c)
G. oblongifolia	0		+				Harvais (1974)
G. pubescens	+++			+			Rasmussen (unpublished)
Platanthera bifolia	0			+	++	++	Downie (1941)
					++		V & D (1986b)
P. hyperborea	+						Harvais (1974)
P. obtusata	+			++			Harvais (1974)
Dactylorhiza purpurella	+						Harvais & Hadley (1967b)

Species							Reference
Dactylorhiza spp.	++						Downie (1941)
	++						Vermeulen (1947)
			+++				V & D (1986b)
Coeloglossum viride	+		++				Downie (1941)
Gymnadenia conopsea	+						Downie (1941)
Orchis spp.	+	+					Eiberg (1969)
			++				Vermeulen (1947)
							V & D (1986b)
Anacamptis pyramidalis	(+)	+	++				Eiberg (1970)
							V & D (1986b)
					(Yeast)		
					++		
Ophrys spp.	+	+					Hadley (1982b)
							Stoutamire (1974)
O. apifera		+	++				Hadley (1982b)
		+	++				V & D (1986b)
					(Fungus,potato extract)		
Corallorhiza trifida	0	0	0	0	+		Downie (1949a)
					(Yeast/amino acids)		
Galeola septentrionalis	(+)				+++		Nakamura (1962)

Notes:

W, water or water agar. M, macronutrient (mineral salts) media. S, water or water agar with sucrose or other soluble carbohydrates. S + M, macronutrients and carbohydrates. OrN, as S + M but with organic nitrogen wholly or partially replacing inorganic nitrogen. Other, plant growth regulators, vitamins, organic supplements.

[a] V & D, Van Waes & Debergh.

The table includes only studies using very simple substrates and studies where two or more directly comparable substrates were used, i.e. substrates where only one factor varies and where single components of complex media have been tested separately. Germination results are indicated as a subjective relative measure, i.e. $0 < + < ++ < +++$.

For those species that need sugars for germination *in vitro* the respiratory substrate for germination in nature is provided from outside, i.e. by the infecting fungus or its substrate; in asymbiotic cultures soluble sugars must be available in the substrate in a form that can be absorbed directly by the seeds. *Cypripedium reginae, Listera ovata, Goodyera repens, G. oblongifolia* and *Platanthera bifolia* are all reported to have germinated on sucrose or glucose media where germination on water agar failed (Table 4.3). Van Waes (1984) obtained a somewhat higher germination percentage for *Orchis morio* with concentrations of 29 mM (59.3%) and 58 mM sucrose (64.9%) than in the control without sucrose (47.9%), but higher concentrations (87 and 116 mM) produced results that were inferior to those of the sugar-free control, as did mannitol controls with a comparable osmolarity. High concentrations of soluble sugars (i.e. a high osmolarity) can thus restrict germination even in seeds that need some external sugar.

In species that germinate in water, germination percentage is unaffected by the level of sucrose, except for the inhibiting effects of very high concentrations which is apparently due to the higher molarity of substrates that are rich in sugar. *Dactylorhiza majalis* germinated equally well in water and in weak solutions of sucrose or an equal molarity of mannitol (asymbiotic germination, Fig. 4.8). However, at a concentration of $16 \, g \, l^{-1}$ sucrose (*c*. 46 mM) germination was significantly reduced, and a mannitol solution with the same molarity also reduced germination. In *D. purpurella* germination was somewhat reduced at concentrations of glucose or sucrose as low as 1% (29 mM) as compared with controls that germinated in distilled water (Harvais, 1972). Mead & Bulard (1979) found that concentrations of 10 and $20 \, g \, l^{-1}$ (29–58 mM) were equally effective in bringing about germination in *Orchis laxiflora*, but that substrates containing $80 \, g \, l^{-1}$ (*c*. 340 mM) had a strong inhibiting effect. Unfortunately they had no controls without sucrose.

During the initial stages of germination, before a photosynthetic apparatus has developed, there are two conceivable sources of carbohydrates available to the embryo, i.e. nutrients stored in the embryo and nutrients acquired from or made accessible by the infecting fungus. Only a few investigations have dealt with the

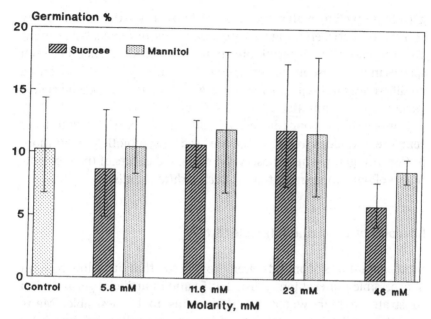

Fig. 4.8. Asymbiotic germination of *Dactylorhiza majalis* in water (control) and in a range of sucrose solutions and mannitol solutions with corresponding molarity. The only significant effect in comparison with the control is the decrease in germination with 46 mM sucrose ($P = 0.024$). Means of 8 replicates with 95% confidence intervals are shown. Original data.

processes of reserve nutrient mobilization (Chapter 6), but mature orchid seeds are usually not starchy, which precludes the possibility of soluble sugars being formed by the hydrolysis of starch. There are two other possible sources of sugars within the nutrient reserve. Small amounts of soluble sugars have been detected in seeds of some South African species (Manning & Van Staden, 1987). Although the species in question did not germinate without external sucrose it is possible that seeds of other species contain adequate amounts of free sugars for them to be independent of external sources during germination. Secondly, the seeds of some species contain a storage of glucoproteins that could release glucose on hydrolysis, but it has not been established at what stage in germination they are mobilized.

If soluble carbohydrates could be made available from the nutrient reserve this might explain why the seeds of some species can germinate well in water, but it is not known whether there is any correlation between a high endogenous content of accessible sugar and the capacity to germinate in water. The difference between species in their response to exogenous soluble sugars could be based on differences in their nutrient reserves which in turn would reflect variation in germination strategies (Chapter 6).

Non-soluble carbohydrates, usually in the form of starch or cellulose, are added to media that are used for symbiotic germination. In Copenhagen we have observed no marked effects of the concentration of oats on germination in *Dactylorhiza majalis*.

Ion concentration and substrate texture

Wild populations of orchids often grow in soil that has a low content of accessible mineral ions (Fast, 1985) and in artificial germination substrates a high water potential seems to be desirable. Many investigators have found that traditional tissue culture media are far too concentrated for orchids, particularly for the temperate species, and should be diluted 2–10 times (Harvais, 1974; Fast, 1980, 1982, 1983; Van Waes, 1984; Mitchell, 1989). Fast (1978) recommended the use of growth media with less than $0.25 \, \text{g} \, \text{l}^{-1}$ of inorganic salts, but Van Waes (1984) observed that in most species germination percentages were reduced when concentrations were as low as $0.05 \, \text{g} \, \text{l}^{-1}$. Anderson (1991) obtained 99% germination in *Spiranthes magnicamporum* on water agar, compared with only 45% on a mineral medium (*c.* $1.2 \, \text{g} \, \text{l}^{-1}$ with 5% potato extract) and 25% on an oat medium (*c.* $0.6 \, \text{g} \, \text{l}^{-1}$).

In propagation work it may be necessary to select a compromise substrate that also satisfies the requirements of the young seedlings, but for the actual germination, the complete absence of mineral salts may often be advantageous. This is probably owing to osmotic effects since none of the mineral ions that are usually added to media are known to inhibit germination to any great extent (see below). The assumption that the seeds are affected by the osmolarity of their substrate is supported by their reactions to high sugar concentrations, as described above.

Observations suggest that mechanical disturbance inhibits germination. When propagating *Disa uniflora* Lindquist (1960) found that it was important not to disturb the seeds during the first 6 weeks after they had been sown in soil, and therefore recommended watering from below. Darnell (1952) noted an improvement in germination in soil when the seed bed was watered by gentle spraying. In the laboratory in Copenhagen we obtained extremely poor germination results for *Dactylorhiza majalis in vitro* in cultures with liquid substrate on a shaking table. The seeds remained viable and germinated about 2 weeks after being removed from the shaker.

It is perhaps for this reason that liquid substrates are not generally used in the propagation of these species. Galunder (1984) recommended as little as 6 g agar per litre, and the use of a soft agar may be practicable as a compromise, since it has a high water potential and at the same time ensures mechanical stability.

Nitrogen compounds

The soils of typical orchid habitats are exceedingly poor in inorganic nitrogen (Stoutamire, 1964; Fast, 1985). In both natural substrates and culture media nitrogen may occur either in the oxidized form as nitrate (NO_3^-), or in reduced forms such as inorganic ammonium salts (NH_4^+) or organic compounds.

In accordance with what has previously been said about mineral salt concentrations in general, germination does not usually benefit from strong concentrations of nitrogen salts. In a number of European species Van Waes (1984) and Van Waes & Debergh (1986b) found that the germination percentage was actually highest when there were no macroelements in the culture medium, and that in some species the presence of inorganic nitrogen prevented germination altogether. In Copenhagen, we obtained a negative correlation between concentration of NO_3^- and germination percentage in *Dactylorhiza majalis* (Fig. 4.9), and Eiberg (1970) noted a steady increase in germination percentages in *Orchis sancta* as the concentration of NH_4NO_3 was decreased, the best result (66%) being obtained when the medium contained none at all. In *Orchis morio*

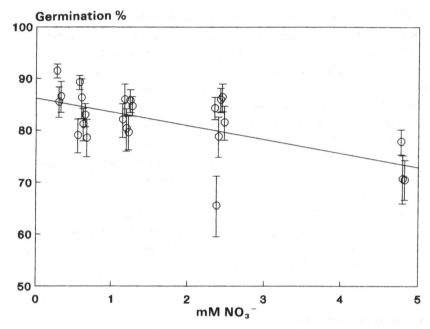

Fig. 4.9. Germination of *Dactylorhiza majalis* on media representing permutations
of Ca^{2+}, K^+, Mg^{2+} and NO_3^-, PO_4^{3-}, SO_4^{2-}, Cl^- within a macroelement
concentration of 9.3–12.6 mM. Basic medium was oat medium
inoculated with a strain of *Tulasnella calospora*. The correlation
coefficient with NO_3^- is −0.621. Germination percentages were
recorded after 6 weeks of incubation in darkness at 20 °C. Means of 16
replicates with 95% confidence intervals are shown (arcsin transformed
data). For clarity, different media with identical NO_3^- concentrations are
slightly out of alignment. Original data.

Dijk (1988*b*) found that the higher the total concentration of
inorganic nitrogen and the proportion of NO_3^-, the lower were the
germination percentages obtained. Seeds of *Bletilla striata* also
seemed to show a preference for low concentrations of inorganic
nitrogen and for nitrogen in reduced form as NH_4^+ (Ichihashi &
Yamashita, 1977).

Asymbiotic germination *in vitro* is often stimulated by exogenous
organic nitrogen (Table 4.3; Fast, 1980), perhaps in part because
nitrogen is less readily absorbed in this form. When organic nitrogen
is added to a recipe it often replaces some of the inorganic nitrogen

otherwise used. Organic nitrogen seems to be tolerated in high concentrations, as shown when yeast extract, which has a high content of amino acids, was used in the germination of *Galeola septentrionalis* in concentrations up to 5% without adverse effects (Nakamura, 1964).

It is often difficult to judge the merits of specific organic nitrogen compounds in the substrate, since autoclaving may alter their composition. For instance, the effect of adding glutamine before autoclaving is simply to increase the level of reduced nitrogen in general since glutamine and several other amino acids cannot withstand temperatures above 100 °C (George *et al.*, 1988). Furthermore, the effects of specific organic nitrogen compounds can be effectively masked when the substrates in addition contain undefined components such as peptone or yeast extract (Dijk, 1988*b*). These supplements are rich in amino acids, and can also contain appreciable amounts of inorganic nitrogen, but the content and the exact composition varies according to brand (typical analyses are given in table A-3 in Arditti, 1982).

To summarize, the effect of nitrate is generally negative, that of organic nitrogen positive or neutral. Compounds containing ammonium seem to yield results that are superior to those with nitrates but inferior to those obtained with organic nitrogen sources. Further investigation into the role of nitrogen compounds is clearly desirable, for the purpose of comparing requirements of the seeds with those of the plant or plant/fungus association. In an experiment aimed at minimizing the media used for symbiotic germination of *Liparis lilifolia* I noticed that on a substrate without yeast extract germination was significantly lower than when yeast was present, and was almost as poor as on a water agar (Rasmussen, unpublished data). This suggests that certain components of the yeast extract, possibly some of the amino acids, were essential for germination and could be synthesized neither by the seedlings nor by the fungus, which is a useful reminder that the seedling/fungus association may obtain organic compounds other than carbohydrates from the organic matter that it feeds on.

High levels of exogenous ammonium stimulate the activity of glutamate dehydrogenase in living cells, thus increasing the build-up of proteins. The reduced forms of nitrogen are also known to stimulate the activity of nitrate reductase enzymes, and in this way make absorbed nitrate usable (Gamborg, 1970). There is evidence that orchid seedlings initially are deficient in enzymes associated with nitrogen metabolism, and that this prevents them from making use of exogenous nitrate (Chapter 9).

Other ions

The role of calcium in germination is of particular interest since many European orchids are confined to calcareous soils, and calcium is known to influence cell membrane properties and to interact with various growth regulators, especially cytokinin (George *et al.*, 1988).

Very little information exists on the reaction of seeds to calcium, however. When modifying orchid media propagators have tended to reduce the amount of calcium nitrate $(Ca(NO_3)_2)$ (from $1000\,mg\,l^{-1}$ in the classic media to less than $100\,mg$ in some of the newer ones; Appendix A). One series of substrates (Lucke, 1971*a* and modifications) completely lacks inorganic calcium salts; Anderson (1990*a*) added some calcium in the form of calcium pantothenate to Lucke's medium, but otherwise calcium is present only as impurities, as a minor component of some of the complex additives or unintentionally included in small amounts with the water. For instance, Riley (1983) indicated that his substrate was mixed in tap water with a pH of 7.5–8.0. This water could contain an appreciable amount of calcium. Perhaps not surprisingly, low-calcium media have been recommended for North American species which are not generally calciphilous, but Lucke (1971*a*) used them for Mediterranean species of *Ophrys* that naturally grow on almost pure calcareous rock.

I found that germinating seeds of *Dactylorhiza majalis* apparently did not react to calcium ions (Ca^{2+}) in concentrations of between 0.25 and 2.5 mM, but in another experiment in which there was *c.* 40 mM Ca^{2+} in the substrate, the germination percentage was significantly lower than in the controls with only about 1.5 mM Ca^{2+} (unpublished data).

Potassium may in some cases have a stimulatory effect on

germination. When seeds of *Galeola septentrionalis* were allowed to imbibe in potassium chloride (1% aqueous solution of KCl) for 24 days and subsequently sown, germination was significantly higher than when the seeds were soaked in water (Nakamura, 1962). The potassium ions (K^+) were probably the active factor, since potassium dihydrogen phosphate (KH_2PO_4) and several other potassium salts produced similar effects in *Galeola*. Nakamura (1962, 1964) also tested other orchid species such as *Bletilla striata, Goodyera* spp. and *Oreorchis patens*, but they did not germinate at all after soaking in KCl, in contrast to the controls where germination was good. In *Dactylorhiza majalis*, I found a weak positive correlation between germination percentage and concentration of K^+ in the substrate (Fig. 4.10).

Potassium nitrate (KNO_3) is an important stimulant in the germination of many plant seeds, but the effect is usually ascribed to the nitrate ion and the physiological background is unclear (Hartmann & Kester, 1983; Hilhorst & Karssen, 1989). Potassium ions are known to act as a cofactor in starch synthesis (Greenwood, 1989). This could be important for the germinating orchid embryo that builds up starch immediately after the hydrolysis of the reserve proteins (Chapter 6).

In *Dactylorhiza majalis* there was a weak positive correlation between the concentration of chloride (Cl^-) in the substrate and the germination percentage ($r = 0.463$: Rasmussen, unpublished data). Several other anions that have been tested with seeds of *Dactylorhiza maculata* proved to have adverse effects, both on germination itself and on rhizoid formation, in concentrations as low as 0.025 M (Eiberg, 1970).

The effects of individual ions on germination seems to vary with the plant species but in general the results obtained are not spectacular, and any effect is usually adverse. When the mineral salt balance in a nutrient substrate is altered by introducing alternative salts, the consequences are complex. The concentration of one ion cannot be changed without inducing increases and decreases in the concentration of other ions, in particular if osmolarity and pH are to be kept constant. It is thus difficult to assess the effects of individual ions independently. If the seeds are pretreated in simple salt solutions

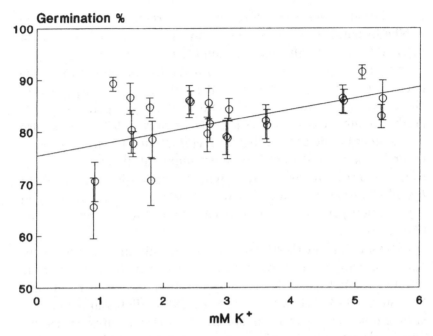

Fig. 4.10. Germination of *Dactylorhiza majalis* on media representing
permutations of Ca^{2+}, K^+, Mg^{2+} and NO_3^-, PO_4^{3-}, SO_4^{2-}, Cl^- within a
macronutrient concentration of 9.3–12.6 mM. The correlation
coefficient with K^+ is 0.515. Germination percentages were recorded
after 6 weeks of incubation in darkness at 20 °C. Means of 16 replicates
with 95% confidence intervals are shown (arcsin transformed data).
For clarity, different media with identical K^+ concentration are slightly
out of alignment. Original data.

before sowing the results can be more easily interpreted, and it is
probably from this type of experiment that we shall obtain clearer
evidence as to the role of mineral ions in germination.

Acidity

Most substrates tend to be somewhat acid (Appendix A), but media
with a neutral or slightly alkaline reaction may be more suitable for
many of the European species, since this conforms better with the
reaction of the soils in their natural habitats. When the culture

substrate has been prepared pH is adjusted to a specified value, but afterwards it is inadvertently changed first by autoclaving and then by the cultured organisms. The absorption of ions affects pH, usually producing a decline (Vacin & Went, 1949), but the interaction with the plant material can be quite complex. For instance, when the living tissue absorbs NH_4^+ it gives off H^+, whereas the absorption of NO_3^- results in the release of OH^- into the substrate (George *et al.*, 1988). The initial proportion of nitrogen ions in the substrate and the capacity of the tissue to utilize them thus affect the direction and extent of changes in the pH during culture.

Changes in pH can presumably be reduced by buffering. Several media in current use include a buffer mixture of basic and acidic phosphates: for instance Vacin & Went medium (Vacin & Went, 1949), modified Burgeff Eg1 (Burgeff, 1936; Borris, 1970), NB 8X medium (Vöth, 1976) and modified Knudson C (Malmgren, 1989a; Anderson, 1990a). However, the effect of this buffering on germination and seedling development has not been analysed.

Some investigators have emphasized the significance of the pH of the substrate for germination in both tropical and extratropical orchid species (e.g. Knudson, 1946; Stoutamire, 1964). The terrestrial species dealt with here certainly vary with respect to pH tolerance and pH preference. *Dactylorhiza maculata* failed to germinate in extremely acid media (pH = 3.0–3.8), optimum germination occurring in media with a pH in the range between 5.0 and 8.0; the related species *D. incarnata* germinated best at a pH of 7.4–8.5 (Eiberg, 1970). I have found that *D. incarnata, D. saccifera, D. elata* and *D. majalis* showed a slight preference for neutral to basic media. According to Arditti and co-workers (1981) the germination of *Epipactis gigantea* was apparently pH tolerant and like other North American species (*Goodyera* spp., *Piperia* and *Platanthera*) could routinely be germinated on media with a pH of 5.0–5.5.

The positive effect of adding activated charcoal to some germination substrates could derive from the fact that it tends to stabilize the pH (Fast, 1980), i.e. to reduce the amount of change in pH that takes place during the period of culture. The addition of charcoal generally also raises the pH according to Eiberg (1970), who also reported that germination in *Anacamptis pyramidalis, Gymnadenia conopsea* and

Orchis militaris in water can be noticeably increased, to 20%, 50% and 60% respectively, if 0.5% activated charcoal is added as compared with 0, 0 and 10% respectively in water that does not contain charcoal. In contrast, on a more complex germination medium, Van Waes (1987*b*) found that activated charcoal consistently reduced germination in a number of species. The negative effects could be caused by adsorption to the coal particles of such media components that might otherwise have stimulated germination. It therefore seems reasonable to expect that the effects of charcoal will be positive in simple and dilute media and the opposite in complex substrates.

Plant growth regulators, vitamins and undefined supplements

The addition of cytokinins such as benzyladenine (BA) or kinetin has been found to improve germination in a number of species that are not usually easy to germinate, for instance *Cypripedium calceolus, C. reginae* and *Epipactis helleborine* (Borris, 1969; Harvais, 1982; Van Waes & Debergh, 1986*b*). The concentrations used range from 0.1 to 0.2 mg BA (Van Waes, 1984) or 0.5 to 5 mg kinetin per litre medium (Borris, 1969; Harvais, 1982; Malmgren, 1989*a*). These species belong to genera where the need for cold stratification has been established (Table 4.2, Fig. 4.4), and it is possible that exogenous cytokinins induce responses resembling those brought on by cold treatment, as they have been shown to do in other plant groups (see the section on cold stratification, p. 47). Cytokinins are also known to counteract the effects of inhibitors, notably abscisic acid (ABA), in many seeds (Hartmann & Kester, 1983) and ABA levels are high at least in *Epipactis helleborine* (van der Kinderen, 1987; see Section 4.2).

Other species such as *Dactylorhiza maculata, Listera ovata* and *Orchis mascula* were found to germinate better in the absence of BA or were unaffected by its presence (Borris & Voigt, 1986; Van Waes & Debergh, 1986*b*).

No positive results have arisen from the addition of GA_3 and other gibberellic acids. GA_3 caused a reduction in germination in *Dactylorhiza maculata* and *Listera ovata*, and produced no effect on *Orchis* mascula, Cypripedium calceolus and Epipactis helleborine (Borris & Voigt, 1986; Van Waes & Debergh, 1986b).

The effect of exogenous ABA can be dramatic; even at concentrations as low as 10^{-7} M it may reduce the germination percentage in Dactylorhiza maculata to half that obtained in the control, and complete failure to germinate occurred at a concentration of 10^{-6} M (Van Waes, 1984).

A survey of the effects of auxins, supplied as indoleacetic acid (IAA), indolebutyric acid (IBA) or naphthaleneacetic acid (NAA), on the seeds of terrestrial orchids (Van Waes, 1984, table II) shows that in eight studies on nine species there was no effect on germination, except in some cases a slight decrease. However, in O. mascula Borris & Voigt (1986) reported a significant increase in germination when 2 ppm NAA was added to the substrate and in Dactylorhiza maculata Eiberg (1970) found that germination was strongly inhibited by 2,3,5-triiodobenzoic acid (TIBA), which has an anti-auxin effect, at concentrations as low as 2 ppm.

Some workers, such as Lucke (1971a) and Fast (1978), have pointed to the importance of the vitamin B complex in orchid substrates, in particular thiamine, nicotinic acid and biotine (Fast, 1983), but this is perhaps mainly because these vitamins influence seedling development (Borris, 1969; Lucke, 1971a). Noggle & Wynd (1943) and Henrikson (1951) indicate that germination in seeds of Cattleya sp. and Thunia sp. was improved when pyridoxine (vitamin B_6) and nicotinic acid were added, although pyridoxine (0.1–0.2 mg l^{-1}) retarded the subsequent growth of the seedlings (Henrikson, 1951). These are, to my knowledge, the only available data on individual vitamins having an effect on germination itself. For instance in Orchis mascula Borris & Voigt (1986) did not observe any reaction to nicotinic acid amide. However, if vitamins are essential to germination the application of yeast extract or peptone to the substrate will amply provide for this. Yeast extract and peptone, both of which are included in many orchid media, contain vitamin B particularly in the form of nicotinic acid (see table A-3 in Arditti 1982). The observation by Nakamura (1964) that peptone is inferior to yeast for germinating Galeola septentrionalis could possibly be connected with the fact that yeast extract has a considerably higher vitamin content than peptone. In addition to nicotinic acid (*c.* 0.03%), yeast extract also contains about 10% amino acids and considerable amounts of phosphorus (*c.* 10%: Arditti, 1982). Yeast extract stimulated germination in *G. septentrionalis* at concentrations up to 5%, but its influence was evident even when the medium contained only 0.1%. A medium with considerable quantities of both peptone and brewer's yeast ('dickbreiiger Brauereihefe') proved successful for germinating the difficult species *Cypripedium calceolus* and *Anacamptis pyramidalis*. On a similar substrate which instead contained extract of yeast there were apparently fewer seedlings, at least as regards *C. calceolus* (Galunder, 1986).

A number of fruit juices and other organic liquids are sometimes added to substrates for orchid culture and an extensive, though hardly complete list is given in Ramin (1983). Unfortunately, as with peptone, yeast extract, oats, etc., these additives are chemically complex and their composition variable. The result, however marked, is difficult to interpret in biological terms. Natural plant juices, which are often added in considerable quantities to orchid substrates, are usually strongly acidic and contain a variety of soluble sugars, some amino acids, vitamins and plant growth regulators (Lucke, 1977) some of which may remain reasonably intact after autoclaving. An interesting property of pineapple juice, which seems to work very well for producing European orchids (Malmgren, 1989a), is that it is strongly reducing; the production of phenolic exudates is therefore very low in such cultures.

5

Fungi

5.1 Isolation

Orchid endophytes can be isolated from naturally occurring protocorms, from roots, and occasionally from rhizomes, tubers or corms.
The advantage of extracting a fungus from protocorm tissue is that
its role in seedling development may almost be taken for granted,
whereas organs of mature plants may house a variety of fungi
(Harvais & Hadley, 1967a; Hadley, 1970b). Furthermore, infection is
extensive in the basal part of the protocorm, and protocorm cells are
often fairly large. In contrast, the cortical cells of roots are long and
narrow and the infection tends to be patchy and irregular. The most
promising places to search for infection in roots are areas of the
cortex below epidermal hairs, especially those close to the root tip
(Bernard, 1909). Infected parts of an orchid root tend to become
faintly yellowish or opaque. The living pelotons, where present, will
be found close to the surface of the organ.

When the rhizome is infected it is usually the outer cortex that is
most suitable for isolation purposes. In *Liparis lilifolia* both the
condensed rhizome and the bases of attached leaves are infected.
Isolation from tubers of orchidoid species has apparently never been
successful, although Fuchs & Ziegenspeck (1925) report that superficial cell layers of globose tubers may be infected as well as the
extremities of palmately divided tubers. Corms are usually without
infection (Chapter 7).

Care must always be taken to remove the soil organisms that are normally found on any underground plant parts. Fungi that occur in the rhizosphere can include strains of *Rhizoctonia* that differ from those that form pelotons inside, and their relevance for the orchid mycorrhiza is not clear (Warcup & Talbot, 1967).

Most investigators surface sterilize the root material to remove contaminants, but with an absorptive structure such as a root there is a risk of killing the endophyte at the same time. Salmia (1988) compared a number of surface sterilization procedures: after 3 minutes in 70% ethanol, pieces of roots were subjected to treatment with either 0.1% mercuric chloride ($HgCl_2$) for 2 minutes, or 1% silver nitrate ($AgNO_3$) for 1 minute, then rinsed in sterile water, after which discs were cut from the mycorrhizal tissue and inoculated on agar. When $HgCl_2$ was used about 60% of the tissue fragments were sterile, but the rest developed colonies of one fungus, presumably an endophyte. $AgNO_3$ completely sterilized more than 90% of the tissue samples, and was thus too effective. Treatment with alcohol alone and with alcohol plus 0.4% sodium hypochlorite (NaOCl) proved, on the other hand, to be so mild that a number of widespread soil organisms appeared as contaminants in the culture dishes. Wolff (1933) sterilized pieces of root or rhizome for 5 seconds in 60% ethanol, and stored them in a sterilized mineral nutrient solution for 24 hours in order to let microorganisms grow out. The procedure was repeated 3 times, after which the mycelium that developed around the plant tissue was assumed to belong to the endophyte.

Although these isolation protocols may eliminate external organisms, there is no guarantee that the isolated fungus has actually formed pelotons in the plant tissue; non-mycorrhizal fungi may also occur within the plant tissues, although they do not form pelotons. Warcup & Talbot (1967) observed hyphae in intercellular spaces that differed from those that formed pelotons in neighbouring cells.

A classic method that my co-workers and I have modified slightly and that works well is simply to rinse the plant material thoroughly, first in tap water with bactericidal soap, then in sterilized water. After that the tissue is cut open under sterile water and the pelotons are teased out of the cells with needles or scalpels and transferred by means of a micropipette through a series of drops of sterile water onto agar (Burgeff, 1909; Costantin & Dufour, 1920; Rasmussen *et al.*, 1990*a*). The technique requires some practice and is time consuming but much labour is saved during later stages since a pure culture is usually obtained at once. A somewhat simpler procedure which requires subsequent transfers so as to obtain a pure culture is to tease out the pelotons in a drop of sterile water in a Petri dish, as above, and then pour over them melted but no longer hot agar medium (Warcup & Talbot, 1967). Mycelia cultivated from individual pelotons will almost certainly constitute a mycorrhizal fungus.

Even when a mycelium is beginning to develop in the culture dishes the final elimination of contaminants may still be a problem. Hadley (1990) recommended dilute substrates for isolation purposes, because the orchid endophytes compete poorly with more aggressive fungi when cultured on nutrient-rich media. Huber (1921) suggested that a brief transfer to a nitrogen-poor substrate would slow down the growth of bacteria. Some investigators have used antibiotics such as streptomycin or novobiocin that have the effect of impeding the development of bacteria so that the hyphae get a chance to grow through the infected patch (Warcup & Talbot, 1967; Clements, 1982a; Rasmussen et al., 1990a; see Appendix A). Novobiocin has the advantage of being autoclavable. Clements et al., (1986) and Mitchell (1989) described a technique in which the inoculate was placed in a window cut in the agar. The hyphae are usually more successful in growing across the bottom of a Petri dish towards the nutrient medium than the bacteria are.

It is probably true of all species of orchids that there are seasons of extensive fungal digestion when it can be difficult to find living hyphae in the tissue, and moreover the mature plants of some species lack infection almost entirely. The living pelotons are loose aggregates of almost transparent hyphae that eventually become digested, forming irregular, often yellowish, clumps. Although the hyphae may appear to be intact when viewed under the microscope, and it may be possible to plasmolyse them, indicating that their membranes are functioning, the excised pelotons often do not grow on agar (Burges, 1939). This problem was first noted by Frank (1891) who interpreted it as the first sign of degeneration and predicted that this modification of the endophyte hyphae would make it impossible to obtain pure cultures from pelotons. Fortunately he has largely been proved wrong, but the problem of extracting viable pelotons remains in a number of orchids, particularly species of Cypripedium (see

Section 13.1), and is possibly caused by a rapid physiological modification of the hyphae once they have entered the host. According to Bernard's recommendations (1909) all pelotons that show the least sign of being digested should be discarded, and a considerable number of pelotons should be extracted from the tissue to ensure successful isolation.

In spite of these difficulties many fungal strains have been isolated from a substantial number of orchids (see Chapter 13), but the status of these strains as endophytes and mycorrhizal fungi should be evaluated in the light of the isolation methods used.

It may be noted that Richardson (1956) investigated a population of *Dactylorhiza fuchsii* growing and proliferating in old clay pits with a very low content of organic matter in the soil (0.12%). The fact that he was not successful in isolating fungi from the plants, and that he observed no endophyte in tissue preparations for the microscope, led him to suggest that no mycorrhiza had been formed. These extraordinary observations have so far not been explained, but they could be related to the season and methods of isolation.

5.2 Identity

Perfect states of *Rhizoctonia*

A diversity of fungi have been found associated with orchids, most of them having been isolated from the roots of adult plants. The largest group, itself a heterogeneous assemblage, comprises mycelia referred to the form genus *Rhizoctonia*, many of which have proved to be compatible with seeds *in vitro*. These mycelia are anamorphic, i.e. sexual sporulation is not present, and they must therefore be distinguished solely on hyphal characters.

These mycelia are characterized by the presence in pure culture of more than one of the following mycelial characters: coloured, septate, relatively wide hyphae; lateral branches arising at an acute angle or at right angles in the distal part of the cell, constricted at the point of origin and with a septum close to the main hypha; the formation of monilioid cells and sclerotia (Fig 5.1*a-b*; Andersen,

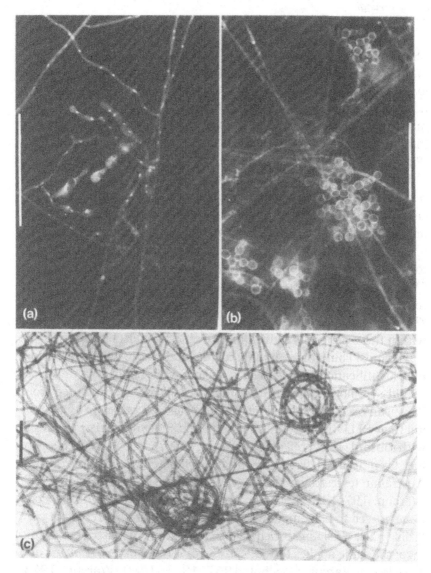

Fig. 5.1. Orchid endophyte structure. (*a*) General aspect showing straight main
hyphae with short lateral branches consisting of monilioid cells.
Endophyte from the root of a young plant of *Listera ovata* found in
Denmark (D 131–1). (*b*) Aggregates of lateral branches with monilioid
cells, the first stage in the formation of sclerotia. Endophyte from the
root of a cultivated specimen of *Orchis sancta* (D 59–2). (*c*) Isolate from
an adult plant of *Galearis spectabilis* growing in Maryland, USA (M26–1)
showing formation of coils in pure culture. Scale bars represent 0.1 mm.
(*a*) and (*b*) by courtesy of T.F. Andersen.

1990*a*). There is usually a clear distinction between the long straight main hyphae and short lateral hyphae that branch repeatedly and form chains of monilioid cells. Sclerotia develop from dense aggregations of the monilioid branch systems. Another characteristic of many of these mycelia is that in pure culture some hyphal tips coil into peloton-like structures almost as they do within the orchid cells (Fig. 5.1*c*; Burgeff, 1909, 1936).

It is usually difficult to match sterile mycelia with corresponding reproductive mycelia (teleomorphs) on which traditional mycological classification is based, but when the formation of reproductive organs is occasionally induced *in vitro* they can be identified as teleomorphic taxa. In this way *Rhizoctonia* strains have been referred to a number of teleomorphic genera (Table 5.1; survey by Andersen, 1990*b*), and among those associated with orchids several have been referred to species within *Tulasnella, Sebacina, Ceratobasidium* and *Thanatephorus*. Key and diagnostic characters to orchid endophytes within these genera are given in Currah & Zelmer (1992).

The most common root and seedling endophyte with a teleomorphic identification is *Tulasnella calospora*, which may be a universal orchid symbiont. Some of the imperfect strains we obtained from European orchids could be referred to *Tulasnella calospora* by matching their RFLP (restriction fragment length polymorphism) patterns with those of perfect reference strains (Andersen, 1990*b*). *T. calospora* has been isolated from *Dactylorhiza purpurella* and a number of Australian terrestrial and epiphytic species (Warcup & Talbot, 1967; Warcup, 1971, 1973, 1981), and strains belonging to this species have proved to be compatible with several holarctic orchids *in vitro* (Table 5.2). *T. asymmetrica* has been found only in Australian orchid taxa; the same applies to *T. allantospora, T. cruciata, T. irregularis* and *T. violacea* (Warcup & Talbot, 1967, 1971, 1980; Warcup, 1973, 1981), but some of them are compatible with seeds *in vitro* and they may prove to be as widespread as *T. calospora* when more of the endophytes from the Northern Hemisphere have been induced to form spores. *T. irregularis* performed reasonably well in symbiosis *in vitro* with seedlings of *Platanthera chlorantha* and *Dactylorhiza majalis* (Fig. 5.3).

Ceratobasidium cornigerum is another common orchid endophyte

Table 5.1. *Distinguishing characters of important groups within the form genus* Rhizoctonia

Anamorph	Character states	Presumed teleomorph
Ascorhizoctonia	Ascomycetous wall; simple pores	*Tricharina*
R. crocorum (type of *Rhizoctonia* s. str.)	Basidiomycetous wall; simple pores	*Helicobasidium purpureum*
Epulorhiza	Basidiomycetous wall; dolipores; parenthesome imperforate; binucleate cells	*Tulasnella + Sebacina*
Ceratorhiza	Basidiomycetous wall; dolipores, parenthesome with 1 or > 1 perforations; binucleate cells	*Ceratobasidium*
Moniliopsis	Basidiomycetous wall; dolipores, parenthesome with 1 or > 1 perforations; multinucleate cells	*Thanatephorus + Waitea*

Notes:
Data from Andersen (1990*b*) and Andersen & Stalpers (1994).
Waitea should probably be excluded from *Moniliopsis* on the basis of sclerotia morphology.

that has been isolated from *Goodyera repens, Dactylorhiza purpurella, Orchis laxiflora* and some Australian orchids (Warcup & Talbot, 1966, 1967; Williamson & Hadley, 1969, 1970; Hadley, 1970*b*; Warcup, 1973, 1981; Muir, 1987). Strains referred to *C. cornigerum* have established seedling mycorrhiza with several orchids *in vitro* (Table 5.2). *C. obscurum* has been isolated from *Amerorchis rotundifolia* as well as from an Australian species (*Cyrtostylis reniformis*), but the compatibility tests *in vitro* have so far been negative (Warcup & Talbot, 1967; Smreciu & Currah, 1989*a*). In a germination test with

Table 5.2. *Seedling mycorrhizas produced in vitro*

Orchid	Symbiont	References
Galeola septentrionalis	*Armillaria mellea*	Terashita (1985)
Spiranthes cernua	*Ceratobasidium* sp.	Warcup (1975)
S. sinensis	*Thanatephorus* sp.	
	Tulasnella sp.	
S. sinensis	*Ceratobasidium cornigerum*	Warcup (1981)
	Thanatephorus cucumeris	
	Tulasnella calospora	
Goodyera repens	*Ceratobasidium cereale* (grass pathogen)	Peterson & Currah (1990)
G. repens	*Ceratobasidium cornigerum*	
	Ceratobasidium sp.	Hadley (1970*b*) (no strains with parasitic tendencies)
	Moniliopsis solani	
	Rhizoctonia sp.	
	Thanatephorus cucumeris	
G. oblongifolia	*Moniliopsis solani* (Rs 10, rice pathogen)	Harvais (1974)
Dactylorhiza incarnata	*Ceratobasidium*	Warcup (1975)
	Thanatephorus sp.	
	Tulasnella sp.	
D. majalis	*Tulasnella calospora*	Rasmussen (unpublished)
D. purpurella	*Ceratobasidium cornigerum*	Williamson & Hadley (1969, 1970)
D. purpurella	*Thanatephorus cucumeris*	Williamson & Hadley (1970)
	Tulasnella calospora	

D. purpurella	*Ceratobasidium* sp. *Moniliopsis solani* *Rhizoctonia* sp. *Thanatephorus sterigmaticus*	} Hadley (1970b) (cf. above)
Coeloglossum viride	*Ceratobasidium* sp. *Thanatephorus cucumeris* *Tulasnella calospora* *Rhizoctonia* sp.	} Hadley (1970b) (cf. above)
Orchis morio	*Ceratobasidium* sp. *Thanatephorus* sp. *Tulasnella* sp.	} Warcup (1975)
O. laxiflora	*Ceratobasidium cornigerum*	Muir (1987)

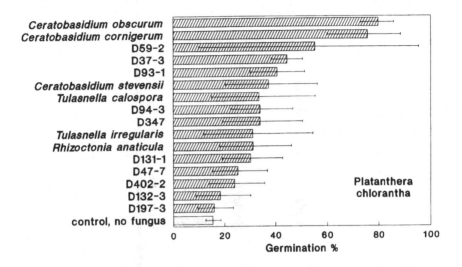

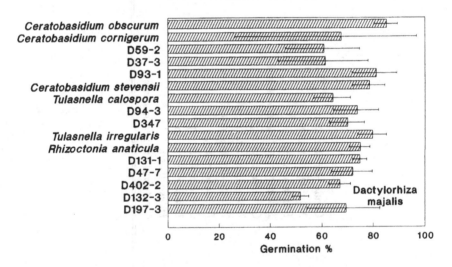

Fig. 5.2. Germination in *Platanthera chlorantha* and *Dactylorhiza majalis* with a range of orchid endophytes and identified reference strains. The choice of strain has a considerable effect on germination percentage in *P. chlorantha*. Germination percentage was recorded after 6 weeks of incubation in darkness at 20 °C on oat medium. Means of 5 replicates are shown, with 95% confidence intervals; arcsin transformed data. Reference strains: *Ceratobasidium obscurum*, UAMH 5443; *C. cornigerum*, CBS 132.82; *C. stevensii*, CBS 477.82; *Tulasnella calospora*, CBS 573.83; *T. irregularis*, CBS 574.83; *Rhizoctonia anaticula*, UAMH 5428. Original data.

Platanthera chlorantha and *Dactylorhiza majalis* the germination percentage was high but the subsequent seedling development was poor (Figs. 5.2 and 5.3). In *Ceratobasidium* there are also some lesser known species so far found only in association with terrestrial and epiphytic Australian orchid species, i.e. *C. angustisporum, C. globisporum, C. sphaerosporum* and *C. papillatum* (Warcup & Talbot, 1971, 1980); a strain of the latter was found to be compatible with seedlings of two tropical epiphytic species (Warcup, 1981). Several isolates from *Dactylorhiza purpurella, Coeloglossum viride* and *Listera ovata* were referred to the genus *Ceratobasidium* without being identified to species level (Warcup & Talbot, 1967).

Most species described in the genus *Thanatephorus* have been associated with orchids. *T. cucumeris* usually occurs as a pathogen, its anamorph being *Rhizoctonia solani*, but has been isolated from orchid roots on several occasions and can enter into a mycorrhizal relationship with *Coeloglossum viride, Dactylorhiza purpurella* and *Goodyera repens* as well as some Australian orchid species (Downie, 1957, 1959*b*; Hadley, 1970*b*; Warcup, 1981). *T. cucumeris* is considered to form only unstable symbiotic relationships with orchids *in vitro* (Hadley & Ong, 1978). *T. gardneri* has been isolated from the Australian orchid *Rhizanthella gardneri* and formed seedling mycorrhiza with this species *in vitro* (Warcup, 1991), and *T. sterigmaticus* found in roots of an Australian terrestrial orchid was compatible with seedlings of *Dactylorhiza purpurella in vitro* (Warcup & Talbot, 1967; Hadley, 1970*b*). A fourth species, *T. orchidicola*, was originally isolated from *Orchis mascula*, but its ability to form seedling mycorrhiza has not been confirmed (Warcup & Talbot, 1966), and this also applies to *T. pennatus*, isolated and described from *Calypso bulbosa* and tested with little success on seeds of several orchid species (Currah, 1987; Smreciu & Currah, 1989*a*).

Sebacina vermifera was first found and described from decaying wood in Germany but has proved to be a widespread orchid endophyte in Australia (Warcup, 1971, 1981). Only on one occasion has it been observed in association with a holarctic orchid, *Platanthera orbiculata*, although the identification was tentative (Currah *et al.*, 1990). *S. vermifera* supports seedling development in Australian species but its activity in the germination of holarctic species has not been established.

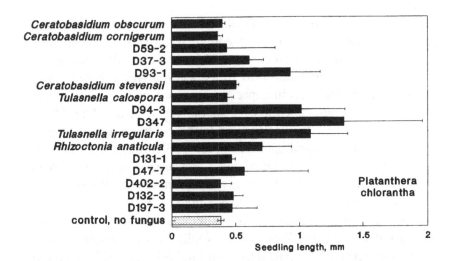

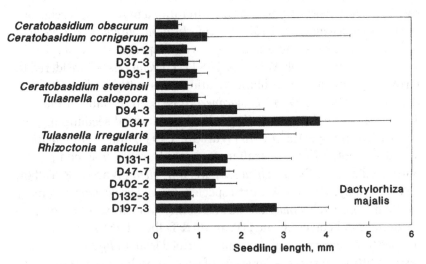

Fig. 5.3. Seedling growth of *Platanthera chlorantha* and *Dactylorhiza majalis* grown
on oat medium with a range of orchid endophytes and identified
reference strains. Maximum length of seedlings attained after incubation
for 6 weeks was recorded. Means of 5 replicates are shown with 95%
confidence intervals. Strains as in Fig. 5.2.

Leptodontidium is a genus of fungi recently described as orchid endophytes. Currah *et al.*, (1990) isolated *L. orchidicola* from *Platanthera orbiculata* and it has previously been found in *Calypso bulbosa*, *Coeloglossum viride*, *Corallorhiza maculata*, *C. trifida*, *Listera borealis* and *Platanthera hyperborea* (Currah *et al.*, 1987*a*). It is thus widespread among orchids, but the exact nature of its relationship with the orchids is not clear.

It is notable that often more than one kind of fungus can be isolated from one orchid species, and some of the fungi that have been isolated from holarctic species are identical with isolates from Australian orchid taxa, suggesting that most species of orchid endophytes have a global distribution.

Sterile mycelia referred to *Rhizoctonia* s.l.

A useful subdivision of *Rhizoctonia* into five groups has been proposed by Moore (1987) based on number of nuclei and hyphal ultrastructure, particularly of septa and their pores. Of these subdivisions, *Ascorhizoctonia* are ascomycetes recognized by the large simple pores and septa consisting of uniformly electron-transparent material, the remainder being basidiomycetes that have three-layered septa. *Rhizoctonia* s. str. including the type species *R. crocorum* also has simple pores whereas *Epulorhiza*, *Ceratorhiza*, and *Moniliopsis* have dolipores. These three genera are distinguished on the number of nuclei and the structure of the parenthesome of the septal pores (Table 5.1).

Ascorhizoctonia includes *R. versicolor*, which was originally isolated from *Himantoglossum hircinum* (CBS 701.82). One of the orchid endophytes we isolated in Copenhagen from *Gymnadenia conopsea* var. *densiflora* (D227–3) had simple pores and was apparently an ascomycete (Andersen, 1990*b*). Although ascomycetous strains apparently can be associated with orchids, it is not known whether they can form seedling mycorrhiza. Most orchid isolates are referable to *Ceratorhiza*, *Epulorhiza* and *Moniliopsis*.

It is possible to match imperfect orchid endophytes with teleomorphic taxa by RFLP analysis, which in combination with a comparison of hyphal ultrastructure has confirmed anamorph/teleomorph correspondence between *Ceratorhiza* and *Ceratobasidium*, between *Epulorhiza* and *Tulasnella* + *Sebacina* and between *Moniliopsis* p.p. and *Thanatephorus* (Andersen, 1990*b*) – the four genera that are of most interest as far as orchid endophytes are concerned. Andersen (1990*b*) suggested that sterile mycelia belonging to *Tulasnella* and *Sebacina* differ from each other in details of the parenthesomes, those of *Tulasnella* often having recurved margins and containing a central electron-transparent layer whereas those of *Sebacina* have straight margins and an electron-dense core. These observations were based on three species of *Tulasnella* and three of *Sebacina* and their diagnostic value was supported by illustrations in all references studied (Andersen, 1990*b*).

Specific distinction within *Rhizoctonia* s. l.

Below the level of the *Rhizoctonia* subdivisions it is extremely difficult to refer sterile mycelia to a taxon. First, the few characters that have been used for distinguishing the published taxa tend to vary somewhat in hyphae within the same mycelium, and the hyphae become modified as the mycelium ages. Moreover, the mycelial structure is influenced by the composition of the substrate (Huber, 1921; Saksena & Vaartaja, 1961; Harvais & Hadley, 1967*a*; Andersen, 1990*a*). Secondly, most specific names, even those that are in frequent use, are illegitimate since there is neither a selected nomenclatural type specimen nor a Latin diagnosis (Andersen & Stalpers, 1994). Of 119 epithets published within *Rhizoctonia* s.l., 48 were found to constitute *nomina nuda*, 30 to belong to taxa that should be excluded from *Rhizoctonia* and 9 to be designated taxa whose identity is uncertain since there is no extant type material and the diagnoses are vague. Of the remaining names, 25 were regarded as taxonomic or nomenclatural synonyms, thus reducing the number of distinguishable taxa with validly published names to 7.

Only two of these accepted taxa are known to have association with orchids: *R. anaticula* and *R. solani* (= *Moniliopsis solani*; although *R. solani* has no extant nomenclatural type specimen it is a well-known and identifiable entity with *Thanatephorus cucumeris* as the corresponding teleomorph). A few other orchid endophyte names could perhaps be saved through neotypification, but in the main the description and illustrations do not even permit a qualified guess as to the original application of the name.

A number of these names have been in frequent use, but may well have been used for different taxonomic entities at different times. Names such as *R. goodyerae-repentis* Costantin & Dufour, *R. moni-lioides* J.T. Curtis and all Burgeff's and Bernard's names lack a nomenclatural basis. For instance, there is no extant type specimen of *Rhizoctonia repens* nor has any authentic culture been preserved and the original description (Bernard, 1909) is inadequate for identifying an anamorphic mycelium (Andersen, 1990a; Andersen & Stalpers, 1994). Thus the repeated association of *R. repens* with the teleomorphic taxon *Tulasnella calospora* is actually based on the tentative identification of mycelia believed to resemble Bernard's original strains.

Since *Rhizoctonia* undoubtedly comprises more than two orchid endophyte taxa, the whole group is greatly in need of taxonomic revision making use of the hyphal characters available with modern methods of analysis. These include ultrastructure, DNA analyses and anastomosis testing, the last having been used successfully for grouping unidentified strains of *Rhizoctonia* by Ramsay *et al.*, (1987). The confused state of the taxonomy should make it clear that any discussion of specificity based on tentatively named strains is futile (Curtis, 1939; Tokunaga & Nakagawa, 1974; Nishikawa & Ui, 1976).

Endophytes outside of *Rhizoctonia*

Orchid tissues, mainly from the non-photosynthetic species, occasio-nally contain endophytes that cannot be included in *Rhizoctonia*. *Galeola septentrionalis* always seems to be infected with hyphae that can be referred to the teleomorphic genus of *Armillaria* (Terashita, 1985; Terashita & Chuman, 1987). The only successful isolation from *Epipogium aphyllum* yielded a mycelium which had clamp connections between the hyphal cells and did not belong in *Rhizoctonia* (Marcuse, 1902, according to Burgeff, 1909).

In orchid species that are chlorophyll deficient the plants often have a complex infection with more than one kind of endophyte. Both clamp-bearing and *Rhizoctonia-* like hyphae have been isolated from *Neottia nidus-avis* and *Corallorhiza* spp. (see Sections 13.5 and 13.34) and in *Limodorum abortivum* two kinds of clampless hyphae have been observed in the same host cells (Riess & Scrugli, 1987). Although *Calypso bulbosa* produces green leaves, this species seems to comply with the infection pattern of the heterotrophic species; Currah and co-workers (1988) found two different mycelia infecting *C. bulbosa* at the same time, one *Rhizoctonia*-like and one clamp-bearing.

Although it is well established that non-rhizoctoneous mycelia associate with the adult, mainly heterotrophic plants, their significance at the seedling stage is less certain. Mycorrhiza in seedlings of the heterotrophic orchid species has been produced only a few times *in vitro*, but *Armillaria mellea* obviously supports the development of seedlings of *Galeola septentrionalis* in culture (Terashita, 1985). The germinating seeds of *Neottia nidus-avis* that were found in the field by Bernard (1899, 1900) were infected with dark hyphae with clamp connections, and we have isolated a mycelium with clamp connections from seedlings of *Corallorhiza odontorhiza* obtained in field sowings (Rasmussen & Whigham, 1993). Similar strains, obtained from adult plants, have been used for symbiotic germination of *C. odontorhiza in vitro*. A clamp-bearing mycelium also stimulated *in vitro* germination in *Corallorhiza trifida* (Downie, 1943a). These examples demonstrate that orchid endophytes other than those referable to *Rhizoctonia* can supply nourishment to orchid seedlings.

5.3 **Properties of fungi in pure culture**

Because of the taxonomic diversity of orchid endophytes their properties can only be outlined in general terms that apply to some entity within *Rhizoctonia* s.l. (special reference will be made to particular features of the endophytes of heterotrophic orchids). Many of the facts presented below are based on fungal material that was only tentatively identified when it was investigated, and for which no voucher specimens are indicated.

Orchid endophytes are aerobic organisms that rarely extend more than *c*. 10 mm below the surface in agar cultures. When orchid fungi are experimentally deprived of oxygen they usually grow only slowly and eventually die (Burgeff, 1909). However, some strains, such as a *Rhizoctonia*-like isolate from *Neottia nidus-avis*, can be observed growing down into the substrate (Fuchs & Ziegenspeck, 1924*b*), apparently preferring slightly anaerobic conditions.

Although the hyphae produce and secrete a number of hydrolytic enzymes, there is no measurable accumulation of low-molecular-weight organic compounds in the substrate (Burgeff, 1936), so the hyphae presumably absorb all available nutrients in solution. Holländer (1932) came to the same conclusion for several orchid fungi kept in sterile compost, but Burges (1936) detected some soluble carbohydrates in sterilized soil about 3 weeks after inoculation with an endophyte from *Dactylorhiza incarnata*.

Glucose, mannose, arabinose, galactose and xylose are all suitable sources of soluble carbohydrate for the orchid fungi (Holländer, 1932). Moreover, many of them produce enzymes for decomposing polysaccharides such as starch, cellulose, hemicelluloses and pectin (Table 5.3). The breakdown of wood can be demonstrated *in vitro* by adding sterilized wood cubes of known dry weight to the agar, then inoculating and after a period of fungal growth reassessing the dry weight of the wood sample (Holländer, 1932). The production of pectic enzymes was demonstrated by Perombelon & Hadley (1965) and Tsutsui & Tomita (1990) succeeded in developing symbiotic seedlings on a culture of *Rhizoctonia* ('R.706') growing on a pectin substrate.

The non-*Rhizoctonia* symbionts of the heterotrophic species of *Galeola* and *Gastrodia* also produce enzymes that decompose lignin (Holländer, 1932). According to Burgeff (1936) those associated with the green orchids tend to not decompose wood. However, the fact that seedlings of *Tipularia discolor* occur in large numbers on woody debris suggests that some green species associate with wood

Table 5.3. *Some physiological properties of orchid endophytes*

Enzyme production
Cellulase (Burgeff, 1909; Smith, 1966[a]; Nieuwdorp, 1972; Barroso *et al.*,
   1986*a*)
Amylase ('diastase', Burgeff, 1909)
β-Fructofuranosidase ('invertase'; Burgeff, 1909)
α-D-Glucosidase ('maltase'; Burgeff, 1909)
β-D-Glucosidase ('emulsin'; Burgeff, 1909)
Pectinase (Nieuwdorp, 1972)
Endopolygalacturonase + endopolymethylgalacturonase (Hadley &
   Perombelon, 1963)
Polyphenoloxidase (Holländer, 1932; Barroso *et al.*, 1986*a*)
Proteolytic enzymes (Burgeff, 1909)

Production of growth regulators and vitamins
IAA + Indole-ethanol (Barroso *et al.*, 1986*b*)
Ethylene (Hanke & Dollwet, 1976[a])
Nicotinic acid or nicotinic acid amide (Harvais & Pekkala, 1975[a])
Thiamine (Harvais & Pekkala, 1975[a])

Translocation
Phosphorus and carbohydrate through the mycelium and into seedlings
   (Smith, 1966, 1967[a], Purves & Hadley, 1975)

Notes:
IAA, indoleacetic acid.
[a]The study mentions voucher specimens.

decomposers (Rasmussen, 1992*a*). Moreover, a *Rhizoctonia* from the
green orchid *Ophrys lutea* has been found to produce polyphenoloxidase (Pais & Barroso, 1983). Wolff (1933) succeeded in growing a
Rhizoctonia-like endophyte of the heterotrophic *Corallorhiza trifida* on
a substrate containing 1–2% tannin, and as a general rule used 0.1%
tannin in his substrates, but neither Holländer (1932) nor Burgeff
(1936) could grow orchid fungi at these concentrations; the highest
tannin levels their isolates tolerated as a sole carbon source were
below 0.05%. Tannins and phenolic acids occur in high concentrations in humus-rich soils where they form chemically resistant

complexes with proteins (Leake & Read, 1990). It is thus possible that these complexes could constitute an important external source of organic nitrogen for the mycorrhiza. If so, differential capabilities of the fungi to break down such compounds would influence their ecological preferences with respect to humus content in the soil.

Proteolytic enzymes are usually produced by orchid endophytes, and the fungi can make use of nitrogen in a complex organic form (Burgeff, 1909). Harvais & Raitsakas (1975) noted an accumulation of amino acids (ninhydrin-positive compounds) in the hyphae, but detected no release to the growth medium.

Wolff (1927, 1933) presented evidence that endophytes from *Corallorhiza* sp. and two tropical orchids increased the total nitrogen content of the substrate, presumably because they assimilated atmospheric nitrogen. Although there are no obvious sources of misinterpretation in the data that Wolff presented, his results were immediately met with scepticism, and attempts to reproduce them have failed (Holländer, 1932; Blumenfeld, 1935). When orchid endophytes were inoculated into nitrogen-free substrates growth ceased within a few days (Burgeff, 1936).

Many orchid fungi have been successfully cultivated on CM_1 medium that contains a mixture of nitrate and organic nitrogen (as yeast extract; see Appendix A). According to Burgeff (1936) some of the fungi isolated from calciphilous European species did not grow at all if nitrate (NO_3^-) was the sole source of nitrogen. Species of *Ceratobasidium* appear to be tolerant with respect to the nitrogen source, whereas a strain of *Tulasnella calospora* did not grow well on a substrate in which ammonium (NH_4^+) was the sole source of nitrogen, doing far better on amino acids and urea (Hadley & Ong, 1978). Two endophytes from *Arundina graminifolia* grew poorly on NH_4^+ and hardly at all on NO_3^- as sole nitrogen source; growth was strongly increased by the addition of yeast extract or other organic sources of nitrogen (Stephen & Fung, 1971a). Apparently the fungi associated with highly mycotrophic orchids are less capable of using inorganic nitrogen than those of the green species, and have a distinct preference for nitrogen given in the form of albumin, peptone or nucleic acids in fairly high concentrations (Holländer, 1932). The fungi tested by Wolff (1927) preferred glycine ('glycokoll') at a concentration of 0.1% to various other sources of organic nitrogen such as albumin, peptone, leucine, haemoglobin or urea.

The vitamins thiamine and nicotinic acid were secreted by a strain of *Moniliopsis solani*, isolated from *Dactylorhiza purpurella*, when it was grown in pure culture in an unshaken macroelement–sucrose solution (Harvais & Pekkala, 1975). Production was highest when the mycelial growth rates were high, when aeration was restricted (as in plant tissue) and when the substrate contained a considerable proportion of NH_4^+ in relation to NO_3^-. Production of nicotinic acid was also noted by Hijner & Arditti (1973) in a strain of *Rhizoctonia* originating from a species of *Cymbidium*.

Like many other mycorrhizal fungi orchid endophytes are capable of producing some indoleacetic acid (IAA: Ek *et al.*, 1983). After a *Rhizoctonia* isolated from *Ophrys lutea* had been grown in pure culture the presence of IAA and indole-ethanol could be detected in both the mycelium and the substrate (Barroso *et al.*, 1986*b*). Symbiotic protocorms of *Dactylorhiza incarnata* contained 10 times as much auxin and cytokinins as asymbiotic protocorms after 10 days in culture; seedlings that were parasitized by the fungi contained even greater amounts of auxin, but less cytokinins (Beyrle *et al.*, 1991).

Burgeff (1936) found that the substrate became distinctly acidified through the activity of the fungi, but neither Clement (1926) nor Downie (1940) detected any decrease in pH. My colleagues and I have noted no consistent changes in pH as a result of growing *Tulasnella calospora* and *Ceratobasidium cornigerum* in fluid oat medium (imperfect strains identified by RFLP analysis). The outcome may largely depend on the initial composition of the substrate. According to Burgeff (1936) fungi obtained from calciphilous orchids do not thrive on substrates with a low pH and I have also observed this; our strain of *Tulasnella calospora* grew a little faster on the solid oat medium if the pH was set within the basic range, up to a pH of 8.0 (Fig. 5.4).

Some orchid endophytes have vitamin requirements; biotin, thiamine ('aneurine'), folic acid and para-aminobenzoic acid in combination or alone have been found to increase hyphal growth in several isolates from orchids (Vermeulen, 1947; Stephen & Fung, 1971*b*; Hijner & Arditti, 1973; Hadley & Ong, 1978). Since yeast extract is

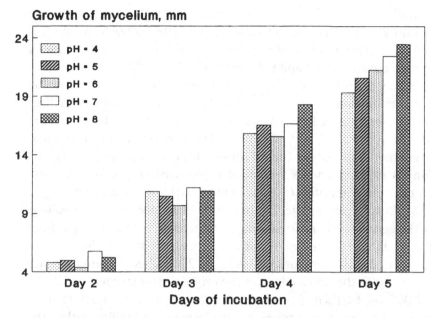

Fig. 5.4. *Tulasnella calospora* (D47–7) growing in pure culture on oat medium adjusted to a range of initial pH. Diameter of mycelium was recorded each day after inoculation. Original data.

often added to substrates used for symbiotic cultures to cater for the possible amino acid or vitamin requirements of the endophytes (see Appendix A; Dijk, 1988*a*), we have little information as to how common such requirements actually are in orchid fungi. However, for endophytes originating from *Corallorhiza, Neottia* and *Limodorum* Blumenfeld (1935) successfully used a medium without any organic additives except glucose. Although some orchid endophytes may have specific vitamin requirements there is no reason to postulate that orchids constitute the source, since such compounds or their precursors could probably be obtained by the fungi from organic debris in their natural substrate (Hadley & Ong, 1978).

In long-term culture of fungi there is often a loss of characters, and the capacity of *Rhizoctonia* spp. to establish orchid mycorrhiza can apparently also be lost or become reduced when endophytes are kept in pure culture (Bernard, 1909; Alexander & Hadley, 1983). Bernard stored his cultures on a salep agar at room temperature in diffuse daylight and found that mycorrhizal capacity was completely lost within 2 to 3 years. Alexander & Hadley (1983) observed a decline within 2 years when cultures were stored at 4 °C on malt agar or potato–dextrose agar and subcultured at 2-monthly intervals.

Bernard (1909) was of the opinion that mycelia with diminishing mycorrhizal activity could be restored to their former capacity if they were co-cultured with seeds, provided of course that they were still able to form some sort of mycorrhiza, and he set up a series of experiments to test his theory. In one experiment, the activity of a fungus isolated from a species of *Paphiopedilum* and kept in pure culture for 6 months (the original strain) was compared with a subculture of similar age that had lived as endophyte with seedlings of the genus *Laelia* during 67 days and had become re-isolated from the seedlings (the re-isolated strain). When seeds of *Laelia* sp. were inoculated with these fungi, 21.4% and 20.5% had germinated after 4 months in the two trials with the original strain, whereas 37.4%, 37.0% and 38.4% had germinated in three trials with the re-isolated strain. Assuming that there is some genetic variability within the mycelium the mycorrhizal seedling will exert selection pressure in favour of genotypes that can survive as endophytes, and it seems possible that growth conditions in a pure culture would tend to favour other genotypes that are better adapted to a life as free saprophytes. However, the existing evidence does not seem convincing, and at least one attempt to repeat Bernard's experiments was unsuccessful (Downie, 1959a).

5.4 Function in germination

The soils of orchid localities are generally characterized as being porous with a well-developed crumb structure, and infection in the plants is often concentrated to roots in the topsoil (see Chapter 13), which suggests that the endophytes prefer a well-aerated soil.

Little is known about the distribution of orchid endophytes in nature, but it is assumed that they are widespread saprophytes, and some are possibly parasites or form other mycorrhizal associations as well. This lack of knowledge is connected with the uncertain identification of many endophytes. When an orchid fungus is induced to sporulate and can be identified as a teleomorph much more information about its occurrence outside orchids becomes accessible. Apparently not only *Rhizoctonia solani (Thanatephorus cucumeris)* but also *Ceratobasidium cornigerum* occur in forms that are pathogenic to a number of plants, as well as forms that are saprophytes, the compatibility with orchids varying. *C. cornigerum* has been found growing on plant debris in the soil of ploughed fields, and as a non-pathogen around the roots of young wheat plants (Downie, 1959*b*; Warcup, 1982). *Tulasnella calospora* has been isolated from mycorrhizal roots of *Pinus banksiana* (Warcup & Talbot, 1967) and *Eucalyptus* roots. Inoculation experiments indicate that it did not form an ectomycorrhiza with the pine tree but probably existed as a rhizosphere organism (Warcup, 1982). The Australian species *Rhizanthella gardneri* is associated with *Thanatephorus gardneri* that forms an ectomycorrhiza on the roots of *Melaleuca uncinata* (Myrtaceae) (Warcup, 1985, 1991). The New Zealand species *Yoania australis*, and *Corallorhiza striata* in North America, also seem to be associated with fungi that form other mycorrhizal relationships with roots of neighbouring woody plants (Campbell, 1970*a,b*).

The production of enzymes by *Rhizoctonia* (Table 5.3 and text above) suggests that they fill a niche as decomposers of a diversity of organic matter. The endophyte of *Goodyera repens* has been found in pine-needle litter even in areas where the orchid did not occur (Downie, 1943*b*). Apparently the distribution of compatible fungi does not restrict the establishment of orchid seedlings to any great extent.

However, locally the distribution of these widespread fungi can probably be quite occasional and patchy. For instance, the endophytes of *Tipularia discolor* presumably only occur in wood debris since seedlings are primarily found on decomposing logs and on stumps (Rasmussen, 1992*a*). Within plots measuring $0.45 \times 0.50 \,\text{m}$ which were used for field sowings of *Goodyera pubescens* and *Corallorhiza odontorhiza* we found that the individual packets of seeds varied greatly with respect to whether mycorrhizal infection was established in the seedlings (Rasmussen & Whigham, 1993), which implies small-scale variation in fungal availability.

It is difficult to evaluate the involvement of fungi in germination in the soil, since germinating seeds have been observed so rarely and since the responses that can be obtained *in vitro* may not be representative for the natural process. Spontaneous seedlings that have been large enough to be found in the soil have always been infected (Ramsbottom, 1929), but whether the association is established at the time of germination or after is not known. Results of our field sowings of five North American species show that the testa of the seeds may rupture, so that by definition germination has occurred, several weeks before the seedlings become infected and before any appreciable growth takes place. Seedlings of *Goodyera pubescens* began to form mycorrhiza about 12 weeks after the onset of germination; in *Galearis spectabilis* we found that there was a delay of at least 21 weeks. In contrast, seeds of *Corallorhiza odontorhiza* apparently became infected at the time of germination or immediately after since no uninfected seedlings were observed (Rasmussen & Whigham, 1993).

While the growth-promoting effect of fungi on the seedlings is an established fact, the effects on germination have not been sufficiently investigated and are often overlooked. Numerous orchid species seem to germinate well *in vitro* both with and without fungi, but a statistical analysis is often required to find out whether the fungi affect the rate or percentage of germination, and parallel symbiotic and asymbiotic experiments should be set up with the same batch of seeds.

In *Goodyera repens* Alexander & Hadley (1983) found no significant differences between the germination percentages obtained with a range of endophytes and those from an asymbiotic control, the germination percentage being high throughout (80–98%). When reviewing his previous observations, Hadley (1990) found no indication that inoculation in itself improved the germination, and this is often the conclusion when the seeds germinate on a nutrient-rich substrate. For instance in the case of *G. repens* the addition of potato extract to the medium yielded about the same germination results as inoculation (Hadley, 1982c).

However, where a dilute substrate or water agar is used the role of fungi in germination is often evident in the sense that the seeds germinate with a symbiotic fungus, but only poorly or not at all in the asymbiotic controls, as in Downie's experiments with *Corallorhiza trifida, Goodyera repens, Listera ovata* and *Platanthera bifolia* using water and simple substrates (Downie, 1940, 1941, 1943a, 1949b). I found that *Liparis lilifolia* did not germinate asymbiotically on any of the simple media that I tested; but when seeds were inoculated, a number of seeds germinated on all the media, even on water agar. With certain species of seeds the stimulatory effect of fungi is also clearly seen on more complex substrates, as exemplified in studies of *Orchis mascula, Platanthera integrilabia, Spiranthes sinensis* and *Epipactis palustris* (Borris & Voigt, 1986; Masuhara & Katsuya, 1989; Zettler & McInnis, 1992; Rasmussen, 1992b). At the opposite end of the spectrum there are species with so few requirements that the germination percentage is high in both asymbiotic and symbiotic culture on water agar, as in *Goodyera pubescens* (Rasmussen, 1994b). Thus in terms of fungal dependency during germination orchid seeds could be characterized according to four criteria:

(a) Germinating well in water or on water agar without infection.
(b) Germinating on water agar only in the presence of a compatible fungus.
(c) Germinating asymbiotically on substrates with specific ingredients.
(d) Germinating symbiotically only if certain substrate requirements are also met.

These criteria are not mutually exclusive, and individual seeds may vary somewhat in their requirements, so that instead of an all-or-none result the effect of more complex germination conditions will be that a greater number of seeds in the batch germinates.

The signals by means of which certain fungi are able to induce or stimulate germination in orchid seeds are not yet known and this presents a challenge to future investigators. In asymbiotic cultures when an increase in germination is achieved by the addition of a chemical factor in the substrate, it is tempting to assume that in nature this substance is supplied by an endophyte. However, the components of a culture medium may initiate processes that in nature would have been activated by other stimuli. For instance, the exogenous plant-growth regulators that are known to affect *in vitro* germination in some orchid species (Chapter 4) could be interfering with processes that in nature are induced by low temperatures and thus be unconnected with the symbiosis.

In seeds that are infected through the suspensor cells, germination may be stimulated during the process of fungal invasion. Downie (1949*a*) observed that non-penetrated embryos of *Goodyera repens* did not germinate. Attempts to identify active substances in fungal extracts or exudates have not met with success. Arditti *et al.*, (1981) were unable to improve germination in *Platanthera stricta* by adding a filtrate from a liquid culture of *Rhizoctonia repens* to the substrate. Similarly, Tsutsui & Tomita (1986) found that neither a macerate nor an extract of a mycelium could induce germination in *Spiranthes sinensis* when added to the growth substrate.

The main problem in detecting an assumed stimulant by means of fungal extracts or macerates lies in the fact that this compound is perhaps produced only as a result of certain outer stimuli, or else is released so slowly from the mycelium that the yield in a macerate does not amount to an effective dose. In an experiment that is more representative of the natural conditions Clements (1988) separated the seeds from the mycelium by inserting a dialysis tube in the culture vessels. Although exudates from the fungi could have passed through this barrier there was no sign that the seeds were affected; only when the dialysis tubes were removed or cut so as to allow the hyphae to approach the seeds did the seeds germinate, implying that the interaction between fungi and seeds requires contact.

The theory that fungi affect germination only indirectly by producing a surplus of soluble sugars and by lowering the pH of the substrate (Knudson, 1925) is untenable since fungi do not usually have this effect on the substrate, as outlined above. Knudson's supposition was partially based on the observation that seeds of *Cattleya* sp. in inoculated cultures did not germinate if they happened to land on the sides of the glass flask instead of on the substrate (Knudson, 1941). He concluded that the fungus did not transmit anything to the embryos which instead were solely dependent on soluble compounds made available by the activity of various micro-organisms. However, Beau (1920*b*) had already succeeded in growing symbiotic seedlings on a glass surface. That the seeds failed to germinate in Knudson's experiment could well have been the result of lack of moisture, as suggested by Smith (1966). It seems doubtful whether the fungus used by Knudson (1929) was compatible and the seedlings in fact were symbiotic; after 2 months the seedlings had attained a maximum length of no more than *c.* 0.5 mm on a 2% starch medium, which represents a rather poor growth rate.

5.5 **Compatibility and specificity**

The question of host specificity is an interesting one in the study of orchid evolution, geographical distribution and ecological competition. Specificity has been the subject of much discussion, but the facts are few because the experimental study of specificity has, until now, been possible only in a number of indirect ways; symbiotic relationships that can be established *in vitro* may have little bearing on what takes place in nature.

Thus it remains to be seen whether the seeds of a given orchid species germinate with the aid of the same fungus or a group of closely related fungi under all environmental conditions and throughout the species' geographical range. Field sowing experiments will in due course tell us which fungi nurture the seedlings in nature, and base the description of the relationship on a broad sample of protocorms from different geographic areas and under different ecological conditions. To date, tests of compatibility have always been carried out on seeds *in vitro* inoculated with cultures of endophytes from adult orchids or spontaneous seedlings. However, there are inherent difficulties with this method as regards the interpretation of results.

Problems related to separating the mycorrhizal fungi from organisms that are found more or less fortuituously in and around the underground structures of adult plants have been dealt with above. Crude isolation procedures such as those where a variety of mycelia are allowed to grow out from large slices of root tissue (e.g. Curtis, 1939) do not provide reliable help in the quest for specific endophytes. Most orchid fungi are difficult to identify, moreover, and there is a risk that phenotypic and genotypic changes will arise in pure culture with time. Finally, as regards the traditional tests for compatibility it remains to be decided how the culture conditions should be defined so as to permit generalization from *in vitro* results to natural conditions. *In vitro* the balance between the organisms can be changed from compatibility to incompatibility simply by altering the composition of the substrate (Burgeff, 1936; Tomita & Tsutsui, 1988; Beyrle *et al.*, 1991) or the temperature (Harvais & Hadley, 1967*b*).

As detailed above the identity of endophytes is especially uncertain in the north-temperate orchids; in germination tests of certain genera of Australian orchids *in vitro* it is possible to demonstrate some degree of preference of some groups for certain fungi (Warcup, 1981; Clements, 1982*a*). When corresponding studies of holarctic taxa were first made the associations were found to be less specific (Table 5.2), as for instance in *Coeloglossum viride, Dactylorhiza incarnata, D. purpurella, Orchis morio* and *Spiranthes cernua*, which all formed seedling mycorrhiza with strains referred to *Tulasnella, Thanatephorus* and *Ceratobasidium* (Hadley, 1970*b*; Warcup, 1975). As more orchid species are studied, however, the picture is changing.

Several investigations have shown that strains differ in their ability to increase the *in vitro* germination in samples from the same seed batch. My colleagues and I have tested seeds of *Dactylorhiza majalis* and *Platanthera chlorantha*, two readily germinating species, with a number of fungi on oat medium and found that the strains varied considerably in their effect. While in the asymbiotic control about 15.5% of the seeds germinated, the symbiotic responses of *Platanthera chlorantha* ranged up to 80%, and many of them differed significantly from the control (Fig. 5.2). The strains performed much more uniformly with the seeds of *Dactylorhiza majalis* on the same medium (Fig. 5.2). This shows that germination is stimulated by various mycorrhizal strains of fungi even in species that are capable of germinating in water, and that the effect is a characteristic of the strain/seed combination, possibly in interaction with factors of the substrate. Warcup (1973) previously reached the same conclusion in a study of Australian taxa. In another study, I inoculated seeds of *Liparis lilifolia* with eight fungal strains isolated from several orchid species all growing within the same forest. The two strains isolated from *L. lilifolia* itself and from *Aplectrum hyemale* were the only ones that increased germination significantly (Rasmussen, in preparation).

When specificity is tested *in vitro* it is also necessary to decide at what point in the course of seedling development the association can be deemed successful. As Warcup (1973) pointed out performance during germination gives no precise indication of the success of the combination later in seedling development. For instance, in our study of *Platanthera* and *Dactylorhiza* the maximum growth rates of the seedlings were by no means reached with those strains that stimulated germination in the greatest numbers of seeds (Figs. 5.2 and 5.3; Rasmussen, unpublished data). Specificity is generally lower at the time of germination than it is during seedling development *in vitro* (Muir, 1989), and seems to decrease again as the plant reaches the photoautotrophic phase when many distantly related fungi may be encountered as endophytes (Warcup, 1981; Alexander & Hadley, 1983). It is the establishment of a fast-growing seedling with a stable seedling mycorrhiza, not merely successful germination, that should be regarded as the decisive criterion of compatibility *in vitro*, but this need not imply that the seedling/fungus combination would also be viable and competitive in the field.

Burgeff (1909) distinguished seven seed/fungus relations that have proved to be adequate for describing the full range of interactions, and that with few changes have been adopted by later investigators (Downie, 1959a; Hadley, 1970b). They are:

(1) No infection.
(2) Fungal hyphae enter the suspensor and a few coils may form in the basal part of the embryos, but no growth occurs.
(3) In seeds that are infected, coils form and a slow but normal development of the seedlings takes place, but the infection is restricted and many seeds remain uninfected.
(4) Normal infection. Most seeds become infected, pelotons form in the basal part of seedlings that develop rapidly.
(5) Infection too heavy. Seedlings are infected and break down pelotons, but they develop slowly and no new non-infected rhizoids appear.

(6) Over-infection. The embryos become extensively infected, maintain only a small non-infected portion at the chalazal end and eventually die.

(7) Instant over-infection. The embryo cells rapidly become filled with hyphae and are broken down. No germination can be said to have occurred.

This classification is obviously based on the assumption that there is a balance between host and endophyte in which either participant organism can be 'strong' or 'weak', in combinations ranging from very strong embryos with a weak fungus (1) to very weak embryos with an extremely virulent fungus (7).

However, since the basis for specificity in the relationship is not understood a one-dimensional scale of compatibility may not hold, and it cannot be unreservedly assumed that embryos have a quantifiable capacity to break down hyphae, and that fungi have a similarly measurable one-dimensional degree of virulence. If this were so, we should be able to rank orchid species and strains of fungi according to 'strength' and hence be able to predict successful (i.e. balanced) combinations.

There are many studies on the production of fungicides in organs of adult orchids, but there is less evidence that seeds and seedlings also produce these compounds. Van Waes (1984) observed that slow-growing seedlings released phenolic compounds *in vitro*, especially when seedling growth ceased through lack of macroelements in the substrate, and that there was also an accompanying unidentified exudate that had a fungistatic effect. In symbiotic cultures of *Dactylorhiza majalis* the micropylar tissue in young seedlings accumulates a substance that reacts positively to tests for tannins (Rasmussen, 1990), and in this tissue invading hyphae do not form pelotons (Fig. 5.5*a*) although infection in other parts of the seedling later produces a regular mycorrhiza. It is notable that hyphae generally do not spread to cells that contain tannins (Ramsbottom, 1923), but the hyphae also seem to avoid a number of other types of cells that have no perceptible tannin content, i.e. epidermal cells, photosynthetic tissue, bundle sheath cells and idioblasts with crystals, raphides and slime (Chapter 8).

5.6 **Pathways of infection**

No chemotaxis has been observed between orchid and hyphae (Burgeff, 1943; Williamson & Hadley, 1970). When a mycelium grows out from an inoculate on an agar surface on which seeds have been sown, there is no apparent distortion of the radial growth pattern that might suggest chemotactic attraction, even when the agar substrate is deficient in nutrients or is a water agar (Rasmussen, unpublished data).

Orchids are infected by means of solitary hyphae that pass through the orchid cell wall by simple penetration, usually with a slight constriction of the hypha (Fig. 5.5*a*, *c*, *d*). Although some fungi associated with orchids will form infection cushions when they invade other plant hosts they do not do so with orchids, and there is very little hyphal growth over the epidermal surface if the infection is compatible (Williamson & Hadley, 1970; see also illustration in Peterson & Currah, 1990). This distinguishes orchid mycorrhiza from other types of endomycorrhiza where close and extensive contact between the host surface and hyphae of the compatible fungus seems to be the rule (Duddridge, 1985). However, orchid seedlings that are grown at temperatures above optimum have a low growth rate and may develop a fungal mantle of densely interwoven hyphae (Rasmussen *et al.*, 1990*a*).

All types of controlled infection require that the host tissue is imbibed (Burgeff, 1936). In other words, the external conditions and the physiological state of the seeds must allow the uptake of water before mycorrhizal infection is possible. Unimbibed embryos will not become infected unless by pathogenic invasion. Such attacks can be observed *in vitro* with fungal strains that are compatible with most of the seeds, so it is possible that the seeds in question are either defective or dead before infection.

When seeds are immersed in a solution of a stain they absorb the liquid entirely through the suspensor (Treub, 1879; Burgeff, 1936), indicating that most of the embryo surface is non-absorbent, the cuticle being continuous except in the region of the suspensor. The cuticle of the rhizoids appears to be fragmented (Fig. 5.5*b*). Since the

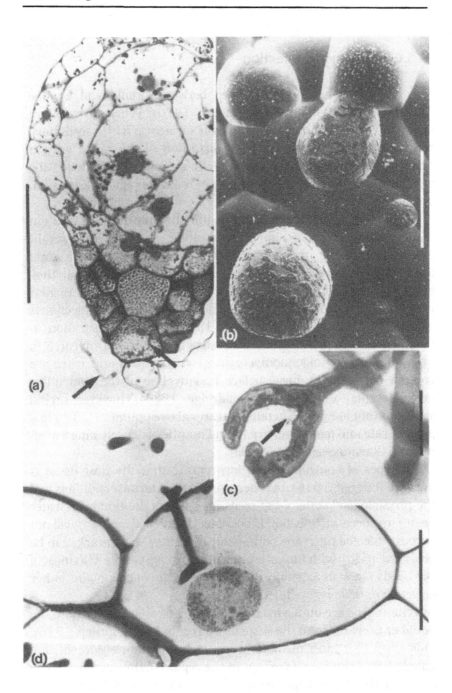

infecting hyphae enter mainly through the suspensor and rhizoids it has been assumed that an absorptive surface with an incomplete cuticle is required if the hyphae are to be able to penetrate (Burgeff, 1936). It should also be noted that cutinases are not among the many enzymes that orchid endophytes have been found to produce (Table 5.3).

Successful infection through the suspensor has been observed a number of times in north-temperate species including *Goodyera repens* (Mollison, 1943) and *Orchis mascula* (Borris & Voigt, 1986). There are also classic observations on tropical species (Bernard, 1904; Ramsbottom, 1929; Burgeff, 1936), and recent observations on species from Australia (Clements, 1988).

In *Bletilla striata* Bernard (1909) observed that infection through the suspensor was checked in the lower part of the embryo. Subsequent vacuolation occurred in the embryo cells, and rhizoids formed, hyphae then invading the seedling through the rhizoids. I observed a similar course of events in *Dactylorhiza majalis* (Rasmussen, 1990), where in some of the seeds hyphae entered through the suspensor, but the embryo cells at the suspensor pole subsequently accumulated tannins which appeared to prevent further invasion (Fig. 5.5*a*; Rasmussen, 1990). The cells of the suspensor region became neither vacuolated nor polyploid, and the hyphae formed no pelotons. The

Fig. 5.5. Infection paths. (*a*) Infection through suspensor (arrows). Note that the cells at the suspensor end of the embryo have not mobilized the reserve nutrients and are not vacuolated; the hyphae do not form pelotons. *Dactylorhiza majalis* in symbiotic culture with *Tulasnella calospora* on oat medium. Semi-thin section stained in toluidine blue. (*b*) Surface view of seedling. Developing rhizoids are seen with fracturing surface layers, possibly the cuticle. Same culture as (*a*), fresh material cryofixed and observed on a cryostage in a JEOL JSM 840A scanning electron microscope. (*c*) Rhizoid tip of *Tipularia discolor* from a plant growing in the field, showing one example of a variety of shapes assumed. The hypha attached (arrow) has probably penetrated the rhizoid wall. Fresh material stained in toluidine blue. (*d*) Surface of *Listera ovata* in symbiotic culture with a compatible *Rhizoctonia* sp. (D 131–1). Hypha penetrates the outer cell wall of the epidermal cells and branches in the vicinity of the host cell nucleus. Semi-thin section, stained with Giemsa. Scale bars represent: (*a*) 0.1 mm, (*b*) 20 μm, (*c*) 50 μm, (*d*) 30 μm.

number of seeds that reacted in this way increased with the length of time of inoculation, but about 40 days after inoculation the seeds tended to lose the visible tannin concentration at the suspensor pole, i.e. at about the time when maximum germination percentage was attained (Rasmussen, unpublished data). Seedlings that had become infected through the suspensor took up water and produced rhizoids above the infected area, and a mycorrhizal infection was eventually established by infection through the rhizoids.

Bernard (1909) suggested that the initial attempt to enter through the suspensor was an important step in the recognition between host and endophyte. However, in the investigation of *Dactylorhiza majalis* described above not all seeds were invaded through the suspensor, but they nevertheless appeared to be equally successful in establishing symbiosis. Seedlings of some species seem to be infected entirely through the rhizoids, for example *Dactylorhiza purpurella* (Williamson & Hadley, 1970), perhaps because they germinate so readily *in vitro* that they have already begun to produce rhizoids when they obtain contact with the fungus, hyphal penetration of the suspensor region apparently not then being necessary. Mutual recognition is more likely to take place during the main infection when the invading hyphae pass close to or through the host nuclei in rhizoid cells and other infected cells (Rasmussen, 1990; Chapter 8).

The rhizoids, each of which is a simple extension of an epidermal cell, are usually formed on the basal part of the protocorm and on roots (Fig. 5.6). On rhizomes and other hirsute plant parts the trichomes are often clustered on hair cushions that develop above an enlarged hypodermal cell with an enlarged nucleus (Burgeff, 1936). This cell divides anticlinally, usually into four daughter cells each of which becomes the basal cell of a unicellular epidermal hair that is often branched. Rhizoids and hairs are mostly simple when they develop *in vitro*, but branching of the tip and distorted shapes can be observed in rhizoids in the soil, as noted on the rhizome in *Corallorhiza trifida* (Jennings & Hanna, 1898) and on seedlings and roots of *Tipularia discolor* (Fig. 5.5c; Clifford, 1899). The rhizoids on *Tipularia* roots are seen to cling to leaf litter and wood debris.

Very little attention has been paid to reports on infection that takes place through the epidermis. In some species the hyphal penetration

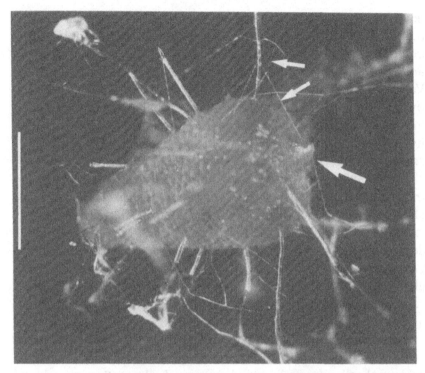

Fig. 5.6. Seedling of *Dactylorhiza majalis*, grown on oat medium with *Tulasnella calospora* (D47–7). Mycotrophic tissue is developing at the suspensor end, the shoot at the other end (large arrow). Hyphae (small arrows) grow along the rhizoids and pass in and out through their surface. Scale bar represents 1 mm. Reproduced from Rasmussen *et al.* (1989) with permission of *Physiologia Plantarum*.

of the epidermis is pathogenic (Williamson & Hadley, 1970). However, my colleagues and I observed that in apparently healthy seedlings of *Listera ovata* growing *in vitro* hyphae had penetrated the epidermis in many places at the base of the protocorms where no rhizoids formed (Fig. 5.5*d*). The fungus was compatible and eventually the infection resulted in the development of leafy plantlets. Species such as *Cypripedium reginae* that are reported to have glabrous protocorms (Harvais, 1973) are perhaps infected through the epidermis in nature, but a smooth surface in seedlings grown *in vitro* is not necessarily a natural phenomenon; rhizoid formation tends to be sparse in asymbiotic cultures, as noted in *Bletilla* (Bernard, 1909) and *Epipactis palustris* (Rasmussen, 1992*b*). The internal pelotons noted in cultivated seedlings of *Epipactis palustris* and *Orchis mascula* before rhizoids had begun to develop (Borris & Voigt, 1986; Rasmussen, 1992*b*) could rely entirely on infection through the suspensor, but it seems more probable that infection also took place through the epidermis. Recently, I have observed that hyphae pass through the epidermis in young seedlings of *Corallorhiza odontorhiza*, *Liparis lilifolia* and *Tipularia discolor* grown *in vitro*.

Neottia nidus-avis, some species of *Platanthera* and *Galeola septentrionalis* are reported to have glabrous roots, and *Epipactis microphylla* almost hairless roots (Irmisch, 1853; Holm, 1904; Niewieczerza-lowna, 1933; Hamada, 1939). Some of these species are highly dependent on mycotrophy. In the roots of *Galeola* hyphae appear to pass through any epidermal cell, and the same may apply to other glabrous roots (see Section 13.8). Roots of *Epipactis helleborine* collected in the field had in fact been infected through the epidermis as well as through the rhizoids (Salmia, 1989*a*).

A common feature of the epidermis in the glabrous roots of *Neottia* and *Platanthera* (Holm, 1904) and seedlings of *Listera* is that the epidermal cells are papillose and are not wholly contiguous, which could create discontinuities in the cuticle that would serve as points of hyphal invasion (Fig. 5.5*d*). In another type of mycorrhiza (VAM) hyphae often penetrate at the junction between two cells in a papillose epidermis (Smith *et al.*, 1990)

Apart from entering through living rhizoids on the roots, rhizomes and protocorms, and in certain instances also through epidermal cells, hyphae can also enter through the secondary surface tissues of roots, i.e. through the dead cells of the velamen and the passage cells of the exodermis.

6

Germination processes

6.1 Structural events in germination

The orchid embryo is a spherical mass of almost uniform cells. In some species the cells at the suspensor pole differ slightly in size from those at the chalazal pole (Fig. 2.1; Ramsbottom, 1929), and a distinct strand of suspensor cells may be attached. The polar regions can also differ somewhat in their content of reserve nutrients, and there can be a structural difference between epidermal cells and internal cells (Harrison, 1977; Manning & Van Staden, 1987; Rasmussen, 1990). Whether or not there is a visible tissue gradient the embryo is apparently physiologically polarized, the suspensor pole invariably differentiating to form mycotrophic tissue and the chalazal pole forming meristematic tissue (Burgeff, 1936).

As the embryo imbibes it increases in circumference rather than in length (Bernard, 1909; Rasmussen, 1990). When germination is observed *in vitro* the testa has been weakened by surface sterilization and it ruptures, usually lengthwise, when the embryo swells. The same appears to happen in the soil, where the testa becomes softened after a long weathering process in which micro-organisms may also be involved (Fig. 6.1a).

Mitotic activity is confined to the cells of the apical meristem at the chalazal end (development of the protocorm: see Chapter 7). Since this meristem gradually broadens, a young seedling is often top-shaped with a tapering mycorrhizal end and a flat meristematic end (Fig. 5.6, p. 111). The meristem first extends the leafless seedling, i.e.

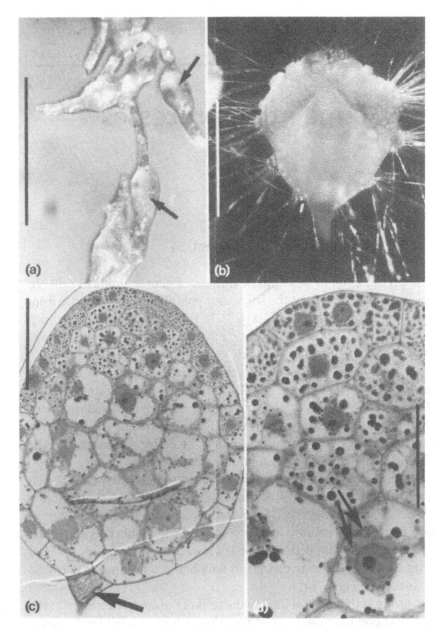

Fig. 6.1. Germination. (*a*) Seeds of *Goodyera pubescens* germinating in the field in May. The testa is rupturing (arrows) but rhizoids have not yet formed. Retrieved after about 6 months in the soil. (*b*) Seedling of *Liparis liliifolia*.

the protocorm, and after some time develops into a traditional shoot meristem with differentiating scale leaves.

The seedling rhizoids are simple extensions of the outer periclinal walls of epidermal cells and the first are usually formed singly, but some seedlings later begin to produce clusters of rhizoids on hair cushions (Fig. 6.1*b*).

6.2 **Histological changes**

While the central cells at the micropylar end of the embryo take up water and vacuolate, their nuclei pass through several endoreduplication cycles reaching ploidy levels of up to 64C in *D. majalis* and 128C in *D. purpurella* (Williamson & Hadley, 1969). Epidermal cells that are to become rhizoid cells also endoreduplicate (Rasmussen, 1990).

Although endoreduplication of DNA and extensive vacuolation are typical features of cells that with time will become infected, these processes are also observed to a lesser degree when seedlings develop without fungi (Alvarez, 1968; Nagl, 1972); in seedlings of *D. purpurella* in asymbiotic cultures, 16C in rhizoids and 64C in cortical cells are the normal maxima as compared with 32C and 128C respectively in symbiotic culture (Williamson & Hadley, 1969).

The seeds germinate and may become infected before any cell

(*6.1 caption cont.*)
The first, succulent scale leaf has been formed and the seedling is covered with hair cushions. Symbiotic culture on oat medium, inoculated with fungus M 31, 9–10 months after sowing. (*c*) Germinating seedling in semi-thin section showing mobilization of stored proteins and vacuolation progressing from the suspensor end of the embryo. As the first sign of infection a hypha is seen within the base of the first rhizoid (arrow). *Dactylorhiza majalis* incubated on oat medium with *Tulasnella calospora*, 11 days after sowing. Stained in PAS–ABB. (*d*) Detail of (*c*) showing coalescing protein vacuoles and newly developed starch grains (arrows), and also the increase in nuclear volume as the cells enlarge. Scale bars represent: (*a*) 1 mm, (*b*) 2 mm, (*c*) 0.1 mm, (*d*) 0.05 mm. (*a*) Reproduced from Rasmussen & Whigham (1993) with permission of the *American Journal of Botany*; (*c*) and (*d*) reproduced from Rasmussen (1990) with permission of *New Phytologist*.

division has taken place (Rasmussen, 1990). When the seedling begins to grow the apical meristem produces new tiers of mycotrophic cells. After a while these cells differentiate so that some, usually the more central ones, become digestion cells while more peripheral and hypodermal layers contain host cells in which the hyphae remain undigested (Burgeff, 1936; Hadley, 1975). The histology and fine structure of infected tissue are known mainly from investigations of roots of adult plants, only rarely of protocorms, and will be dealt with in Chapter 8.

6.3 Reserves and early nutrition

The nutrient reserve consists mainly of protein and lipids (Fig. 1.2*c*, p. 9). In mature seeds there can be small amounts of soluble sugars, sucrose, mannose and a few other kinds of sugars having been detected in South African species (Manning & Van Staden, 1987). Starch is usually absent, which seems to be a characteristic common to most orchids, both holarctic species (Carlson, 1940; Harvais, 1974; Rasmussen, 1990; Richardson *et al.*, 1992) and species from other regions (Knudson, 1929; Nakamura, 1964; Harrison, 1977; Manning & Van Staden, 1987), but there are a few exceptions, such as *Calypso bulbosa* that has prominent starch deposits in the mature embryo cells (Yeung & Law, 1992). The seeds of this species are also unusual in that they lack distinct protein bodies. Starch deposits are apparently also present in mature seeds of *Bletilla striata* and *Cymbidium ensifolium* (Mei-sheng *et al.*, 1985; Shun-xing & Jin-tang, 1990).

The distribution of the nutrient reserves in the mature seed is usually fairly uniform, although epidermal cells may differ slightly from the central cells (Rasmussen, 1990). In some species such as *Cattleya aurantiaca* and × *Laeliocattleya* the nutrients have a polarized distribution, the protein bodies being concentrated to the chalazal end of the embryo which is the last part to take up water (Burgeff, 1936; Harrison, 1977).

The first change to be observed in the nutrient reserve during imbibition is the hydrolysis of protein bodies (Harrison, 1977;

Manning & Van Staden, 1987; Rasmussen, 1990). Under the light microscope the first signs of nutrient mobilization in embryos of *Dactylorhiza majalis* were seen about 10 days after incubation at 20 °C. The content of the proteins bodies first dissolved at the suspensor end, and in a matter of 1–2 days the process gradually extended towards the other pole (Fig. 6.1*c*; Rasmussen, 1990). The cells at the suspensor end enlarged dramatically while vacuolating and accumulating starch (Fig. 6.1*d*).

Whereas protein mobilization is a water-dependent process that occurs in imbibing seeds, the breakdown of the lipid reserve seems to entail a higher level of activity in the embryo cells, and lipolysis has been regarded as a crucial step in the germination of orchid seeds (Manning & Van Staden, 1987). There are apparently no glyoxy-somes present at the onset of germination (Harrison, 1977; Manning & Van Staden, 1987). In the seeds investigated by Manning & Van Staden, external energy in the form of sucrose was required before the embryos could initiate the lipolysis. Glyoxysomes appeared about 4 days later when sucrose was provided. In the absence of external sucrose, the lipid droplets merely coalesced; protein dissolved slowly with little vacuolation of the cells though a few starch grains occasionally appeared.

In contrast, during asymbiotic germination in *Cattleya aurantiaca*, lipolysis was essentially the same whether there was sucrose in the substrate or not, but was slower when no sucrose had been added (Harrison, 1977). Glyoxysomes were never observed, but the lipid reserves became depleted within 27 days on a substrate containing sucrose, and within 65 days on a sucrose-free substrate. The breakdown was apparently related to peculiar cup-shaped mito-chondria that partially enclosed the lipid bodies, suggesting that the products released by lipolysis were respired instead of being directed towards carbohydrate synthesis (Harrison, 1977).

Nevertheless, in *Goodyera oblongifolia* the dissolution of lipids seemed to be associated with the accumulation of starch, and in imbibed seeds that failed to germinate the lipids were gradually exhausted while no starch was deposited (Harvais, 1974).

The accumulation of starch after germination is characteristic of orchids, taking place in particular in asymbiotic seedlings provided with external sources of sugars. In symbiotic seedlings of *Goodyera repens* cultivated by Purves & Hadley (1976) starch was deposited in smaller quantities than in asymbiotic seedlings while the growth rate was considerably higher.

The species showing the most extreme dependence on mycotrophy need an external source of energy before they can even begin to mobilize their own nutrient reserves, so that germination does not take place without such a source. Less extremely specialized species can germinate asymbiotically without the help of external sucrose and are gradually able to use up their nutrient reserve, sometimes even the lipids, and produce small amounts of starch during germination (Burgeff, 1936), but the nutritional status of orchid seedlings immediately after germination differs from species to species. Although the seedlings can remain alive for a shorter or longer time in a state of starvation they do not develop beyond a certain point without an external source of nutrients. Some of the seedlings of *Goodyera pubescens* that we observed in the ground lingered without beginning to grow for up to 36 weeks, when the first indications of seedling mortality became visible (Rasmussen & Whigham, 1993). Cultivated seedlings usually do not reach the stage of foliage leaf development without the addition of nutrients (Harvais, 1974).

The protocorm and shoot tip of epiphytic orchids and some terrestrial species turn green when illuminated, but for several weeks after germination it is impossible to grow them *in vitro* without an external supply of carbohydrates, apparently because photosynthesis does not function (Knudson, 1924; Harrison & Arditti, 1978). In spite of a visible content of chlorophyll, and cultivation in light, asymbiotic seedlings cease to grow and ultimately die if the substrate does not contain some of the simple sugars that they can utilize (Ernst *et al.*, 1970). Initially the seedlings are thus heterotrophic, although a photosynthetic apparatus is eventually established and assimilation of carbon dioxide can make a smaller or larger contribution to seedling nutrition.

The supplement of primary photosynthetic products and mineral salts alone is not always sufficient to sustain a culture of asymbiotic seedlings (Harvais & Hadley, 1967b); many species, for instance seedlings of *Dactylorhiza purpurella*, show a continued requirement for certain amino acids even as green plantlets (Harvais & Raitsakas, 1975 and others; see Chapter 9). When seedlings were supplied with glutamic acid, arginine or ornithine from the substrate they seemed able to produce the remaining amino acids independently. A number of vitamins, such as pantothenic acid, thiamine and pyridoxine, are also often required even when the seedlings are green and are illuminated (Chapter 9; Burgeff, 1934; Harvais, 1973). Not only is photosynthesis inadequate as a source of energy for the seedling, but a number of other essential anabolic pathways prove to be inadequate or blocked while the seedlings are young.

The diversity of mechanisms displayed by these few studies indicates that further investigation is necessary before the mobilization of orchid seed reserves is fully understood and we can explain the role of external nutrients, not only of soluble sugars but also of other organic compounds that are essential.

It is clear that young orchid seedlings, even green ones, are heterotrophic and only occasionally and to a lesser extent make use of photosynthesis when they are subjected to light. Absorption of compounds from the soil or breakdown of the endotrophic hyphae seem to be only two ways in which the seedlings in nature can meet their requirements for external nutrients. Seedlings of *Goodyera pubescens* that we observed in the soil did not develop beyond the point when the testa ruptured unless they had formed mycorrhiza; those of *Galearis spectabilis* did not become infected and remained the size of newly emerged seedlings (Rasmussen & Whigham, 1993; Rasmussen, 1994*a*). In neither case could the activity of surrounding micro-organisms in the soil, or the contents of the soil itself support development of the seedlings.

It can never be proved that under no circumstances are seedlings able to establish in nature without infection. The opposite could be effectively demonstrated if uninfected seedlings in various stages of development could be found outside the laboratory. Pending such evidence the only feasible assumption is that fungi form the main nutritional source in the early seedling stages.

7

Underground organs

All underground parts of terrestrial orchids must either accommodate the endophyte or actively reject it. The mycotrophy of the protocorm is obligate, and most roots are also mycotrophic in varying degrees, but in many species the rhizome loses its mycotrophic function while the plant is still young. Storage organs such as root-stem tubers and corms are usually not infected.

An apparently important aspect of the mycotrophic organs is their longevity; although in the tuberous species some roots function for only about 9 months, orchid roots typically continue to function for a number of years.

7.1 Protocorm

The germinating embryo develops into the protocorm. Unlike other angiosperm seedlings it has no radicle, since the suspensor end of the embryo becomes specialized to form mycotrophic tissue. Neither root cap nor meristem is formed, and the suspensor end of the embryo remains stationary in the soil (Fig. 5.6, p. 111; Fig. 6.1c, p. 114; Fabre, 1856; Stojanow, 1916). As far as is known all orchid seedlings are modified to form mycotrophic tissue in the basal part, even in *Bletilla* which has the most highly differentiated embryo yet observed (Bernard, 1904).

The opposite, i.e. chalazal, end with the functional meristem is solely responsible for all increase in width and length in the seedling;

from this end a condensed or elongated mycorhizome develops which allows the shoot tip with leaf primordia to progress in the soil.

Strictly speaking the protocorm comprises only that part of the seedling axis that develops below the lowermost leafy appendage and corresponds to the radicle and hypocotyl in seedlings of other plants. The boundary between protocorm and mycorhizome is indistinct, however, since the first scale leaves are often inconspicuous and occasionally vestigial. The protocorm stage may be defined in practice as the stage from germination until the seedling has a shoot tip with primordial leaves but no roots, the mycorhizome stage being initiated when the apical meristem elongates and the first roots develop.

Even seedlings that are grown asymbiotically *in vitro* develop most of the histological features pertaining to mycotrophy; a large-celled tissue is formed at the suspensor end and DNA endoreduplication occurs to some extent in the nuclei of these cells (Alvarez, 1968; Williamson & Hadley, 1969; see Chapter 6). There are generally few conducting elements in the protocorm and most elements are phloem. A well-developed endodermis with Casparian strips surrounds the small stele (Fuchs & Ziegenspeck, 1925).

The surface is usually evenly covered with simple rhizoids, but some protocorms are only sparsely hairy or even glabrous. At the upper end of the protocorm and on the mycorhizome the rhizoids often form in clusters, usually from raised cushions on the surface of the protocorm. Seedlings such as those of *Liparis lilifolia* and *Corallorhiza odontorhiza* are glabrous basally and begin to form clusters of rhizoids after a certain amount of growth has taken place (Fig. 6.1b). Simple rhizoids are generally characteristic of roots and the root-like end of the protocorm, most clustered rhizoids being produced on the mycorhizome and stems.

In vitro germination shows that there is individual variation in shape among seedlings in a single batch, but particular characteristics appear to be shared by larger groups of species or genera. Mitchell (1989) noted that protocorms of *Orchis* and *Ophrys* species are usually rounded and isodiametric like onions, in contrast to the elongated, conical turnip-shaped protocorms of *Dactylorhiza* spp. (Fig. 5.6, p. 111). These differences in habit are apparently also seen in protocorms that have been found in soil. Protocorms of *Tipularia discolor* from natural habitats were unusual in being broader than long and more or less lobed owing to the division of the meristem below the lowermost visible scale leaf (Fig. 7.1a, b). The result is a fairly large, broad structure consisting of mycotrophic tissue except at the growing points. One or more of these meristems eventually produces a leafy shoot with one or more roots at its base (Fig. 7.1b). Similar habits possibly occur in the protocorms of *Aplectrum hyemale* and *Calypso bulbosa* (see Sections 13.32 and 13.33), according to published illustrations and descriptions of 'coralloid offsets' (Gillman, 1876; MacDougal, 1899a, b). Asymbiotic protocorms of these species also develop into lobed structures *in vitro*, but seedlings of *Tipularia* grown under the same conditions (Stoutamire, 1983) did not assume the habit that I observed in field situations, so the appearance *in vitro* may not always be typical. There is a report on similar seedling structure in *Govenia liliacea* from Mexico which in several respects resembles *Aplectrum* (Dressler, 1965). Bernard (1909, table IV) also showed illustrations of seedlings of the genus *Vanda* that, in cultivation with certain fungi ('*Rhizoctonia lanuginosa*'), became lobed, and cited Prillieux & Rivière (1856) as observing similar seedlings in *Eulophidium maculatum*.

In other species, branching takes place in the axils of scale leaves, in actual fact from the mycorhizome. In *Galearis spectabilis* the protocorms/mycorhizomes found in the field were long and slender, usually with small lateral shoots of determinate growth (Fig. 7.5a, p. 138), the seedlings differing entirely from those of the European species of *Orchis*. Seedlings from field sowings of *Corallorhiza odontorhiza* (Rasmussen & Whigham, 1993) began to branch from the lowermost scale leaves. Extensive branching of this type gives rise to structures like the mycorhizome system known from this genus and from *Epipogium aphyllum* (see Section 7.3).

The duration of the protocorm and mycorhizome stages in nature is not known for certain in any species. Field sowing experiments indicate that typical protocorms of *Goodyera pubescens* may persist for at least 6 months after germination (Rasmussen & Whigham, 1993). Protocorms/mycorhizomes found in the field sometimes show one or more constrictions; this could be an indication that several phases of

growth are involved, but whether or not the constrictions correspond to annual increments is not known. Like orchid roots the young
seedling is contractile, so that if adjustments of the position in the soil
have taken place there will be transverse folds on the protocorm
surface. Some investigators, primarily Fuchs & Ziegenspeck, have
assumed that the protocorm stage is perennial, but larger protocorms could easily be mistaken for some of the more advanced
underground stages, in which a tuber has replaced the protocorm
(Fig. 7.1c; see also Chapter 10).

7.2 **Root**

A general characteristic of mycorrhizal plants is that they have thick,
sparsely branched roots (Brundrett & Kendrick, 1988). Terrestrial
orchids have an extremely simple root system, each plant developing
only a few, usually unbranched roots (Fig. 7.2a; Niewieczerzalowna,
1933) that typically are fleshy, thick and brittle owing to the strongly
developed cortex and small amount of supporting tissue (Fuchs &
Ziegenspeck, 1925). The roots serve a variety of functions apart from
mycotrophy, i.e. absorption, storage and anchorage; often they are
contractile (Beer, 1863). Anatomically they are usually monostelic,
but more than one stele has been found in the roots of *Platanthera
hachijoensis, P. minor, P. platycorys, P. takedai* and *P. tipuloides* (Ogura,
1953).

Since no primary root is formed, all roots are adventitious. The first
root develops close to the top of the protocorm at the node of one of
the first scale leaves and is exogenous, i.e. contiguous with the cortex
of the protocorm (Fig. 7.1c). Since the fungus can easily colonize the
cortex of the young root from the protocorm cortex, it seems very
likely that the original fungus is passed on to the new root, although
according to Hadley (1982a) the new root is often invaded from
outside.

As a rule a number of the subsequently formed roots also develop
exogenously, but roots produced on older seedlings develop endogenously (Fuchs & Ziegenspeck, 1925), originating from the uninfected
tissues in the centre of the rhizome; they are shielded by their

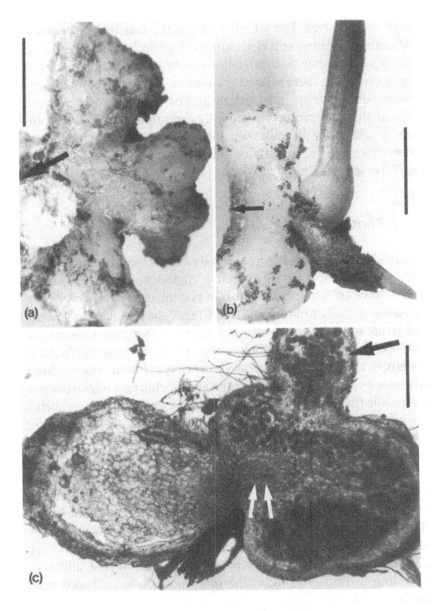

Fig. 7.1. Protocorms and seedlings. (*a*) Lobed and leafless protocorm of *Tipularia discolor*, found in September in Maryland, USA. There are several growth points; the morphological base is at the left (arrow). (*b*) Protocorm of *Tipularia discolor* with three growth points (morphological base at the left: arrow) which has produced bud and root from the

epidermis while they grow out through the cortex. There is thus no continuity of cortical (mycotrophic) tissue between the rhizome and the new root, since epidermal cells usually do not provide a pathway for infection. There is a greater chance that the new root will be colonized by a secondary infection from the soil. The possibility of secondary infection could explain why fungi that are isolated from roots of more mature plants are often unsuitable for germinating the seeds.

The first roots to be formed typically develop before the plant has any leaves above ground and hence absorption of water is not an important function. These roots are usually thicker than the later roots, and have a wide cortex consisting of large cells reminiscent of those of the inner tissue of the protocorm, which is probably an adaptation to mycotrophy. Endogenous roots that appear later have the elongated cortical cells that would normally be found in roots, but infection can nevertheless be extensive (Fig. 7.2b).

Starch is stored in the cortex before infection, but these deposits disappear in the cells that have become infected (Dangeard & Armand, 1898; Burgeff, 1936; Borris et al., 1971). Fuchs & Ziegenspeck (1925) distinguished between roots that are mycotrophic from the beginning, and roots that become mycotrophic after first functioning as storage organs for some time. During my own work with the isolation of endophytes I have often seen tissues containing both starch and dead pelotons, indicating that the storage function can be resumed after the breakdown of the hyphae. The seasonal dynamics of infection and storage in orchid roots is in need of further investigation.

> *(7.1 caption cont.)*
> middle growth point. Found during the same season as (*a*). (*c*) Leafless underground seedling of *Orchis militaris* found in early June on Öland, Sweden. The old part (left) is probably an almost exhausted tuber that has sprouted forming a mycorhizome with one exogenous root (black arrow). Infection with living pelotons is heavy in the younger part. Conducting tissue runs from the old tuber into the mycorhizome with a branch to the root, but it has only been grazed by the section at the base of the younger tissue (white arrows). A new tuber is developing close to the tip of the mycorhizome (not shown). This stage corresponds to Fig. 10.2(*c*), third summer. Hand section, stained in toluidine blue. Scale bars represent: (*a*) and (*b*) 5 mm, (*c*) 1 mm.

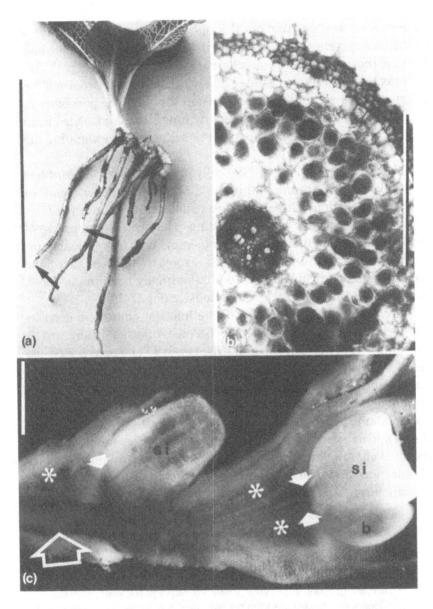

Fig. 7.2. Roots. (*a*) *Goodyera pubescens* showing the entire root system in October with few and unbranched roots. Only the two most recent ones have an active meristem (arrows); the rest are from previous seasons but are still alive. (*b*) Transverse section through root of *Tipularia discolor* showing extensive infection in the cortex, but fungus-free zones around the stele,

As regards the transport of water, Fuchs & Ziegenspeck (1925) considered this function as antagonistic to mycotrophy. They suggested that when the cortical cells are filled with pelotons and their remains the volume of the symplast is considerably reduced, thus interfering with the symplastic transport of water. Furthermore they advanced the theory that a major part of the conducting capacity of the stele is phloem which is required for transporting products from the digestion of the hyphae. These views were formed in the course of an extensive comparative study of European species from a variety of habitats, and are based mainly on a description of the anatomy. In plant roots in general there is a preponderance of xylem over phloem in the stele; in orchid roots the proportions of xylem and phloem vary considerably, but the phloem groups are often well developed or even dominant. In species such as *Listera cordata, Pogonia ophioglossoides, Isotria verticillata, Cleistes divaricata* and *Cephalanthera austinae* there is an abundance of phloem in the roots of mature plants (Holm, 1904). Niewieczerzalowna (1933) estimated the relative amounts of xylem and phloem in roots of European species and found particularly high proportions of phloem in highly mycotrophic species such as *Neottia nidus-avis* (Table 7.1). Fuchs & Ziegenspeck (1925) further noted that in species of *Platanthera, Orchis, Himanthoglossum* and several other genera, the highly mycotrophic first root on the seedling developed considerable amounts of phloem but practically no contiguous xylem. Roots that appeared later, which were less infected, contained more xylem. These observations are consistent with a functional involvement of the phloem in mycotrophy, and indicate that the variation in the xylem/phloem ratios is not merely due to specific differences.

(7.2 caption cont.)
in the multiple epidermis and in hypodermis. March, root 4–6 months old, hand section, stained in safranine. (*c*) Longitudinal section of the rhizome of *Hammarbya paludosa* showing an internal root (large arrow) emerging at the base of the rhizome segment and piercing the previous segment. Every segment is terminated by a fungus-free swollen internode (si) bounded by lignified tissue (white arrows) towards the rest of the rhizome. The renewal bud (*b*) has not yet produced internal root. Mycotrophic tissue is marked with asterisks. Hand section, unstained. Scale bars represent: (*a*) 5 mm, (*b*) 0.4 mm, (*c*) 2 mm.

Table 7.1. *Relationship between degree of infection and amount of xylem in the stele*

Species	Extent of mycorrhiza	No. of xylem elements	Relative area of xylem (%)
Cypripedium calceolus	(+)	49	15.95
Cephalanthera rubra	+ +	56	11.77
Epipactis helleborine	(+)	28	11.54
E. palustris	(+)	24	10.85
E. atrorubens	+	26	10.8
Listera ovata	(+)	24	9.6
L. cordata	+ +	14	7.62
Neottia nidus-avis	+ + +	10	4.93
Goodyera repens	+ + +	32	4.2
Corallorhiza trifida[a]	+ + +	14	4.16
Epipogium aphyllum[a]	+ + +	0	0
Orchis ustulata	+	18	12.66
O. morio	+	23	8.49
O. militaris	+	30	6.73
Coeloglossum viride	+ + +	23	10.83
Traunsteinera globosa	(−)	33	9.64
Gymnadenia conopsea	(+)	44	6.5
Platanthera bifolia	+ + +	27	5.45
P. chlorantha	+ + +	40	6.5
Dactylorhiza maculata	+	28	6
D. majalis	+	26	5.56
D. incarnata	+	22	5.48

Notes:
Data from Niewiezerzalowna (1933, tables V, VI and I combined).
The degree of mycorrhiza was subjectively assessed in the original source as 'rare', 'assez rare/assez faible', 'assez souvent', 'souvent' and 'regulier', here translated to a scale ranging from (−) to + + +. The number of xylem elements was counted in cross-sections of the root stele, and the area of xylem in relation to area of stele was calculated.
[a]Rhizome stele measured in these species.

Mycotrophic roots are nutritionally independent of the rest of the plant, which probably implies that root fragments can remain alive for a long time in the soil. Bud production on roots, which would enable detached roots to give rise to new plants, have been observed in many species (see Chapter 13). Moreover, roots of *Neottia nidus-avis*, *Listera cordata*, *Pogonia ophioglossoides* and possibly *Isotria verticillata* are unusual in that the root tip meristem itself transforms directly into a shoot meristem (Warming, 1874; Holm, 1900; Champagnat, 1971; Rasmussen, 1986). The root cap is shed when a shoot meristem with leaf primordia has been formed underneath, and new roots arise at the nodes of the shoot. If the short and densely infected side roots in *Cephalanthera rubra* become detached by the decay of the long root they can live on and produce a bud at the tip that will grow into a new rhizome and eventually produce leaves. When after flowering the rhizome of *Neottia* dies numerous mycotrophic roots are released, all capable of an independent existence and of producing new plantlets from the root tips. Vegetative reproduction by means of root fragments may be a fairly common phenomenon, since roots often show a strong tendency to break. However, there are few actual observations of this type of vegetative reproduction.

The role of roots in a mycotrophic lifeform can be better evaluated by examining the modifications that characterize those orchid species with little chlorophyll or in which the above-ground structures are of brief duration. The roots are often specialized for different functions, for example in the West Indian species *Wullschlaegelia aphylla*, in which some roots are heavily infected while others are not (Johow, 1885). Other species have an extensive system of long roots (*Galeola* spp.) with short mycotrophic lateral roots, the role of the long roots apparently being to ensure that the lateral roots reach pockets of high fungal activity, and provide communication between these centres of mycophagy (Hamada, 1939). Similar types of specialization in roots can be found in certain autotrophic species, such as species of *Cephalanthera* (Fuchs & Ziegenspeck, 1924b, 1925; see Section 13.3).

Other chlorophyll-deficient species such as *Corallorhiza* spp. and *Epipogium aphyllum* form no roots at all. Apparently the subterranean rhizome system and the hyphae attached to it provide sufficient water for transpiration when the inflorescence eventually emerges above ground. The Australian species *Gastrodia sesamoides* has a cormous rhizome, also devoid of roots (Pate & Dixon, 1982), while *Rhizanthella gardneri* represents further modification in that it produces a rootless cormous rhizome system which displays its flowers in the uppermost levels of the soil, barely protruding above ground (Dixon, 1991).

Species that have foliage leaves and photosynthesize to some extent apparently never entirely omit the development of roots, but the bog-inhabiting species of *Liparis* and its closest relatives develop only one or a few roots annually, a growth pattern which is probably connected with the wet habitat. One of these roots is 'internal', i.e. it grows down into older parts of the rhizome, apparently resorbing nutrients and water from the decaying tissue and carrying over infection into the new rhizome segment (see Section 13.31). In *Hammarbya paludosa* the internal root is the only root developing (Fig. 7.2c).

Roots that arise from a slender rhizome are probably always perennial as in species of *Cypripedium, Epipactis, Listera, Galeola, Goodyera* and *Cleistes* (Harvais, 1974; Fuchs & Ziegenspeck, 1925; Hamada, 1939; Gregg, 1989, Stoutamire, 1990). The number of scars left on the rhizome by the erect part of the annual shoot allows its age and history to be determined, the age of the root in turn corresponding to the age of the rhizome segment to which it is attached. The normal life span of roots is up to 3 years but the bases of damaged roots may persist longer (Möller, 1968, 1987a). Roots on certain cormous rhizomes as in species of *Bletilla* (Masuhara *et al.*, 1988) and *Tipularia* are also perennial. Judging from various illustrations (Correll, 1950) the same applies to *Aplectrum*, but not to *Arethusa, Calopogon* and *Calypso*.

In the species of *Spiranthes* the roots are mycotrophic as well as having a storage function. Reports on the occurrence of fibrous roots in *Spiranthes* may have been based on observations of old roots, since the roots become more slender when the store of nutrients begins to be depleted. The old roots persist for a few years. In contrast, orchidoid species such as *Orchis, Platanthera* and *Dactylorhiza* show pronounced root dimorphism, developing root-tubers as well as slender, mycotrophic roots. The latter are produced in either autumn or spring, but disappear the following summer when the leaves die down, the life span thus being less than 1 year. The tubers have a special anatomy and ontogeny (Section 7.4).

Often terrestrial orchid roots grow horizontally or even upwards towards the soil surface, as shown in analyses of *Dactylorhiza majalis*, *Gymnadenia conopsea* and *Nigritella nigra* (Kutschera & Lichtenegger, 1982). In marsh-inhabiting species such as *Dactylorhiza majalis* and *D. incarnata* the roots and extensions of the tubers curve upwards (Lichtenegger & Kutschera-Mitter, 1991), which could be the result of seasonal fluctuations in soil moisture (Fuchs & Ziegenspeck, 1925) and aeration. Roots of *Orchis papilionacea* and *O. morio* grow horizontally when young and eventually curve downwards in the distal parts (Kutschera & Lichtenegger, 1982).

It has been assumed that the orientation of orchid roots in the soil reflects diversity of function, the roots that grow towards the mineral soil being mainly water conducting whereas the roots that grow horizontally or upwards within the humus and into the leaf litter are most likely to be strongly mycotrophic. In *Tipularia discolor* roots emerge all round the horizontal rhizome, and are oriented in all directions from vertically downward towards the mineral soil to obliquely upwards into the leaf litter, and this species was used for testing whether there is a correlation between orientation and mycotrophic activity (M. Bosela, unpublished data). Contrary to expectation no such correlation was found, the infection intensity varied in an unpredictable manner.

7.3 Rhizome

The protocorm meristem produces a mycorhizome that may be elongated to varying degrees according to the species (Fig. 7.3). Although the term 'mycorhizome' seems to suggest an organ distinct from the rhizome it actually denotes the earliest and most heavily infected part of the rhizome. Particularly in species with a prominent horizontal rhizome, such as in the genera *Cypripedium*, *Cephalanthera*, *Epipactis* and *Goodyera*, there is a gradual transition to a less

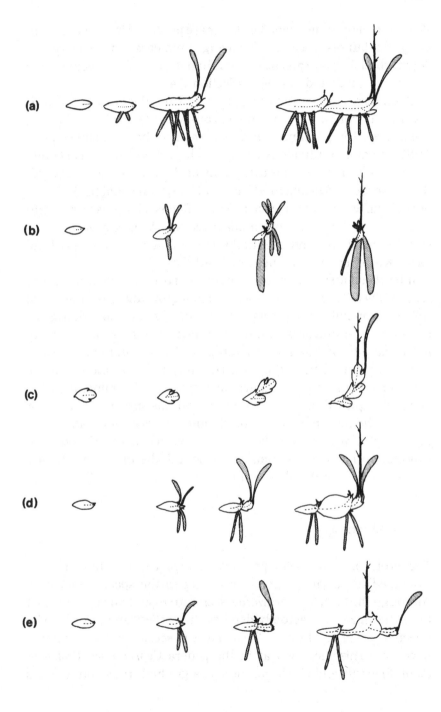

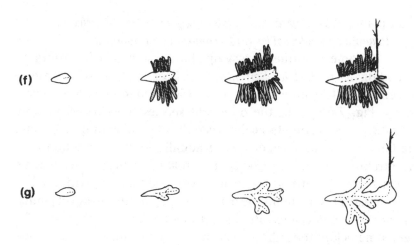

Fig. 7.3. Variations in development of the rhizome, semi-diagrammatic. (*a*) Creeping rhizome with fairly long slender segments, each terminated by an aerial shoot, growth continuing from axillary bud. (*b*) Basically as (*a*), but rhizome segments short, ascending, making roots and leaf rosettes emerge in a cluster. (*c*) Short, ascending rhizome, each segment having a succulent terminal internode below which branching occurs. (*d*) As (*a*) but each rhizome segment showing cormous development. (*e*) As (*d*) but the proximal part of each segment slender. When flowering is separated in time from the leafy season the inflorescence is often produced from one of the axillary buds on the rhizome instead of from the terminal shoot. (*f*) Rhizome with densely clustered roots growing monopodially until flowering after which it withers, since none of the axillary buds develops. (*g*). Rhizome begins to branch while terminal bud is still growing. Examples are given in the text.

heavily infected rhizome or one that is entirely free from infection. In these plants the rhizome remains unbranched until a considerable number of nodes have formed while at the same time it produces adventitious roots (Fig. 7.3*a*). As the development of the roots changes from exogenous to endogenous there is an accompanying decrease in the degree of infection in the cortex of the rhizome (Fuchs & Ziegenspeck, 1925). After the mycotrophic function is transferred from the rhizome to the roots, the general extent of mycotrophic tissue subsequently decreases. However, infection can be maintained in the rhizome for a considerable time, as Holm (1900) observed in adult plants of *Goodyera* spp., *Listera* spp., *Cleistes divaricata*, *Isotria* spp., *Pogonia ophioglossoides* and *Triphora trianthophora*.

The rhizome continues as a sympodium that can be horizontal or ascending as in *Isotria*, *Spiranthes*, *Cleistes*, and possibly *Pogonia* according to illustrations in Correll (1950), and more or less conspicuous (Fig. 7.3*a–e*). In the orchidoid species the rhizome merely constitutes connecting tissue between the old and young tubers and the base of the leafy shoot, and gradually loses its function as a mycotrophic organ. In *Liparis* and related genera the rhizome remains infected throughout its life span, and together with the leaf bases constitutes the main site of mycotrophy in this group of plants. The protocorm develops into a short mycorhizome that persists for some time before the rhizome becomes a sympodium. The rhizome branches below the uppermost internode which is a swollen storage organ (Fig. 7.3*c*) protected from infection from the rhizome below by a barrier of lignified tissue. Infection is then carried from one rhizome segment to the next by means of a root that grows from the younger segments into the older (Fig. 7.2*c*; see Section 13.31). Usually only the two youngest rhizome segments, representing 2 years' growth, are alive at one time.

Many species produce a cormous rhizome. In most species the whole rhizome segment is swollen to form a corm (Fig. 7.3*d*; *Arethusa*, *Calopogon*, *Calypso*, *Tipularia*), but in *Aplectrum hyemale* only the distal part of each segment is cormous (Fig. 7.3*e*). The corms serve mainly as storage organs for nutrients, in particular carbohydrates and water, the reserves being drawn on for shoot initiation and seed set, and to a lesser extent for vegetative reproduction and the replacement of parts damaged by herbivory and disease (Zimmerman & Whigham, 1992). Only rarely can any infection be observed in cormous rhizomes, which implies that infection cannot be transferred from old roots to new ones through the cortex of the rhizome. MacDougal (1899*a*) has described how this transfer is accomplished in *Aplectrum hyemale* (Chapter 10), but apart from this species new roots in the cormous orchids are presumably always infected from the soil.

Many cormous species lose the older corm when the younger one becomes fully developed, so that there is never more than one full-sized corm present on a plant at one time (plates in Correll, 1950). However, in *Tipularia* corms remain alive for at least two seasons and occasionally up to four (Whigham, 1984), and in *Aplectrum* they survive for 1–3 years (MacDougal, 1899a).

The rhizome of the heterotrophic *Neottia nidus-avis* is covered with short roots in which the infection is concentrated (Fig. 7.3f). In contrast, the rhizomes in *Corallorhiza* and *Epipogium* are heavily and permanently infected. These plants never develop any roots but they are densely branched, each rhizome branch being supported by a small scale leaf (Fig. 7.3g). Since the main axis is often bent at the points where lateral shoots emerge, the branching pattern appears to be dichotomous or 'coralloid' (Reinke, 1873). In *Epipogium aphyllum* certain branches become slender and elongated, functioning as stolons (Reinke, 1873). The genus *Hexalectris* appears to have roughly the same habit though the rhizome is less branched (illustrations in Correll, 1950). In accordance with their mycotrophic function the rhizomes of all three genera are characterized by an abundance of phloem compared with xylem, in *Hexalectris* occasionally forming a complete ring around the xylem (Table 7.1; MacDougal, 1899b).

Scale leaves developing on the rhizome are mainly fibrous and have no mycotrophic function. In species of *Liparis*, *Malaxis* and *Hammerbya*, however, the scale leaves are fleshy and their bases heavily infected, while the upper part functions as a storage tissue. The swollen rhizome ensheathed by the thick leaf bases constitutes a bulb in the morphological sense.

7.4 **Tuber**

Tubers are essentially modified roots that have nutrient storage as their overriding function. In species of *Spiranthes* all roots are tuberous when young; these tubers arise like other adventitious roots below a node, as roots generally do in orchids (Rasmussen, 1986). The anatomy resembles that of other roots and in addition to serving as storage organs they are usually heavily infected (Holm, 1904).

In contrast, the tubers formed in the orchidoid orchids such as

Orchis, Platanthera, Dactylorhiza and *Ophrys* have a different anatomy and are almost entirely storage organs, usually being infected only in peripheral extensions and in superficial tissue. This kind of tuber originates as an axillary shoot that produces a basal extension resembling an adventitious root (*Orchis mascula*: Sharman, 1939). It is thus always formed in the axil of a leaf or scale leaf and has an apical bud. Some authors use the term root-stem tuber (or 'tuberoid') on account of the unique anatomy (Dressler, 1981).

Root-stem tubers vary considerably in habit. In the genera *Orchis, Ophrys, Serapias, Anacamptis* and *Herminium* they are globose, whereas in *Gymnadenia, Dactylorhiza, Coeloglossum* and *Nigritella* they are flattened and palmately divided to somewhat resemble a hand. The species of *Platanthera* have spindle-shaped tubers. Tapering extensions develop the anatomy and functions of roots, including mycotrophy (Fuchs & Ziegenspeck, 1925), but the main part is usually not infected. In species such as *Galearis spectabilis, Amerorchis rotundifolia* and many species of *Platanthera* the tubers are slender enough to have a superficial resemblance to roots.

The young root-stem tuber is often stalked, which serves both to adjust the depth of the plant and to colonize new soil at some distance from the parent plant. The ontogeny of the stalk varies, being formed either by an expansion of the basal tissue between the mother stem and the bud (Figs. 7.4*a* and 7.5*a*) or by the bud meristem becoming concave to form a deep cup at the bottom of which the youngest leaf primordia of the bud are located (Figs. 7.4*b* and 7.5*d*). The stalk can also develop by a combination of these two processes. While the basal extension generates a solid stalk and a tuber with an exposed bud (Fig. 7.5*b*), the cup-shaped development of the bud meristem forms a hollow stalk, the bud being in a protected position in the bottom of the tube (Fig. 7.5*e*). Solid stalks are produced in *Dactylorhiza, Platanthera* and *Galearis spectabilis*, tubes in *Aceras anthropophorum, Himantoglossum hircinum, Ophrys holoserica, Orchis mascula, O. militaris, O. pallens* and *Serapias* spp. (among others listed by Fuchs & Ziegenspeck, 1927*c*; Ogura, 1953; Kumazawa, 1958; Rasmussen, unpublished data).

The stalk has been variously named by different authors, some using the term attachment-tube (obviously only in reference to the

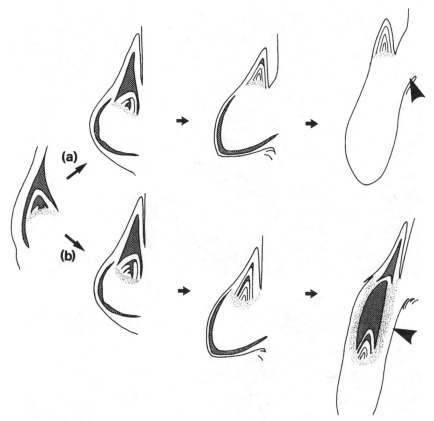

Fig. 7.4. Development of stalk on root-stem tubers, semi-diagrammatic. Tuber develops from an axillary bud which produces an adventitious root. (*a*) The meristem of the bud remains shallowly conical (lightly shaded) and the connecting tissue between bud and supporting axis extends into a massive stalk (arrowhead). (*b*) As above, but the meristematic receptacle develops into a deep cup (lightly shaded) forming a hollow stalk (arrowhead). Drawn to a decreasing scale of magnification from left to right. Based on anatomical analyses in the literature, particularly Kumazawa (1958) and Sharman (1939).

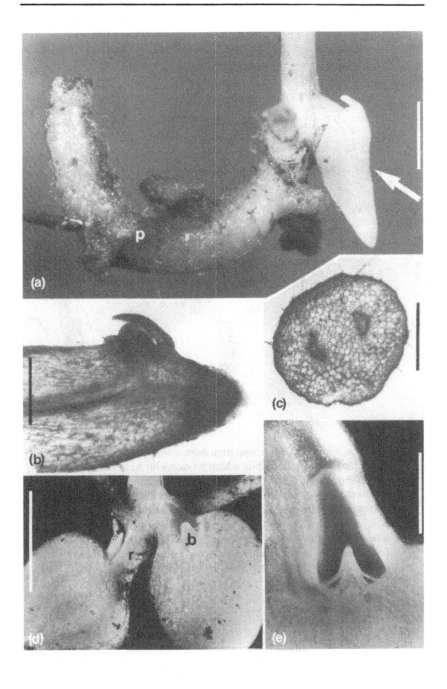

hollow type) or sinker (e.g. Vermeulen, 1947) whereas others, such as Sharman (1939), used the term sinker to denote the tuber as a whole. 'Dropper', 'stolon' and 'root-like outgrowths' are other expressions used (Pate & Dixon, 1982), probably all in reference to the various morphological modifications of stem tissue connecting the tuber with the mother plant.

An unusual anatomical characteristic of the root-stem tuber is that there is usually more than one stele. Slender tubers such as those formed by *Galearis spectabilis* probably never contain more than one or two (Fig. 7.5*c*; White, 1907), but some species can form up to about 50 steles (Stojanow, 1916). There is no direct correlation between the number of steles and the number of finger-like extensions in the palmately shaped tubers (Arber, 1925). Fuchs & Ziegenspeck (1925) presumed that the first tuber to be formed in the young plant contained only one stele, and that more steles would be added annually in those species that develop polystelic tubers, so that the age of the plant could be estimated on this basis. There is no evidence to support this hypothesis. On the contrary, when Stojanow (1916) compared young and old tubers in *Orchis morio*, only 59 of 100 plants had more steles in the daughter tuber than in the parent tuber; in 9 plants the number of steles remained the same, whereas in the remaining 32 plants it decreased from one year to the next. The number of steles could increase by as many as 17 from one year to another, but the likelihood that the number would increase tended to diminish the greater the number of steles in the parent tuber (Fig. 7.6). It seems that when the number of steles in the parent tuber

Fig. 7.5. Root-stem tubers. (*a*) Young tuber (arrow) of *Galearis spectabilis* developing at the base of a leafy shoot in May (note the elongated protocorm (p) with a number of short lateral branches). (*b*) Longitudinal section of a very young tuber with exposed bud as on the right in Fig. 7.4(*a*). (*c*) Transverse section of the distal part of a tuber showing two steles. (*d*) Longitudinal section through a small plant of *Orchis militaris*. The older tuber to the left has produced a short rhizome (r) that has grown into a leafy shoot (not shown) with the young tuber emerging at the base (right). The terminal bud (b) on the young tuber is concealed at the base of a tube, as to the right in Fig. 7.4(*b*). (*e*) Detail of a similar tuber from a larger plant, showing bud and hollow stalk in longitudinal section. Hand sections, (*b*) and (*c*) stained in toluidine blue. Scale bars represent: (*a*) 2 mm, (*b*) 1 mm, (*c*) 2 mm, (*d*) and (*e*) 5 mm.

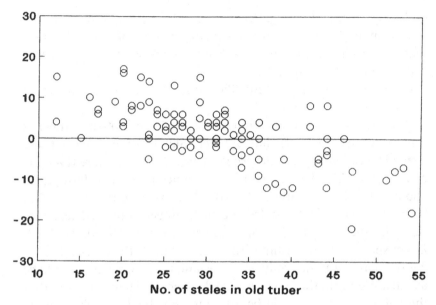

Fig. 7.6. Relation between the number of steles in old and young tuber recorded in 100 plants of *Orchis morio*. Data from Stojanow (1916).

exceeded a maximum of *c.* 35, a smaller daughter tuber was more likely to be produced than one of the same size or larger.

In general no infection is observed in orchidoid tubers, raising the question as to how the plant is able to retain the infection from year to year, since it has no roots during the resting period. Fuchs & Ziegenspeck (1925) observed hyphae in the outermost layers of globose tubers and in the root-like extensions of palmately divided tubers. As regards the latter type the observation was confirmed by Vermeulen (1947), although he did not mention living hyphae in the superficial layers of ovoid tubers from *Orchis* spp. In an analysis of tubers from several species of the Australian genus *Diuris* infection was also observed in the surface layers. Hyphae occurred on, between and within epidermal cells, and in some tubers there were also coils to be found in the outer cortex. These hyphae were identified after sporulation as the well-known orchid symbiont *Tulasnella calospora* (Warcup, 1971).

Almost all investigations on the fungicidal effects of orchid tissue, beginning with Bernard's classic experiments with *Himantoglossum hircinum* (1911), have been carried out on globose tubers (Chapter 8) and have shown that they have a high phytoalexin content. However, the fungicidal effects have always been demonstrated in tissue fragments from the internal parts of the tubers and not from the superficial layers, so that survival of hyphae in the superficial layers is not impossible.

The tubers contain salep, a polysaccharide which has been used medicinally, nowadays mainly for its demulcent and nutritive properties. Salep consists mainly of mucilage and starch (Lawler, 1984); the mucilage is both water retentive and reduces the freezing point of the tissue, thus rendering tubers fairly drought and frost resistant (Jaretzky & Bereck, 1938; Borsos, 1983).

8

Orchid mycorrhiza

There are several early nineteenth century observations of disorganized clumps of material in the cortex of orchid roots, but Reissek
(1847) may have been the first to realize that these clumps were the
remains of fungal hyphae coiled inside the plant cells. He studied the
roots of *Neottia nidus-avis* and observed tissues that were so young
that living hyphae were still present. Reissek also reported these coils
from roots of other European orchid species belonging to *Orchis,
Ophrys* and *Gymnadenia* and from several tropical orchids. Some
years later, 'une matière granulaire' originating from decomposed
hyphal coils was observed in the basal cells of orchid protocorms
(Prillieux & Rivière, 1856). These observations stimulated interest in
orchid mycorrhiza, particularly as regards investigating the species
that display marked chlorophyll deficiency (Prillieux, 1856; Johow,
1885, 1889; Groom, 1895; Janse, 1897; MacDougal, 1899*a, b*;
Magnus, 1900).

In a survey of more than 500 cultivated orchid species, Wahrlich
(1886) managed to find the characteristic fungal infections in every
one of them, and Frank (1891) concluded from observations of the
infected tissue that the hyphae became lysed, and that the nuclear
modification in the infected cells was a sign of the active participation
of the host in this process.

A more complete understanding of the functions of fungi in the life
history of orchids arose from Bernard's germination studies during
the years 1899 to 1909, and from Burgeff's study that primarily dealt
with the fungi (1909). Both of these investigators succeeded in

germinating orchids *in vitro* and in observing the process of infection in embryos and seedlings. Among the early investigators opinions differed as to the nature of the relationship. Bernard (1904, 1909) regarded the dissolution of hyphae within the plant tissues as a defence reaction to keep a parasitic infection at bay. He considered that the benefits for the orchids were indirect, presumably an activation of metabolic pathways that would remain blocked without infection. He gained some adherents, and as late as 1936 Burges described the relationship as a 'controlled pathogenic invasion'. In contrast, Frank (1891) and Burgeff (1909) spoke of a 'digestion' (Verdauung) and 'digestion cells', stating explicitly that there was a nutritional relationship, either mutualistic or beneficial to the orchid alone. 'We [I] postulate a causality between the uptake of carbon compounds and the mycorrhiza' ('Wir postulieren also einen ursächlichen Zusammenhang zwischen der Aufnahme der Kohlenstoffverbindungen und der Verpilzung': Burgeff, 1909, p. 185). That the orchid could live as a parasite on the fungus was also the obvious explanation to most investigators who were studying the chlorophyll-deficient species (e.g. MacDougal, 1899b).

These studies and experiments such as Beau's (1920b), in which seedlings were fed entirely through hyphal connections, supported the view that orchids consume their endophytes. The fact that the seedlings died when the hyphal connection with the nutrient substrate was severed showed that development could not be based on the nutrient reserve of the seed alone. The only possible conclusion was that the mycelium acquired nutrients from its environment and transferred them to the seedling where they were taken up by the plant through lysis of the internal hyphae. This view seems to have been widely accepted until a method was developed for growing orchid seedlings to maturity without fungi on nutrient agar *in vitro* (Knudson, 1922), which initiated new speculations as to the role of the fungi.

Burgeff (1909, p. 78) had succeeded in growing seedlings of × *Laeliocattleya* without fungus on a substrate with 0.33% sucrose and essential minerals. The seedlings grew to about twice the length that could be achieved if they were kept on a substrate without sucrose; in contrast to seedlings without sucrose, those provided with sucrose remained alive, but they did not develop further. This may have been because of the low sucrose concentration, since Knudson (1922) succeeded in keeping seedlings growing on a substrate containing 2% sucrose.

On the basis of his experiments, Knudson concluded that seedlings did not have an absolute requirement for infection; they were clearly able to take up nutrients directly from the substrate and any effect of inoculation could be based merely on the capability of fungi to decompose complex nutrient compounds and thereby make them accessible for absorption. This point of view gave rise to a discussion, at times heated, within scientific circles on the general importance of mycotrophy in the development of orchids (Arditti, 1990). Knudson's approach and that of his followers (e.g. Burges, 1936) was strictly *physiological* and focused on what was technically possible, while most of his opponents were mainly concerned with the *ecological* requirements for seedling development as it might occur in nature (Ramsbottom, 1929 and others). The disagreement was thus based entirely on differences in perspective and is now mainly of historical interest.

The case against mycophagy was unconvincing because the only alternative hypothesis that could explain the nutrition in underground seedlings and in plants with chlorophyll deficiency was that they take up nutrients direct from the soil. However, if protocorms and other underground structures of orchid are to compete effectively with the soil microorganisms for these nutrients, they should be provided with a large absorptive surface. This is not so; rather the opposite is true (Chapter 7). Roots of most other angiosperms seem better equipped with respect to absorptive surface, and yet it has probably never been postulated that direct absorption of soluble carbohydrates from the soil is essential to these plants, even if they can take up soluble sugars through the roots when grown in hydroponic culture.

With reference to the species with chlorophyll-deficient adult stages, MacDougal & Dufrenoy (1944, p. 462) aptly stated that 'parasitic fungi may infect the roots and stems of some orchids, but when the association follows a regulated pattern accompanied by a loss of absorbing organs and photosynthetic apparatus, the assumption that the fungus yields nutritive material . . . is not only allowable but inevitable'.

One of the most convincing discoveries was that carbohydrates labelled with radioactive ^{14}C could be traced *in vitro* from the point and time they were fed as soluble sugars to a mycelium, until they appeared as structural polysaccharides in the symbiotic seedlings (Smith, 1966; Table 5.3, p. 94). Smith showed that nutrients were translocated through hyphae towards infected seedlings. The passage of phosphorus compounds has since been traced by radioactive isotopes from the site of feeding through the external mycelium to the endophytic hyphae (Smith, 1966, 1967; Alexander *et al.*, 1984; Alexander & Hadley, 1985). The transfer of ions and carbohydrates from the pelotons and into the actual seedling tissues is difficult to demonstrate by these methods of analysis but can be assumed, since the infected seedlings increase to such an extent in size and biomass. In an analysis of symbiotic seedlings whose fungi had been fed with radioactive carbon (Smith, 1967), 50% of the carbon was found 160 hours later in the symbiotic seedlings in the insoluble fraction which mainly comprises cell wall material and proteins. This strongly suggests that orchid tissue is built up, whereas translocation in the opposite direction, towards the fungus, has not been demonstrated in spite of several experiments set up to provide such evidence (Hadley & Purves, 1974; Purves & Hadley, 1975). For instance, radioactive carbon fed to the leaves of *Dactylorhiza purpurella* as $^{14}CO_2$ could not be traced to the external mycelium that grew out when the labelled plantlets were transferred to a fresh medium, but their roots leaked small amounts of tracer into the medium, as did asymbiotic controls. The fact that hyphae can grow out from the infected plant on a nutrient-poor substrate has been seen by some investigators as evidence that nutrient transfer can take place from orchid to fungus (Burgeff, 1959; Smith, 1967), but the same effect could be achieved if there were an allocation of resources within the system of living hyphae.

The interpretation of the orchid mycorrhiza that has thus become prevalent is that it represents a source of energy for the plant which either supplements, replaces or alternates with phototrophic nutrition. However, the introduction of asymbiotic propagation methods in the 1920s meant that studies of the mycorrhizal relationship lost their technical purpose and received much less attention than previously. A few investigators have since successfully advanced this field of research, but there has been a widespread disregard of orchid mycorrhiza as insignificant, even among many professionals. This neglect has been a scientific setback which, in spite of the undisputed usefulness of asymbiotic propagation, may also have delayed further development of culture techniques.

8.1 Histology of infected tissues

As a rule mycotrophic orchid tissues show no signs of cell death or accumulation of defence substances. Two histological types of orchid mycorrhiza are recognized: *tolypophagy*, found in the great majority of species, and *ptyophagy*, noted only in a number of highly mycotrophic species such as *Lecanorchis javanica* (Janse, 1897) and several species of *Gastrodia* (Kusano, 1911; Burgeff, 1932).

Ptyophagy seems to be rare outside the tropics and has received little attention in recent years. Jonsson & Nylund (1979) cultivated seedlings of *Bletilla striata in vitro* with a strain of *Favolaschia dybowskyana*, Aphyllophorales, taken from an African orchid and analysed the ultrastructure of the association, which seemed to be ptyophagic. There was a considerable amount of host cell death, however, so the mycorrhizal association may have verged on parasitism. Ptyophagy is characterized by deformation and lysis of the intracellular hyphal tips through which the fungal cell contents are released (Burgeff, 1936), presumably into the interface between the host plasmalemma and the endophyte cell walls. Ptyophagy is presumed to be a continuous process in which the leaking hyphal tips make new materials accessible to the host.

In tolypophagy the endophyte forms well-defined hyphal coils, known as pelotons, in the infected cells before any lysis takes place. Tolypophagy is characterized by successive waves of peloton formation, lysis and reinfection. The hyphal cytoplasm is at all times separated from that of the host cells by an interface consisting of hyphal plasmalemma, hyphal cell wall, interfacial matrix and host plasmalemma.

The mycotrophic tissue is found mainly in the cortex of protocorms, roots and rhizomes; other organs, such as leaf sheaths and tubers, are occasionally infected (Chapter 7). The hyphae are confined to certain cell types with a raised DNA content (Williamson, 1970), apparently being unable to spread to any diploid or mitotic cells or cells containing tannins, raphides, crystals or mucus. Cells with chlorophyll are usually also avoided (Ramsbottom, 1923). This applies, for instance, to the green parts of the roots of epiphytes. Burges (1939) induced greening in roots of *Dactylorhiza incarnata* by exposing one side of them to light, and noted that the green side did not become infected.

A sterile cell sheath is maintained around conducting elements (Figs. 7.2*b* (p. 126) and 8.1). Epidermal cells are also generally free from infection except where the hyphae pass through them in connection with the entry and exit of hyphae, but a more widespread infection has been reported in the epidermis of some species, as in the roots and rhizomes of *Cleistes, Isotria, Pogonia* and *Triphora* (Holm, 1900).

In many studies a clear distinction is observed in the cortical tissue between host cells usually in the outer cortex where the hyphae can survive, and digestion cells where they proliferate and sooner or later are dissolved (e.g. Dangeard & Armand, 1898), but this distinction does not always seem clearcut. It is possible that the zonation is seasonal or varies according to growth conditions, or that it mainly occurs in organs and species with a prolonged mycotrophic function, as suggested by Fuchs & Ziegenspeck (1927*a*). The vacuoles of cells in which lysis takes place have a higher pH than the host cells and epidermal cells (Burgeff, 1936). The entrance of hyphae into a cell is followed by the disappearance of starch from the cell (Dangeard & Armand, 1898; Burgeff, 1909). Behind the meristematic root tip there is a maturation gradient of tissue, beginning with a zone of starch-filled storage cells, then a zone of vacuolated cells without starch where living pelotons occupy the cells, and finally a zone where lysis takes place (Fig. 8.1). The central core consists of the stele

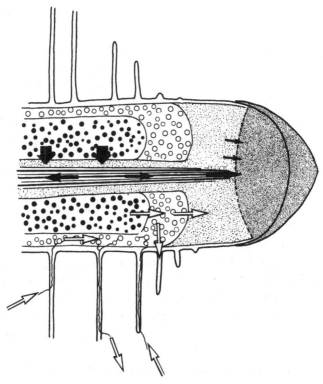

Fig. 8.1. Diagrammatic longitudinal section of a growing mycotrophic root tip
(the tip of a protocorm would be similar except for the structure of the
meristem itself). Behind the meristem (dark shading) is a tissue gradient
of increasing age, the youngest consisting of starch-containing cells
(light shading). Cortical cells contain living pelotons (open circles);
further back fungi are being lysed (filled circles) except in 'host' cells
below the epidermis. The vascular cylinder is free from infection; the
epidermis is not infected except where hyphae pass in and out. Dark
arrows show presumed movement of carbohydrates in solution from
lysed pelotons to deposits in the bundle sheath or to the stele where they
stream towards the meristem or towards the rest of the plant. The
starch deposits in the region immediately behind the meristem may also
contribute to new growth. Open arrows show some infection routes:
from outside through the rhizoid and epidermis into permanently
infected outer cortex and into digestion layers of the inner cortex; from
infected tissue towards the uninfected part of the cortex and towards
rhizoids where they leave the plant. No quantitative information is
intended. Based partially on a description and illustration in Clements
(1988).

and a bundle sheath of storage cells that are not infected. As Burgeff (1936) pointed out a front of starch accumulation and starch mobilization runs ahead of the hyphal front, as the infection spreads from mature tissues towards the young tissue below the root tip. Products from the digestion zone must pass through the zone of newly infected tissue, probably within the conducting tissue, if they are to be utilized for new growth at the tip.

Infecting hyphae pass directly from cell to cell without occupying the intercellular spaces to any great extent. They break through the plant cell wall without fracturing it and are usually slightly constricted where they pass through the wall, often making use of existing pit fields (Fig. 8.2a; Dangeard & Armand, 1898; Burgeff, 1936; Barroso, 1988). A local degradation of the wall around the point of penetration has been noted by Peterson & Currah (1990). The fungi are able to synthesize various wall-degrading enzymes in pure culture (Table 5.3), which suggests that these enzymes could be responsible for the breakdown of host walls (Pais & Barroso, 1990).

Fine structure

Tolypophagic infections have been studied in detail as regards structure as seen in the electron microscope (EM) and cytochemistry, and the picture has proved to be quite similar in heterotrophic orchids and species with green leaves. The investigations have mainly concerned terrestrial European species such as *Dactylorhiza maculata* (Strullu & Gourret, 1974), *D. purpurella* (Hadley *et al.*, 1971; Hadley, 1975), *Goodyera repens, Listera cordata, Epipactis helleborine* (Nieuwdorp, 1972), *Ophrys lutea* (Barroso, 1988; Barroso *et al.*, 1986a), *O. insectifera* (Hofsten, 1973), *Corallorhiza* spp., *Limodorum abortivum* and *Neottia nidus-avis* (Dörr & Kollmann, 1969; Nieuwdorp, 1972; Barmicheva, 1990). A few tropical species, such as the epiphytic *Epidendrum ibaguense* (Dexheimer & Serrigny, 1983) and the terrestrial *Habenaria dentata* (Borris *et al.*, 1971), have also been investigated, but with respect to histological detail no fundamental differences in the the mycorrhiza of different lifeforms and habitats have been pointed out.

Most studies are based on root material from adult plants found in the field, some on cultivated and inoculated seedlings, but little attention has been paid to the dynamics in the timing and location of infection. Thus little is known about the progressive changes in the interaction at the ultrastructural level, nor is it known to what extent rhizoid cells and cortical host cells differ ultrastructurally from digestion cells. Most of the micrographs in the literature presumably show digestion cells only, and little information is available as to how digestion products are exported to the conducting tissue, or by what means neighbouring cells remain free from infection. Barmicheva (1990) described an aggregation of electron-dense material in the periplasmic space of storage cells that could be an indication of a local defence reaction to infection, and in the endodermal cells of infected roots she noted ingrowths of the radial wall close to the Casparian strip, an observation that suggests great transport activity between apoplast and symplast.

Digestion cell membrane system

The plasmalemma of the plant cell recedes in front of the invading hypha, forming a vast sequestration membrane as the hypha branches and curls up to form a dense peloton that may almost fill the cell lumen, traversing the plant cytoplasm and the central vacuole (Dörr & Kollmann, 1969; Hadley *et al.*, 1971; Nieuwdorp, 1972). In general the contact surface between hyphae and plant cell membrane is extensive. Each hypha is sheathed by a tube of the host plasmalemma, which in turn is covered by a layer of plant cytoplasm. Where the hyphal strands traverse the vacuole, they are further sheathed by an extension of the tonoplast, making the distance between the hyphal wall and the lumen of the vacuole of the plant cell as short as possible (Fig. 8.2*b*).

Thin sections may give the impression that the central vacuole is split into smaller vacuoles, the sectioned coil of hyphae giving the appearance of an abundance of yeast like single cells (Hofsten, 1973). In a three-dimensional view, however, the apparently separate hyphal segments are in fact connected and most of the small vacuoles are probably connected in a labyrinthine system into which the central vacuole is transformed, although some of them seem to derive from dilated portions of the endoplasmatic reticulum (ER) (Borris *et al.*, 1971).

Evidently the formation of a new peloton is associated with a high growth rate in the hyphae as well as in the membrane system of the plant cell. This should appear in EM as a profusion of cortical vesicles both within the tip of the growing hypha ('Spitzenkörper') and inside the sheathing host plasmalemma, but this developmental stage has apparently not been observed in any of the EM studies.

The plant plasmalemma facing the mature peloton appears wavy and irregular (Borris *et al.*, 1971; Nieuwdorp, 1972; Strullu & Gourret, 1974), which may indicate that a large number of vesicles either fuse with the membrane or are in the process of being formed there. This irregularity can, however, also be an artifact resulting from aldehyde fixation. Barroso & Pais (1987) observed numerous coated pits along the host plasmalemma, and large numbers of coated vesicles and partially coated ER were found in the cortical cytoplasm, but it was not possible to determine whether the vesicles were moving in the direction of the plasmalemma or into the cell. The first possibility would imply a secretion from the plant cell, perhaps of enzymes that could take part in the breakdown of the hyphae, the second would represent an uptake of material from the intercellular space.

Digestion cell cytoplasm

Cells that are not infected contain more dictyosomes than the infected cells and large ethioplasts (Borris *et al.*, 1971). Infected cells have poorly developed plastids but contain numerous mitochondria, ribosomes and profusely developed ER, in some places with dilated cisternae (Dörr & Kollmann, 1969; Barroso *et al.*, 1986a). The ER is associated with an intensive production of hydrolytic enzymes, in particular acid phosphatases that increase in concentration during infection (Williamson, 1973; Barroso, 1988). Acid phosphatases occur in the lumen of the rough ER and show a high concentration

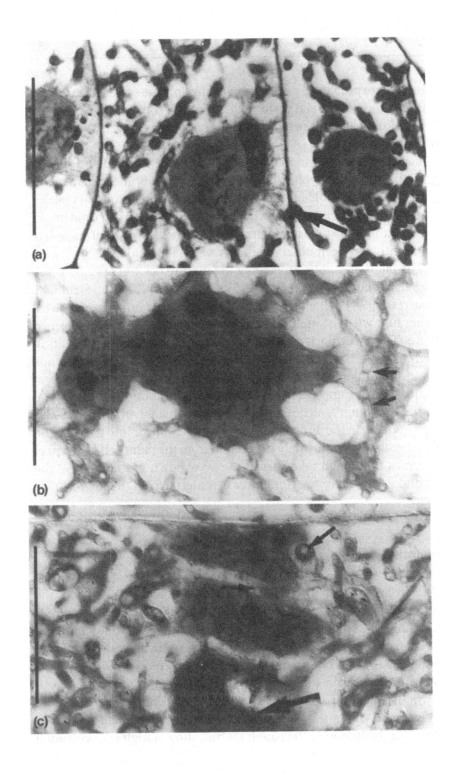

immediately outside the host plasmalemma and in the interfacial space while the hyphae are still alive; when the hyphae have begun to degenerate these enzymes can also be detected inside the hyphal walls (Dexheimer & Serrigny, 1983; Serrigny & Dexheimer, 1986). The occurrence of acid phosphatases in the lumen of decaying hyphae could indicate that there is an element of autolysis in the breakdown of the endophyte, but it is also possible that exogenous enzymes can enter the hyphae when the degradation of membranes is advanced. The enzymes 1,3–β-glucanase endochitinase and N-acetylglucosaminidase that can break down hyphal walls in orchid endophytes have been detected in mycorrhizal seedlings of *Dactylorhiza majalis* and even, in lower concentrations, in asymbiotic seedlings (Pegg & Breadmore, unpublished data, cited in Hadley & Pegg, 1989).

There are numerous microbodies in the digestion cell cytoplasm, more than in non-infected cells, and they usually contain a protein crystal or twin crystals (Borris *et al.*, 1971; Barroso *et al.*, 1986*a*, 1988; Richardson *et al.*, 1992). Uricase and catalase activity has been traced to these microbodies that are considered to be involved in either amino acid metabolism or recycling of DNA and RNA (Barroso *et al.*, 1988).

Fig. 8.2. Infected tissue of seedling. (*a*) The hypodermal cell on the right is the source of infection of the other two cells. A hypha has passed through the host cell wall (arrow) close to the nucleus of the middle cell which is enlarged and attached to the cell wall by cytoplasmic strands. Infection is recent, especially in the left cell; no fungal breakdown is evident. (*b*) Digestion cell with hypertrophied and amoeboid nucleus. Hyphae, still alive but vacuolated (arrows), are confined to the cytoplasma around the nucleus and cytoplasmic strands that traverse the vacuole. (*c*) Some hyphae are dilated and filled with glycogen; some are seen in the immediate vicinity of lobes and constrictions in the nuclear membrane (small arrows). The clump (large arrow) below the bilobed nucleus represents the remains of previous infection. *Dactylorhiza majalis* in symbiotic culture with *Tulasnella calospora* (D47–7) on oat medium, cortical cells of seedling. Semi-thin sections stained in toluidine blue. Scale bars represent: 50 μm. (*a*) and (*c*) Reproduced from Rasmussen (1990) with permission of *New Phytologist*.

Digestion cell nucleus

The nuclei of digestion cells undergo dramatic changes, which are regarded as a sign of high metabolic activity. Shortly before the cell is infected the nucleus becomes hypertrophied and irregular in outline, and the ploidy level rises to as much as 128 C (Williamson & Hadley, 1969). Before infection the cell usually has one nucleolus (rarely two nucleoli), and no nuclear bodies; after infection there are up to four nucleoli that are often vacuolated, and nuclear bodies occur, usually adjacent to either nucleoli or the nuclear membrane. Since the nucleus must produce considerable amounts of RNA for enzyme synthesis, it is possible that these bodies consist of messenger RNA temporarily accumulated before being exported to the cytoplasm (Barroso & Pais, 1990).

The invading hypha appears to establish close contact with the nucleus in each of the cells that it passes through, i.e. the rhizoid cell, the hypodermal cell and the cortical cells. Shortly before invasion the plant nucleus moves to the side of the cell adjacent to the infected neighbouring cell (Fig. 8.2*a*; Burgeff, 1909; Rasmussen, 1990). The outer surface of the large and amorphous nucleus increases greatly and numerous pores are observed in the nuclear membrane. Portions of the cytoplasm and hyphae often become almost enclosed in cavities on the nuclear surface (Fig. 8.2*c*; Dangeard & Armand, 1898; Burgeff, 1909; Barroso & Pais, 1990).

Fungal wall and interface matrix

The hyphal wall of typical basidiomycetes consists mainly of polysaccharides such as mannans, bound in glucoproteins, β-glucans and chitin (Ruiz-Herrera, 1992). Hadley and co-workers (1971) observed that in pure culture the fungal wall appeared in EM as an electron-dense inner layer and a thin transparent outer layer; inside the orchid an outer layer was distinctly visible. The outer layer sometimes appeared compact, while at other times had a loose structure and did not form part of the septal walls (Strullu & Gourret,

1974). Its true origin is uncertain since it is not altogether clear whether it forms part of the hyphal wall and whether it is produced by the fungus. Some investigators have interpreted it as being an interface matrix derived from the plant (Dexheimer & Serrigny, 1983; Barroso & Pais, 1985). Judging from some published micrographs (Barroso, 1988; Peterson & Currah, 1990; Richardson *et al.*, 1992) the endophyte walls including the outer layer measure from 0.06 to 0.66 μm in thickness, while the inner layer alone appears to be only 0.015–0.05 μm thick. This is thin compared with the inner layer of hyphae in pure culture which are 0.06–0.1 μm thick, rising to 0.32 μm in monilioid cells. Even if the outer layer is included, the endophyte wall only occasionally exceeds the usual thickness of hyphal walls in pure culture. Either the width of the fungus wall becomes considerably reduced when the hyphae grow intracellularly or else the outer layer seen around the endophytic hyphae is of fungal origin.

Whether this layer represents the outer endophyte wall or an interface matrix, it reaches its highest development in the later stages of infection (Nieuwdorp, 1972; Barroso *et al.*, 1986*a*), and seems to increase at the same time as the glycogen reserves in the hyphae become depleted. A positive aniline blue reaction in the interface suggests that callose is a component of this material (Richardson *et al.*, 1992); treatment with cellulase removes a large part of it and it reacts to the PATAg stain. These tests both suggest that cellulose is a major component, but the substance does not appear to be fibrillar and neither a cellulase test nor the PATAg staining are entirely specific. Cellulase is able to break down a variety of polysaccharides and oligosaccharides provided they have β-1,4-glycosidic bonds. A fibrillar network that is resistant to cellulase was observed adjacent to the (inner) endophyte wall (Barroso, 1988), and extraction with pectinase or proteinase also affected the density of the material, indicating that it also contains pectic compounds and proteins (Barroso & Pais, 1985).

The content of vesicles both dispersed in the host cytoplasm and close to the plasmalemma showed a PATAg reaction similar to that of the material in the interface, as did invaginations in the plasmalemma. Provided this material is identical with that of the interface the observation suggests production within the host (Barroso & Pais, 1985).

The inner part of the endophyte wall can develop protuberances which are fairly large bulges on the wall surface (Hadley *et al.*, 1971). Many of the published micrographs of endophyte walls show that they tend to flake into layers (e.g. Borris *et al.*, 1971; Barroso, 1988). Both the swelling and the fracturing could be indications of the beginning of a breakdown of the hyphal wall.

When the hyphae have become completely lysed a residue is left that reacts strongly with acriflavine-HCl, indicating the presence of polysaccharides. Peterson & Currah (1990) suggested that this substance could originate from chitin, which is known to occur in the hyphal wall, and which breaks down into products that the seedlings do not seem to be able to assimilate (Harvais & Raitsakas, 1975).

Fungal membrane system and cytoplasm

Invasion is accompanied by a high polyphenol oxidase activity in the hyphae (Pais & Barroso, 1983; Barroso *et al.*, 1986a). The plasmalemma of live peloton hyphae indicates that substances are secreted by means of vesicles fusing with the membrane, since the plasmalemma is extremely irregular and wavy and surplus membrane is apparently guided back towards the vacuole (Hadley *et al.*, 1971). Polyphenol oxidase can be demonstrated first in the cytoplasm of the fungi, later in the interfacial space (Barroso, 1988). Blakeman *et al.*, (1976) found that there was about a fivefold increase in polyphenol oxidase, ascorbic acid oxidase, peroxidase and catalase 1 hour after the seedlings had become infected, but this analysis could not tell whether the enzymes originated from plant or endophyte.

The cytoplasm of the hyphae becomes loaded with glycogen both in pure culture and when the hyphae grow inside the plant (Hofsten, 1973; Strullu & Gourret, 1974; Barroso *et al.*, 1986a). The accumulation of glycogen is particularly noticeable in dilated hyphal cells resembling the monilioid cells formed in pure culture (Fig. 8.2c; Rasmussen, 1990).

In the early stages of lysis the hyphal cells vacuolate, and the dolipore septa begin to disintegrate (Richardson *et al.*, 1992). During digestion the intracellular contents become depleted; some of the membrane systems persist when most other elements of the cytoplasm have disappeared and the hyphal cell walls begin to collapse. At this stage spherical electron-dense bodies that contain phosphorus compounds (as shown by energy-dispersive spectroscopy) appear to be released from the hyphae. When the hyphal clumps collapse the adjacent plant cytoplasm is found to contain a number of lipid bodies, as well as microbodies with protein crystals (Richardson *et al.*, 1992) which could represent the accumulation of products from the breakdown of the peloton. The periplasmic canals in which the hyphae lie presumably then fuse, since the flattened hyphal remains are observed to become aligned in a dense pattern of parallel 'shells'. The space between them decreases progressively while the remaining wall materials appear to become reduced as well as condensed. It is not known whether the hyphal remains eventually disappear completely.

The stages of mycolysis can be briefly outlined as follows (Strullu & Gourret, 1974):

(1) Hyphae lie in separate periplasmic canals that traverse the host cytoplasm. The cytoplasm and hyphal wall are structurally intact and the hyphae contain glycogen.
(2) Hyphae become compressed and their contents disorganized.
(3) Hyphal walls collapse.
(4) Wall material is lysed.

Discussion of the ultrastructural evidence

When interpreting the electron microscopic observations it is useful to take into account that the host receives the net benefits of the relationship. The existence of holomycotrophic species of orchids contradicts any notion of balanced mutualism or a net benefit for the fungus. If transfer of any material takes place from the plant to the fungus it is bound to be limited in time and amount as compared with the contribution made by the fungus to the plant. There is some discrepancy between this fact and the way in which much of the ultrastructural evidence has nevertheless been interpreted as though it concerns secretion from the plant.

The prevailing view is that the material in the interface is produced by the plant. According to Nieuwdorp (1972) it is produced as a defence reaction and continually becomes dissolved by exudates from the endophyte, only to accumulate in greater quantity when the fungal growth eventually declines before lysis. This would entail the uptake of host material by the endophyte fungi in the initial phase of infection. If the plant is to have a net gain there would either have to be a subsequent resorption or transport of different substances into the plant that exceeded the cost of energy represented by the interface matrix.

According to Barroso and co-workers the observation of PATAg-positive compounds in intracellular vesicles in the cortical cytoplasm of the host, at the plasmalemma and in great amounts in the interface matrix, provides evidence that the interface material is produced within the plant protoplast and exported in vesicles that release their contents by fusing with the plasmalemma. Viewed in isolation this seems convincing but, again considering the nutritional function of the hyphae, it seems more probable that materials move towards the plant. The observed distribution of materials would also be consistent with the movement of material from interface into the host cell, since it is difficult to determine the direction of moving objects in an image produced in fixation. Furthermore, although the contents of the interface and vesicles have been identified as mainly cellulose this cannot be regarded as conclusive, given that the tests were not highly specific. When the hyphal wall is being broken down a diversity of more or less complex carbohydrates could be imagined to arise and many of these compounds could probably produce the enzyme reactions observed in the interface material. Certain structural modifications of the endophyte wall suggest that it undergoes some decomposition while the fungal cell is still alive.

The transport from fungus to plant could either occur after the death of the hyphae (necrotrophic) or begin while the intracellular hyphae are still alive (biotrophic) and be completed after their death. Proof of the second possibility would require evidence of both a biotrophic and a necrotrophic transport of material.

There is currently no clear evidence of a biotrophic transfer; although Hadley & Williamson (1971) and Mollison (1943) succeeded in showing some growth stimulus in the seedlings of *Dactylorhiza purpurella* apparently before the breakdown of pelotons begins, it has been technically difficult to confirm this result under other conditions because the lysis begins as early as 48 hours after contact between the mycelium and the plant tissue (Purves & Hadley, 1976). Biotrophic transfer appears to be dominant in the mycorrhizal systems of other plant groups, but in contrast to orchid mycorrhiza these have mostly proved to be two-way exchanges (Harley, 1984). However, they resemble the orchid mycorrhiza in producing an interface matrix, which is considered to constitute an extremely important element in nutrient transfer in these systems, particularly since cell wall precursors comprise one of the most likely sources of carbohydrate that are available to the carbon-demanding partner (Duddridge, 1985).

8.2 Regulatory mechanisms, phytoalexin

In 1911, Bernard showed that the internal tissue of tubers of *Himantoglossum hircinum* contained a strongly fungicidal substance that diffused into agar from cut surfaces, and that this substance decomposed at 55 °C. Magrou (1924) confirmed Bernard's observations and extended them when he found that a fungus inoculate placed on an agar surface with a fragment of tuber grew until the hyphal front reached a certain distance from the tissue. Growth then stopped, the hyphae began to degenerate and within a matter of a few weeks the whole mycelium was dead. Nobecourt (1923) investigated whether the fungicide was produced only in the presence of fungi or whether it was permanently present in the tissue. By killing the cells either with chloroform or by freezing he assumed that any pre-existing fungicide would remain active, so when no fungicidal effects could be detected in the dead tissue he concluded that the fungicide was only produced in living tissue when fungi were present. This conclusion was confirmed by Boller *et al.*, (1957). The production of orchinol begins when the tissue of a tuber has been in contact with the fungus for about 36 hours, and reaches a peak within 8 days. The reaction spreads to the whole tuber but does not reach concentrations that protect the tuber entirely against new infection. In the vinicity of the initial attack, however, there is a zone of high orchinol activity (Gäumann & Hohl, 1960).

The first compound to be described was orchinol, which was isolated from *Orchis militaris* (Gäumann & Kern, 1959); chemically related substances, called loroglossol (= 'substanz A') and hircinol, have since been described (Urech *et al.*, 1963). Chemically these compounds are all phenanthrenes, and by virtue of their antifungal properties they are referred to the varied group of compounds known as phytoalexins (Stoessl & Arditti, 1984). While both orchinol and hircinol have fungicidal properties, loroglossol has been considered inactive and is apparently an intermediary metabolite; however, in tests carried out by Ward *et al.*, (1975) loroglossol also showed antifungal activity.

The capacity to react to an infection by producing phytoalexins is widespread among European orchids. For most of the species that have been tested tissue from mature plants from natural habitats has been used to measure fungicidal activity and all have reacted positively, which suggests that they had previously been in contact with the appropriate fungi. Of the 24 species that were examined by Gäumann and colleagues (1960) a group of 16 produced orchinol, a few others produced related but as yet unidentified substances, and a few species, such as *Orchis militaris*, produced a mixture of compounds. There was no obvious chemo-systematic pattern since closely related orchids did not necessarily belong to the same group as regards fungicidal activity.

A number of fungal strains isolated from various orchids induced antifungal reactions in *Orchis militaris*, but when a group of unspecific saprophytic and semi-parasitic fungi from soil (24 strains) were tested none of them induced a phytoalexin reaction in the orchid tissue (Gäumann *et al.*, 1960). On the other hand a number of different fungi, including several ascomycetes, that were found growing as endophytes in *Himantoglossum hircinum*, all induced a defence reaction in fragments of tubers (Gäumann *et al.*, 1961).

The antifungal protection in orchids is, in the main, non-specific. Forty-three strains of endophytes and common soil fungi that were tested by Gäumann and co-workers (1960) were all inhibited in their growth, but bacteria were not affected. Since the production of phytoalexins is apparently provoked only by symbiotic infection, the establishment of a mycorrhiza is required to provide the plant with this broad-spectrum protection against parasites and pathogens. Wounding seems to be another way of inducing phytoalexin synthesis in orchid tissue (Gehlert & Kindl, 1991).

Most tests have been carried out on tissue from tubers, with little attention being paid to the other underground organs. However, Gehlert & Kindl (1991) induced the production of a number of phytoalexin precursors in the rhizome of *Epipactis palustris* by infecting it with a strain of *Rhizoctonia*. Gäumann and co-workers (1960) found low concentrations of orchinol in roots – about 1% of that measured in tubers. When comparing extracts made from tubers, stems, leaves and roots of *Dactylorhiza incarnata* Burges (1939) found that the fungicidal activity was much higher in extracts derived from stem and tuber than in those from roots.

Even less attention has been paid to the mechanisms by which the spread of infection is regulated within the tissues. Since many workers describe tolypophagy as a rhythmic alternation between fungal proliferation and lysis, it is possible to conceive of a fluctuating balance between two antagonistic physiological processes, the one being conducive to fungal growth, the other promoting the breakdown of hyphae. The seasonal changes in infection that some species experience suggest that the balance between peloton formation and digestion is influenced by external factors such as temperature and moisture, and perhaps by the prevailing conditions governing the growth of the external mycelium.

Like other fungi orchid endophytes are inhibited by phytoalexins, but fairly high tolerance levels (orchinol concentrations up to $150\,\mu\mathrm{g}\,\mathrm{l}^{-1}$) have been observed in strains of *Ceratobasidium cornigerum* and *Thanatephorus cucumeris*, two fungal species that often occur in association with orchids (Pegg & Breadmore, unpublished data, according to Hadley & Pegg, 1989).

8.3 Physiological effects

The substances assimilated by orchids through their mycorrhiza are, briefly, water, mineral salts, carbohydrates and other organic compounds, the last arising mainly from the hydrolysis of fungal glycogen, lipids and proteins (Burgeff, 1936). Fuchs & Ziegenspeck (1924b) found that the nitrogen content in leafless plants of *Dactylorhiza* sp. from a natural habitat increased by about 43% during the winter months, stored carbohydrate increased by 25%, and structural polysaccharides, fibres, etc., increased by 11% during the time when the plants were provided for only by fungi.

As Hadley (1982c) expressed it, there is 'no sound evidence that the symbiotic growth stimulus can be replaced by [the addition of] nutrients alone'. It is difficult to see how in asymbiotic culture a substrate, however closely it conforms to the nutritional needs of the seedling, could wholly substitute the effects of a well-balanced mycorrhiza in terms of growth rate. When the hyphal pelotons are established within the plant cells, the external hyphal network forms an effective system of absorption. For instance, according to Hadley (1984) the rate of [^{14}C]glucose uptake was considerably higher in symbiotic seedlings of *Dactylorhiza purpurella* and *Goodyera repens* than in asymbiotic seedlings. Beyrle and co-workers (1985) found that symbiotic seedlings that were planted on a water agar but connected by hyphae with a nutrient agar, grew at the same rate as those that were planted on the nutrient agar, which implies that the amount of nutrients absorbed directly by the seedling from the agar was negligible as compared with the absorption occurring through the hyphae. Split-plate cultures of other orchid species have confirmed this conclusion (Tsutsui & Tomita, 1990).

Burgeff (1936) noted many communication hyphae emanating from rhizoids in parts of the root where the pelotons were still alive, and estimated that an adult plant of *Platanthera chlorantha* had roughly 11 000 rhizoids which would provide it with about three times as many hyphal connections to the soil. When the hyphae began to become lysed the number of hyphal connections to the soil was decreasing and some of the observed connections were dead hyphae. At the time of flowering and fruiting the lysis was so widespread that few connections remain.

Not only does the mycelium have a large surface area through which nutrients can be absorbed from the growth medium, but the internal pelotons also form an extensive interface with the plasmalemma of the orchid cells (see the section on the digestion cell membrane system above). The absorbing surface in a symbiotic seedling must be orders of magnitude larger than that of an asymbiotic seedling. It is characteristic that those substrates that are recommended for symbiotic culture are much less concentrated than those used for asymbiotic culture (see Appendix A).

In addition to the mainly symplastic pathway leading from the hyphal cytoplasm across the interface and into the host cytoplasm, the apoplastic pathway along the hyphal and plant cell walls should also be taken into account. Apoplastic transport of water and ions along hyphae may be of crucial importance in older roots where most exodermal walls are suberized. Hyphal passages that existed before suberization took place, and hyphae that penetrate the passage cells, would remain a potentially functional apoplastic pathway as suggested in other mycorrhizal systems (Smith *et al.*, 1990).

Since the symbiotic seedling is able to exploit a rapid proliferation of the mycelium in response to the local or sudden appearance of food sources, it shares some of the advantages innate to microorganisms, and since the fungi exude enzymes that break down complex compounds in the soil the orchid will have access to resources that would not otherwise have been available.

Water

Within the mycelium water is distributed by cytoplasmic streaming, and liquid is sometimes observed to exude from free hyphal tips (Burgeff, 1936; Vöth, 1980b). Streaming is maintained with the help of local water stress in the mycelium or, in various mycorrhizal fungi, of the transpiration stream of the host plant (Lucas, 1977).

It has previously been suggested that when the starch in the orchid tissue dissolves after infection the osmotic potential becomes more negative, which could give rise to an increase in the absorption of water from the environment (Bernard, 1904). This could happen either directly or through a promotion of the uptake from hyphae within the plant. Hydrolysis of glycogen in the pelotons could function in a similar manner in relation to the mycelium outside.

When Beau (1920b) cultivated seedlings on a glass plate connected to the nutrient agar by means of a mycelium only, he had to water the seedlings. Apparently the moisture obtained through the hyphae was insufficient to replace the amount lost in transpiration, but in this experimental environment the water loss from the seedlings may have been greater than it would be in soil. In agreement with what is observed during the propagation of orchids *in vitro*, the seedlings can be regarded as hygrophytes with little capacity to withstand water stress.

Ions

The transportation of mineral salts in the mycelium accompanies the movements of water. It has been demonstrated that phosphorus (^{32}P) can be transported into the tissues of infected seedlings of both *Dactylorhiza purpurella* and *Goodyera repens*, the uptake in *G. repens* being 100 times greater than in the asymbiotic control seedlings (Smith, 1966; Alexander *et al.*, 1984). In symbiotic seedlings of *Cymbidium* sp. Holländer (1932) found a very high nitrogen content: almost 5% of dry matter as compared with only 1.66% in the asymbiotic seedlings after 5 months of cultivation on a nutrient agar. This indicates that there is an active import of nitrogen into the infected tissues reaching concentrations that considerably exceed those usually found in cultivated plants (Salisbury & Ross, 1985). This effective uptake could be of competitive advantage in the usually mineral-poor soils typical of the habitats of natural orchid populations. Dijk (1990) reported that the positive effects of infection on seedling growth occurred at low nitrogen concentrations, presumably when nitrogen was the limiting factor for asymbiotic growth.

Organic compounds

Mature hyphae growing in a substrate secrete enzymes to break down complex substances in their surroundings into soluble compounds that they can absorb. Enzyme production by the invading hyphae may explain the manner in which they penetrate plant cell walls, but otherwise there is no evidence that the plant tissue is broken down. It is not known whether intracellular hyphae in the mature pelotons secrete any enzymes into the interface, but this appears not to be the case with acid phosphatases which are secreted by hyphae in pure culture (Dexheimer & Serrigny, 1983).

Within a mycelium the net translocation of nutrients is directed from source to sink, typically from old hyphae towards the growing tips, but the movement of soluble compounds is also dependent on the general direction of water flux. In the young stages of peloton formation in orchid mycorrhiza the intracellular hyphae grow vigorously and therefore constitute a sink for both nutrients and water that could either be replenished from the host or be drawn from the established mycelium outside the plant. Since the result is ultimately a net gain for the plant it seems more likely that the outside mycelium contributes to the building up of pelotons, including their glycogen deposits. The starch that disappears in the infected host cell could be metabolized locally, for instance during the extension of the host membranes and the production of defence substances such as phytoalexins and hydrolytic enzymes.

When the hyphae grow out from the plant (Burgeff, 1936; Williamson & Hadley, 1970) they encounter new resources in the environment, but usually at much lower concentrations than those available within the plant. However, cytoplasmic streaming in the external hyphae towards the plant would be favoured if a constant sink situation were maintained within the internal hyphae, i.e. if the plant were able continually to deplete the intracellular hyphae of essential compounds that would have to be replenished from the external mycelium. Another possibility is that the plant can induce the internal hyphae to develop storage structures so that the accumulation of glycogen in the peloton constitutes the sink that directs a stream of nutrients from the outside. The first hypothesis presupposes a biotrophic transfer, whereas the second could work irrespective of whether the eventual transfer were biotrophic or necrotrophic.

Several investigators have suggested that the fungi produce activating substances that are beneficial to the host in more than the purely nutritive sense (Burgeff, 1934; Downie, 1940, 1949a). Burgeff (1934, 1936) increased growth rate and development of chlorophyll in asymbiotic seedlings of *Vanda* sp. by adding an alcoholic extract of fungi, and suggested that the effect resulted from the presence of vitamins. The reaction was not specific since extracts of both compatible and incompatible strains of *Rhizoctonia* could stimulate seedlings of a given species and the effects resembled those obtained with extract of yeast and other fungi. Burgeff's (1936) attempts to identify an active substance were unsuccessful, although he found that the activity was preserved in aqueous extracts of the fungi, did not come into solution in ether, and was fairly resistant to heating. Downie (1949a) also observed that an aqueous homogenate of fungal mycelium had a certain stimulatory effect, but found that heating destroyed this effect and that under none of the tested conditions was the activity as strong as that of the living fungus. Active components were found in both the alcohol and water fractions of fungal extracts.

These results are not surprising, since extracts of a fungal mycelium would probably contain not only soluble sugars but also organic nitrogen, which often stimulates germination and growth (Chapters 4 and 9). On the other hand, a chemical extract from a fungal mycelium cannot be expected to contain the same compounds and concentrations as the intact hyphae could produce and secrete over a period of time.

8.4 Developmental effects

When seedlings are infected with a compatible fungus their growth rate increases both in darkness and in light, provided that the seedlings can withstand irradiation. This applies both to green seedlings and to those that are devoid of chlorophyll (Hadley, 1983). It is thus difficult to make valid comparisons between seedlings grown with and without fungi. What are perceived as characteristic traits of the symbiotic seedlings may actually reflect the fact that their development is more advanced than that of their asymbiotically grown siblings.

In general, however, inoculation of seedlings *in vitro* leads to an increase in the number of rhizoids or hair cushions (Burgeff, 1909; Hayes, 1969). Bernard (1909) found that in asymbiotic culture of *Bletilla striata* the apical meristem of the embryo developed into an elongated shoot before roots appeared. When inoculated with a weak strain of fungus, seedlings merely established mycorrhizal infections in patches along the shoot, below tufts of rhizoids, but when the seedlings were infected by a fast-growing strain the stem remained short, no leaves developed and roots emerged, the stem and root tissues becoming densely infected. These seedlings became much larger and developed roots earlier than those raised with a weak strain of fungus or asymbiotically.

Purves & Hadley (1976) emphasized the correlation in time between peloton formation, mobilization of starch and accelerated growth in seedlings. Asymbiotic seedlings on a substrate containing soluble carbohydrates grew slowly while accumulating starch; these deposits could be mobilized either by moving the seedlings to a substrate without accessible carbohydrate or by infecting them, which indicates that infection entails a physiological change that stimulates the consumption of starch reserves.

8.5 **The mycotrophy/phototrophy balance**

Although the heterotrophic nutrition of the seedlings is now beyond debate, it remains to be settled how large a contribution, if any, the fungi make to the mature plant. In orchid species that develop green leaves there have been few attempts to demonstrate the relative importance of the two nutritional systems. Different approaches have been used in preventing photosynthesis and mycotrophy, respectively, in order to evaluate the outcome.

When mycorrhizal and leaf-bearing plantlets of *Goodyera repens* were grown on agar with the addition of fungicide, which stopped the growth of the external mycelium, the growth rate of the plants, as well as their phosphorus and nitrogen contents, were reduced compared with untreated controls and the root/shoot ratio increased (Alexander & Hadley, 1984). However, when working with larger plants of *Goodyera repens* Alexander & Hadley (1985) could detect no transfer of ^{14}C from mycelium to plant nor in the opposite direction. Since the transfer from mycelium to seedlings of the same species was evident, the result suggests that *G. repens* gradually lost its dependency on carbohydrates from the mycelium. On the other hand, adult individuals obtained appreciable amounts of ^{32}P and under low phosphorus conditions the growth rate of the plants could be reduced by fungicide treatment (Alexander *et al.*, 1984). In *Dactylorhiza purpurella* the growth rate of plants and their organogenesis remained about the same as in symbiotic control plants in spite of fungicide treatment (Hadley & Pegg, 1989).

There are too few such studies to obtain a general picture of the importance of mycotrophy in adult orchids, but it may be characteristic that both *Dactylorhiza purpurella* and *Goodyera repens* have well-developed foliage and the latter is evergreen as a mature plant.

The effect of a lack of photosynthesis can be studied in the leafless season, or by removing leaves or excluding light to the leaves. In species of *Dactylorhiza* Fuchs & Ziegenspeck (1924*b*) reported an increase of 11% in structural polysaccharides in the leafless season from autumn to spring. The rise in storage polysaccharides in the same period was 25%. *Tipularia discolor* is the only green orchid species that has been subjected to a full year-round biomass analysis (Whigham, 1984). This species, which produces a single leaf in autumn and is photosynthetic only during the winter season, showed a slight increase in total biomass after the leaf was shed in May, reaching a maximum in July. This increase may reflect mycotrophic activity in early summer when the plant is devoid of chlorophyll. When photosynthesis was totally prevented in *Tipularia* by removal of the leaf when it emerged in autumn the result was a decrease in leaf and corm biomass the following year, so apparently mycotrophy could not provide adequate compensation. However,

Table 8.1. *Albino mutants found in the field*

Species	References
Epipactis helleborine	Renner (1938), Griesbach (1979), Jorgensen (1982), Salmia (1986), Case (1987)
Cephalanthera damasonium	Renner (1938)
C. rubra	Burgeff (1954)
Platanthera hyperborea	Light & MacConaill (1989)
Triphora trianthophora	Keenan (1988)
Orchis militaris	Vöth (1966)

there was only about 20% mortality in the plants that had their leaf cut off 3 years in succession.

Even if adult mycorrhiza may be of minor significance in some species it remains a fact that albino mutants have been reported in a number of species (Table 8.1) and their presence in a population can be stable. Salmia (1989a) reported that no obvious morphological or anatomical differences were found between roots of green and chlorophyll-deficient individuals of *Epipactis helleborine* and the infection patterns were similar. The endophytes also seemed identical. Apart from these occasional chlorophyll mutants, some species, including *Neottia nidus-avis* and *Epipogium aphyllum*, are mainly mycotrophic throughout their life cycle. Species of orchids will probably be found with all imaginable modifications ranging from those that have a long-lived and plentiful foliage and rely almost entirely on photosynthesis as mature plants, to those with little chlorophyll and an extreme dependency on mycotrophy.

The disappearance of above-ground parts and the subsequent reappearance of full-sized photosynthetic plants has been documented in a number of studies with marked plants that have been monitored over several years (Table 8.2). However, there are few indications as to the cause, nor has the habit of the plants while underground been ascertained, and it is possible that these stages, if occasionally found, could be confused with some of the more advanced seedling stages.

Table 8.2. *Sojourn below ground: length of time below ground recorded in terrestrial species*

	No. of years underground	References
Cypripedium acaule	1–5	Gill 1989
C. acaule	12	Gill (personal communication)
C. calceolus var. *parviflorum*	N.i.	Sheviak (1974)
C. candidum	N.i.	Sheviak (1974)
C. reginae	N.i.	Sheviak (1974)
Epipactis helleborine	3	Light & MacConaill (1991)
Limodorum	1[a]	Bernard (1902)
Neottia nidus-avis	1[a]	Bernard (1899)
Listera ovata	1–2	Inghe & Tamm (1988)
Spiranthes cernua	N.i.	Sheviak (1974)
S. magnicamporum	N.i.	Sheviak (1974)
S. spiralis	1	Wells (1967)
Dactylorhiza sambucina	1	Inghe & Tamm (1988)
Orchis mascula	1	Inghe & Tamm (1988)
O. pallens	N.i.	Fuchs & Ziegenspeck (1927c)
O. simia	1–2	Willems (1982)
Ophrys apifera	1–3	Wells & Cox (1991)
O. sphegodes	1–2	Hutchings (1987a)
Aplectrum hyemale	N.i.	Sheviak (1974)
Cleistes divaricata var. *bifaria*	4	Gregg (1991)

Notes:
N.i., an underground interlude is noted but the duration is not recorded.
[a]Cleistogamous underground flowering.
There are also records of Australian taxa such as *Caladenia* spp. and *Leporella fimbriata* carrying out seasonal replacement of tubers without producing a shoot above-ground (Dixon, 1991).

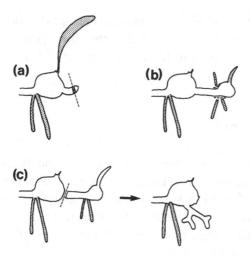

Fig. 8.3. Organogenetic experiment in *Aplectrum hyemale*. (*a*) Tip of a young shoot removed in spring resulted in no regeneration from axillary buds. (*b*) Roots of a young shoot removed in autumn when the young leaf appeared; new roots subsequently formed to replace them. (*c*) Young rhizome segment separated from the old segment; it developed normally but the old segment produced a branched 'coralloid' rhizome from one of the axillary buds. Diagram based on description in MacDougal (1899*a*).

The foliage bud can fail to develop if mechanical damage, frost or disease has destroyed the apical bud; Tamm (1972) considered the activity of slugs and other leaf predators to be a frequent cause of leaf disappearance. In this event the plant may lack a photosynthetic apparatus for a whole season, since it apparently has a very limited potential for immediately activating reserve buds. In an experiment with *Aplectrum*, MacDougal (1899*a*) severed the tip of the renewal shoot and noted that no axillary buds were activated to produce a new leaf (Fig. 8.3*a*). Likewise, in Whigham's (1990) experiment when the functional leaf was removed from *Tipularia discolor* the plants remained leafless for a whole season.

Reversion to pure mycotrophy could be associated with reversion to juvenile morphological features as well. MacDougal (1899*a*) showed that when the youngest rhizome segment in *Aplectrum hyemale* had been removed in its entirety, including the only functional leaf, the old corm could send out mycotrophic 'coralloid' shoots, i.e. repeatedly branched and hairy rhizome systems with traces of scale leaves, resembling the mycorhizome of this species (Fig. 8.3c).

8.6 Symbiotic systems

The orchid/fungus association viewed as a compound organism presumably relies on organic debris for its subsistence. Since symbiotic culture of a range of orchid species has been achieved *in vitro* on organic substrates that contained cellulose, starch, proteins, etc., there is a considerable evidence that orchid fungi can subsist as saprophytes on dead organic matter.

It is uncertain to what extent orchid mycorrhiza in natural habitats can live in mutualistic or parasitic associations with autotrophic living plants or other microorganisms. Burgeff (1954) attempted to establish some kind of symbiosis between isolates of orchid fungi and a number of herbaceous plants typically occurring in the same habitats as orchids, but did not succeed. Since none of these plants became infected orchid mycorrhiza was apparently not involved in chain-symbiosis.

MacDougal (1899a), however, observed a close contact between the roots of *Aplectrum hyemale* and those of oak or maple trees. Hyphae that formed a complete mantle on the surface of the tree roots apparently invaded the orchid roots at certain points. The orchid root cortex was filled with endotrophic hyphae and showed typical signs of orchid mycotrophy, but it is unclear whether all the hyphae observed belonged to the same fungus. An endophyte extending from the rhizomes of *Corallorhiza striata* was observed to form a sheath and Hartig net with roots of *Pinus strobus* growing nearby (Campbell, 1970a) and Correll (1950) cited a plant collector as having seen orchids, in a species of *Hexalectris*, in close association with tree roots. Other examples of orchid mycorrhiza in mycorrhizal association with trees have been mentioned earlier (Section 5.7).

In *Corallorhiza maculata* an association with the parasitic fungus *Armillaria mellea* was suspected (Campbell, 1970a), and *Galeola septentrionalis* has been shown to form seedling mycorrhiza with this tree parasite (Terashita, 1985). On several occasions orchids have been found to be infected with strains of *Rhizoctonia solani*, which mainly occurs as a plant parasite (Chapter 5).

It is thus likely that orchid mycorrhiza in some cases subsist on living, instead of dead, biomass, either in a mycorrhizal relationship with a living plant or as parasites on it. Since it is usually possible to cultivate holarctic species in pots where there are no foreign living roots available, this cannot be the general rule. However, the possibility of such complex relationships is worth considering when certain orchids species appear to be particularly difficult to grow in saprophytic culture.

Other, still more complex relationships may exist with other fungi and bacteria. A fairly frequent occurrence of bacteria in mycorrhizal tissues of Australian orchids has been established by Wilkinson and co-workers (1989), who also noted some effect of these bacteria during germination and early seedling stages.

Strains referred to *Rhizoctonia* sp. have been observed on several occasions in pot cultures of vesicular-arbuscular mycorrhizal fungi and seemed to be in an intimate symbiosis with these. When vesicles from plant tissues with V-A mycorrhiza were isolated and cultivated hyphae of *Rhizoctonia* grew out on the agar, and inoculation with these *Rhizoctonia* in conjunction with V-A mycorrhizal fungi resulted in most cases in a more rapid onset of the vigorous growth response of the host plants. The same strains of *Rhizoctonia* proved compatible with orchid seedlings (Williams, 1985), and since V-A mycorrhiza is the most widespread form of mycorrhiza in higher plants these observations open new perspectives as to the general role of *Rhizoctonia* in the vegetation. If orchid endophytes have access to resources flowing in other mycorrhizal systems this might strongly influence the competitiveness of orchids in natural habitats and their chances of establishing as seedlings.

9

Abiotic factors in growth and development

There is a considerable body of observations on the development of orchid seedlings *in vitro* in relation to various physical and chemical factors. Provided that conditions *in vitro* support reasonably healthy growth, such observations can give us a fairly accurate idea as to how development is regulated in nature, in various soil types and during the different seasons. However, little has yet been done to supplement the laboratory evidence by setting up explantation experiments in the field.

As regards physical factors, their effect can be studied by subjecting cultures to varying regimes of temperature, light and atmosphere. *In vitro* cultures are not suitable for observation of the effects of moisture conditions since humidity *in vitro* is consistently high. The effects of chemical factors have been studied in numerous *in vitro* experiments with substrates of varying composition, but these observation may or may not be applicable to natural conditions. Some of the requirements of asymbiotic seedlings reflect the absence of both mycotrophy and fully functional photosynthesis, all nourishment being obtained by direct absorption from the substrate. The external requirements of symbiotic seedlings may yield some information on the needs of the orchid/fungus association in a natural substrate.

Seedlings grown *in vitro* pass through some developmental stages that appear to be critical. The first occurs immediately after germination when many seedlings die before a shoot tip has differentiated (Stoutamire, 1974; Van Waes, 1984). This is understandable since many species have less complex chemical requirements for germination than they have for asymbiotic growth, so that germination in some cases takes place readily on a substrate that is insufficient for seedling development (Chapter 4). Typically, the first symptom of poor seedling growth is browning which begins at the surface of contact with the medium and spreads until the seedling is dead (Van Waes, 1984).

The second critical stage coincides with the formation of a shoot and the first root (Stoutamire, 1974). Van Waes (1984) observed a high mortality rate in species of *Dactylorhiza*, *Gymnadenia*, *Platanthera* and *Spiranthes* after the first root had emerged and begun to grow down into the substrate. The root began to turn brown and the seedlings eventually died, apparently in response to either nitrate or cytokinins in the substrate.

9.1 Temperature and humidity

Under natural conditions while the seedling is in a purely heterotrophic life stage it is entirely surrounded by the substrate – either soil or plant debris, or even living plant material such as a moss cushion. It can be assumed that the temperature regime is roughly the same as in the air, though with somewhat less extreme fluctuations, and the availability of water will usually be much more stable and predictable than at the surface of the substrate. A few observers have reported that seedlings have died in shallow soils as a result of severe drought in summer (Chapter 1) and I have observed how seedlings of *Tipularia discolor* withered after germinating in exposed places on logs. Fuchs & Ziegenspeck (1926a) noted the collapse of rhizoids in seedlings that were periodically dried out. Considering the significance of the rhizoids for contact with the external mycelium, this condition is equivalent to starvation for the heterotrophic seedlings.

Temperatures below 0 °C could probably also dry out seedlings, although there are no reports of such incidents. However, among more mature plants of *Orchis simia* and *Ophrys apifera* growing in shallow soil the mortality rate is high during the severe winters that they encounter at the northern limits of their distribution (Willems & Bik, 1991; Wells & Cox, 1991).

The optimum temperature for growth *in vitro* is usually around 20–25 °C. As regards seedlings of *Galeola septentrionalis* there is a narrow optimum of 22–24 °C (Nakamura, 1976) and for *Dactylorhiza majalis* this lies between 23 and 24.5 °C (Rasmussen *et al.*, 1990*a*). Above the optimum (i.e. above *c.* 26 °C) the growth rate of symbiotic seedlings of *D. majalis* became significantly lower and growth virtually ceased at 29 °C. This confirms the previous observation by Harvais & Hadley (1967*b*) that protocorms of *D. purpurella* grown at 29 °C were significantly smaller than those grown at 23 °C. The most striking feature of plants grown at above-optimum temperatures is that very few new rhizoids develop and existing rhizoids tend to collapse, although humidity in the culture dishes remains high. Thus rhizoids suffer damage not only from drought but also from temperatures that are too high. Although it was evident that these seedlings did not thrive at the high temperatures there was no indication that the fungus became parasitic, but the fact that there were few rhizoids may account for the infrequency of infection (Harvais & Hadley, 1967*b*; Rasmussen *et al.*, 1990*a*).

At optimum temperatures the protocorms of *Dactylorhiza majalis* became larger before the bud differentiated, apparently because the chronological age was more important for the onset of bud development than was seedling size (Rasmussen *et al.*, 1990*a*). If this information can be generalized to natural conditions it means that a seedling that experiences optimum growth temperatures in summer will require a shorter chilling period in winter for the above-ground shoot to develop, since the necessary cold treatment decreases with increasing seedling size (see below).

Below the optimum, down to 12–15 °C, growth was slower but development in every respect normal (Rasmussen *et al.*, 1990*a*). Some species even develop better *in vitro* at cool temperatures, with a lower mortality, according to Malmgren (1989*a*, 1993) who recommended subjecting small seedlings of most European species to low temperatures (5–10 °C) for a few months. The growth curve for symbiotic *Dactylorhiza majalis* (Jørgensen, n.d.) shows that the bud, in particular, elongates at a temperature of about 5 °C though the leaves do not unfold until the seedlings are transferred to higher temperatures (see below).

When seeds require cold treatment for germination, the chilling can also affect subsequent seedling development. This was clearly seen in *Epipactis palustris*, in which growth was unsatisfactory in the few seedlings that developed in the controls without cold stratification. Chilling apparently activated the metabolic systems that are responsible for organogenesis in the seedlings and the formation of a stable symbiotic association with the fungus (Rasmussen, 1992*b*).

Thermoperiodicity

Breaking bud dormancy by chilling was first described by Curtis (1943) for *Cypripedium*, and then by Borris (1969) for tuberous European orchids, but their findings seem to have been overlooked for a number of years and it was only recently that the requirement for chilling was brought to notice with the works of Riley (1983), Malmgren (1988), Mitchell (1989), Anderson (1990*a*), Yanetti (1990), Stewart & Mitchell (1991) and Mrkvicka (1992).

Most investigators suggest a period of 3 months at about 5 °C for effective vernalization. Jørgensen (n.d.) found that in *Dactylorhiza majalis* bud dormancy can be broken in considerably less time if the seedlings are fairly large when the cold treatment is begun. For example, about 50% of seedlings that are 5–8 mm long break bud dormancy after only 8 weeks of cold treatment; more than 90% of the seedlings that are longer react to the same amount of chilling (Fig. 9.1).

Vernalization has the effect of synchronizing the unfolding of the leaves and appearance of the inflorescence of summergreen species. Without chilling a similar response can be induced in tuberous orchids by increasing the length of the photoperiod, but the response is delayed, not synchronous, and the plants produce smaller tubers and are less likely to flower the following year (Mrkvicka, 1992).

After vernalization the leaves elongate and unfold, initiating a photosynthetic phase. In young seedlings of *Cypripedium* chilling may lead to yet another heterotrophic phase during which a new segment of the rhizome develops. Leaf development is induced by a second cold treatment (Curtis, 1943) suggesting that *Cypripedium*

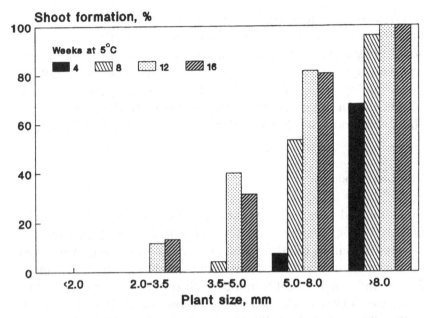

Fig. 9.1. Shoot induction in *Dactylorhiza majalis* by chilling in relation to initial seedling length, tested at four durations of cold treatment. Length classes include lower limits. Symbiotic culture on oat with fungus D47–7, grown in darkness, at 20 °C until cold treatment. Shoot development was recorded 20 days after return to 20 °C. Data from Jørgensen (n.d.).

can remain underground for two growing seasons. Other cultures of *Cypripedium* can develop a leafy shoot with a single long period of chilling (10–12 weeks at 4–5 °C: Stewart & Mitchell, 1991). Such differences in response could be related to the size of the seedlings at the beginning of the cold treatment.

The evergreen *Goodyera repens* is another rhizomatous species whose seedlings react to chilling by shoot elongation (Rasmussen, unpublished data), but this species can also be grown without vernalization (e.g. Alexander & Hadley, 1984). However, corms of *Calopogon tuberosus* remain dormant if they are kept continuously at room temperature (Anderson, 1990a).

Species of *Dactylorhiza* and many other European tuberous and summergreen species generally need a period of chilling in order to expand their leaves, but the mainly southern European wintergreen species such as *Ophrys* spp. and *Serapias* spp. which in nature produce a rosette of leaves during autumn do not need vernalization. The genus *Orchis* is divided into species with and without chilling requirements, apparently according to their leaf phenology in nature (Mitchell, 1989). Mrkvicka (1992) indicated that decreasing autumn temperatures break bud dormancy in wintergreen species.

9.2 Light and tropism

Newly emerged seedlings of most terrestrial species are extremely light sensitive and investigators recommend complete darkness for the first few months after germination (Harvais, 1973; Van Waes, 1984; Mitchell, 1989). Van Waes (1984) reported a seedling mortality above 85% in *Orchis morio* if the cultures were subjected to bright illumination within 24 weeks after germination, although they had already developed shoots that were *c.* 20 mm long. Lower light intensities was tolerated better, and when the seedlings were about 14 weeks old they had become less susceptible (Fig. 9.2). Australian species are apparently not adversely affected by light, since it is customary to grow the seedlings in 20 hour photoperiods (McIntyre *et al.*, 1974). Some North American species have been raised in dim light (Knudson, 1941; Stoutamire, 1964); this means that they will tolerate light but there is no indication that they require light for successful cultivation at this stage. The seedlings of terrestrial species often remain without visible chlorophyll even when subjected to light.

According to Henrich and co-workers (1981) seedlings grown in darkness do not elongate abnormally. I have also noticed this in connection with numerous symbiotic seedlings I have raised, and etiolation is not usually mentioned as a problem by those investigators who favour cultivation in darkness. Ichihashi (1990) did observe some elongation of internodes in *Bletilla striata* grown in darkness, but even so the height of the plants was roughly the same in light and in darkness. The fact that seedlings of terrestrial orchids do not elongate in the dark, or do so only to a slight degree, is yet another indication of the obligate nature of the seedling mycotrophy.

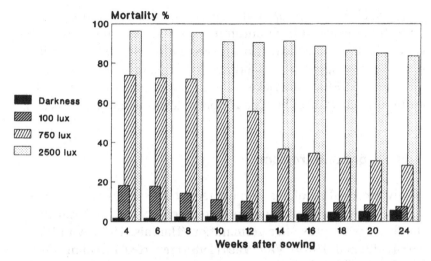

Fig. 9.2. Seedling mortality of *Orchis morio* in light. Cultures were brought into the light at the time indicated on the abscissa. Asymbiotic culture on BM1 medium at 20–25 °C. Mortality was recorded 8 weeks after transfer to the light. Data from Van Waes (1984, table XXIX).

The elongation response of seedlings in general is a mechanism that can help the leaves to reach above ground; such a response is clearly inappropriate to seedlings with a long heterotrophic life stage in the soil.

Darkness promotes rhizoid formation on seedlings (Mitchell, 1989; Ichihashi, 1990), and rhizomes that are not completely covered by the soil have rhizoids only on the underground part (Fuchs & Ziegenspeck, 1927c). Darkness is required for cultures where protocorm proliferation is artificially induced by wounding (Fast, 1982), which is to be expected since in the terrestrial species the natural vegetative reproduction by budding usually occurs below ground. In tissue culture of a species of *Aceras* Ponchet and co-workers (1985) achieved proliferation of bud tissue *in vitro* in 8 hour photoperiods, but in 16 hour photoperiods the tissue browned and withered. The investigators interpreted this as an indication of summer dormancy brought on by long-day photoperiods; this fits the normal annual growth cycle of *Aceras* but it would be desirable to have experimental data from intact plants.

Seedlings of epiphytic species usually turn green when exposed to light, as do those of some terrestrial species; other terrestrial species remain pale except for the shoot tip (Knudson, 1941; Stoutamire, 1964). In spite of visible chlorophyll it is doubtful whether the seedlings can immediately photosynthesize (Chapter 6). Nevertheless there are indications that the heterotrophic life stage is shortened by exposure to light, as Burgeff (1936) has shown in a tropical *Cymbidium* hybrid. While darkness stimulates the development of the protocorm and mycorhizome it inhibits the formation of the shoot (Fast, 1982) as well as the formation of roots, both of which are required for the change-over to the photosynthetic stage (Burgeff, 1936; Ueda & Torikata, 1972; Ichihashi, 1990). The same reaction is noted in seedlings of *Calopogon tuberosus* (Anderson, 1990a). Even at low light intensities (*c.* $0.6\,W\,m^{-2}$) root development and elongation of leaves was induced in *Bletilla striata*, whereas no roots developed in darkness (Ichihashi, 1990). In *Cymbidium* sp. Burgeff (1936) observed that when seedlings were illuminated a leaf unfolded before the root developed, but in holarctic species the first root usually develops while the leafy shoot is still in bud (Chapter 10).

The results of Werckmeister (1971) indicate that polarity in seedlings of *Cymbidium* can be established by using a substrate that is darkened with charcoal and illuminating the cultures; this caused the roots to grow down into the medium. Stoutamire (1974) observed positive geotropism in mycorhizomes of several species of Australian terrestrial orchids, which may have been due to the fact that the cultures were illuminated; after some months of cultivation, when the bud developed, the shoot tips turned upwards. In my cultures *in vitro* the seedlings are usually horizontally oriented when they are very young. When the apical bud is well developed it is directed vertically upwards. Since the cultures are kept in darkness this is probably a response to gravity, but we did not set up any experiments to ascertain this and the cultures received a certain amount of irradiation during observation and manipulation. In nature, shoot tips close to the surface could orient their growth in relation to the light that penetrates the uppermost levels of the soil, but seedlings that germinate deeper must react to other stimuli, such as gravity.

When they have reached the adult stage most terrestrial orchids can utilize natural light for photosynthesis, but some remain highly mycotrophic, some wholly mycotrophic. *Neottia nidus-avis* is extremely adapted to heterotrophy; chloroplasts and chlorophyll a develop but there is no chlorophyll *b*, and no photosynthetic activity can be detected (Montfort, 1940). In the species of *Corallorhiza* the inflorescences are usually visibly green, but since the foliage is reduced to scale leaves the photosynthetic surfaces are extremely small and the amount of carbon dioxide assimilated is only barely sufficient to balance the respiration of the above-ground inflorescence in the few weeks of flowering (Montfort, 1940).

9.3 **Atmosphere**

Although the soil environment creates a local atmosphere that has a different composition from that of air (Chapter 4) very few experiments have been set up to show how the ambient atmosphere affects seedling development *in vitro* or in the soil. Since they are heterotrophic, the seedlings have a net production of carbon dioxide (CO_2) while exhausting the available oxygen (O_2). Seedlings *in vitro* have greater demands for O_2 when they are in symbiotic culture than when grown on an asymbiotic substrate, consuming up to 4 times more (Blakeman *et al.*, 1976); this may be attributed in part to the higher growth rate. If ventilation in the culture vessels is insufficient, low O_2 levels and high CO_2 levels are generated.

The growth rate of asymbiotic seedlings of *Galeola septentrionalis* is highest at about 8–11% O_2, which is considerably less than that in atmospheric air; they grew poorly at 3–5% but there was less tendency to browning of the tissues at the low O_2 concentrations. The optimum concentration of CO_2 was about 2–3%, which is a level that could be found in soil (Chapter 4). Higher concentrations of CO_2 (6–10%) inhibited the development of scale leaves, but on the other hand there was no growth below 1% CO_2. The older the seedlings, the broader the tolerance range for atmospheric conditions (Nakamura, 1976).

In one experiment seedlings *in vitro* were found to grow faster in tightly closed culture vessels than in containers with a certain amount of ventilation (*Cattleya aurantiaca*: Hailes & Seaton, 1989), but usually the reverse is true. Kano (1968) reported a high seedling mortality in tightly capped cultures (*Cymbidium goeringii*, *Cypripedium acaule*), best seedling growth being obtained in vials that were ventilated by a glass tube with sterile cotton wool inserted in the stopper. My experience has been that symbiotic seedlings of *Dactylorhiza majalis* became significantly longer when the Petri dishes were not sealed with plastic film (unpublished data). These studies cannot, of course, be directly compared since the resulting concentrations of gases in the culture containers were not measured.

Burgeff (1909) compared the development *in vitro* of two-leaved seedlings of × *Laeliocattleya* both when CO_2 was allowed to accumulate in the vessels and when it was absorbed by sodium or calcium hydroxide. On a substrate that was devoid of carbohydrates seedlings required CO_2 as well as light if they were to develop normally. If the substrate contained 0.33% sucrose the seedlings developed normally even in the absence of CO_2, and if in addition they were inoculated then development was more rapid and they turned an intense green when exposed to light. This experiment showed that the seedlings needed soluble carbohydrates at this point in their development and were capable of producing them from assimilated CO_2, but that absorption from the substrate could adequately replace photosynthesis, especially with the assistance of fungi.

In *Dactylorhiza majalis* I found that the seedling increased in length in the presence of ethylene. Elongation of cells and organs in response to ethylene is well known in other systems, where ethylene is also known to affect the density of root hairs (Smith & Russell, 1969) and possibly to cause negative geotropism of root growth in waterlogged soil (Jackson, 1985). In orchids with hand-shaped tubers the root-like extension tends to bend upwards in wet soils (Fuchs & Ziegenspeck, 1925), but it is not known whether ethylene is involved in this reaction.

These scattered observations cannot be generalized, but they do suggest that the concentration of various gases is of considerable importance. Further studies on the influence of the soil atmosphere on seedling and plant development would be welcomed.

9.4 Growth medium

Carbohydrates

Soluble carbohydrates are always added to growth media even when asymbiotic seedlings are cultivated in light (e.g. McIntyre *et al.*, 1974), on the assumption that photosynthesis does not function at a sufficient level to support development (Knudson, 1924). Asymbiotic seedlings placed on substrates without any kind of sugar do not grow; the seedling may just reach beyond the stage of germination (Knudson, 1922; LaGarde, 1929). Photosynthetic activity, indicated by gross oxygen evolution per milligram chlorophyll, was found to be relatively slight in the green seedlings of *Dendrobium* sp. at the protocorm stage (1–3 mm), but increased considerably during later stages of seedling development (Hew & Khoo, 1980).

Soluble sugar in asymbiotic substrates is usually added in concentrations of between 10 and 20 g l^{-1} (i.e. 29–58 mM: Van Waes, 1984) and is generally supplied in the form of glucose, fructose or sucrose (Knudson, 1943). Sucrose is partially decomposed into glucose and fructose if the substrate is heat sterilized during preparation. In media for symbiotic cultures the mono- and disaccharides are partially or wholly replaced by polysaccharides such as starch (as in oats) or cellulose, which are often supplied at considerably lower concentrations (Appendix A).

Asymbiotic seedlings are not able to use all types of soluble carbohydrate equally well. Glucose, fructose, and oligosaccharides containing only these components are satisfactory for seedling growth (*Cattleya* sp., *Phalaenopsis* sp.: LaGarde, 1929; Ernst *et al.*, 1971). Mannitol is used by the seedlings of some species (*Galeola septentrionalis*: Nakamura, 1982), but usually it is neither utilized nor toxic (Fig. 9.3; Smith, 1973; Purves & Hadley, 1976; Van Waes, 1984) and can therefore serve as an osmoticum in experiments when it is necessary to distinguish between nutritive and osmotic

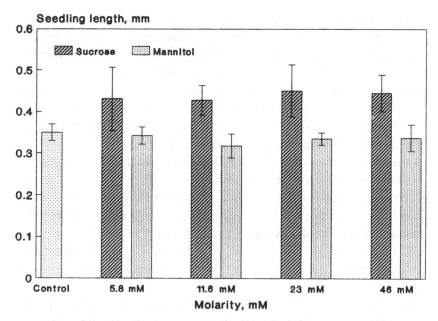

Fig. 9.3. Asymbiotic seedling development in *Dactylorhiza majalis* in water (control), and in a range of sucrose solutions and mannitol solutions with corresponding molarity. The increase in seedling length in response to sucrose was significant in comparison with that in water culture and the mannitol controls ($P \leq 0.05$). Means of 8 replicates with 95% confidence intervals are shown. Original data.

effects of sugars. Another fungal sugar, trehalose, is suitable for asymbiotic development in seedlings (Smith, 1973).

Ernst and co-workers (1971) found that galactose was toxic to the seedlings and that oligosaccharides containing one or more galactose unit (melibiose, lactose, raffinose, melezitose and stachyose) were clearly inferior to other sugars, although the seedlings are able to break down oligosaccharides to obtain the beneficial units such as glucose or fructose. Since breakdown products of the oligosaccharides were detected in the medium after culture the seedlings probably secreted enzymes (invertase) that decomposed the oligosaccharides extracellularly (Ernst *et al.*, 1971).

The capacity of orchid seedlings to exude hydrolytic enzymes across plasmalemma could be relevant for the digestion of hyphae within the plant. Components of the hyphal walls must constitute a significant part of the nutrients that the plant gains through mycotrophy; during lysis the endophyte walls are structurally altered but it is not known with certainty which wall components are preserved in the 'clumps' remaining in the digestion cells. When a basidiomycete wall is lysed products such as mannose, glucose and N-acetylglucosamine could be formed (Chapter 8). Both mannose and glucose can support seedling growth *in vitro* (Ernst *et al.*, 1970), although mannose is less effective than glucose in the development of seedlings of *Cymbidium* sp. (Fonnesbech, 1972*b*). However, N-acetyl-glucosamine was not utilized by asymbiotic seedlings when it was added to the substrate (Harvais & Raitsakas, 1975), and since this compound is a breakdown product of chitin it is possible that the chitin component in the endophyte walls cannot be utilized and is included in the waste products that agglomerate in the digestion cells.

Polysaccharides that cannot be used by the seedlings directly are accessible in symbiotic culture since they are metabolized by the fungi, which can also use monosaccharides and disaccharides. Starch produces about the same growth rate in symbiotic seedlings as a corresponding amount of a simple sugar, such as glucose or sucrose (*Dactylorhiza incarnata*: Penningsfeld, 1990*b*). Cellulose is apparently somewhat less suitable than starch for symbiotic culture (Beyrle *et al.*, 1991), but Hadley (1969) used 1% cellulose with good results. Although they are toxic to the seedlings, galactose and galacturonic acid can both be used to support seedling growth in symbiotic culture if the seedlings are connected with the substrate by means of hyphae and are not in direct contact with it (Tsutsui & Tomita, 1990).

Increasing concentrations of carbohydrate produce progressively higher growth rates in the seedlings (Fig. 9.4; Beyrle *et al.*, 1991) up to a point (depending on the orchid/fungus combination) above which seedling mortality increases or growth rate is reduced (Harvais & Hadley, 1967*b*; Tomita & Tsutsui, 1988). Concentration of available carbohydrates also affects organogenesis. In asymbiotic seedlings of epiphytic orchids root development increased with increasing concentration of sucrose, both in terms of length and

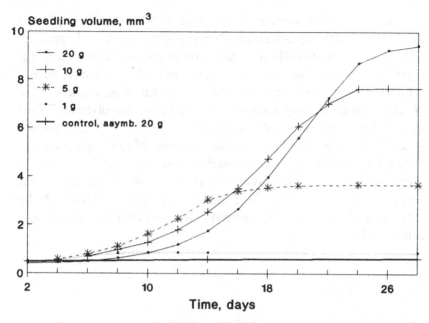

Fig. 9.4. Influence of sucrose concentrations on growth of symbiotic seedlings of
Dactylorhiza incarnata grown with a strain of *Rhizoctonia* and asymbiotic
control. Symbiotic medium contained macroelements in addition to
varying concentrations of sucrose. Asymbiotic medium in addition
contained peptone ($1 \, g \, l^{-1}$), microelements and birch sap. Means of 50
protocorms (abscissa: days after transfer from asymbiotic germination
culture) are shown. Seedling volume was assessed according to the
method of Hadley & Williamson (1971). Data from Beyrle *et al.*, (1991).

number of roots, while the length of the shoot was reduced (Yates &
Curtis, 1949), which suggests that the change-over to photosynthetic nutrition was delayed as long as sufficient carbohydrate was
available in the substrate. In seedlings of *Cymbidium* sp. leaf, root and
chlorophyll development were all inhibited by high concentrations of
sucrose (above *c.* 1.6%: Homes & Vanséveren-Van Espen, 1973;
Vanséveren-Van Espen, 1973). This could also explain why very
little chlorophyll developed in callus cultures of a cultivar of × *Aranda*
when the substrate contained abundant carbohydrate and the initial
C/N ratio was high (Chia *et al.*, 1988).

Species that react to illumination by turning green are often easier to cultivate than those that do not, and which probably represent species that naturally have a long heterotrophic phase underground. The onset of photosynthesis apparently helps asymbiotic seedlings through a critical stage in culture (Stoutamire, 1964). On substrates with $20\,g\,l^{-1}$ sucrose, e.g. on media such as Knudson C and Burgeff N_3f, Stoutamire (1964) noted an increase in the rate of development of asymbiotic seedlings of *Platanthera* spp. after they had turned green. This suggests either that the seedlings of some species make only poor use of exogenous sucrose or that they had requirements other than primary photosynthetic products which were not met through absorption from the substrate and which could only be satisfied when all processes connected with photosynthesis had come into function.

Ion concentration

The soils of European orchid habitats are often poor in inorganic salts (*c.* 0.02%, according to Möller, 1985*a*; Fast, 1985). The ion concentrations that have been found to be suitable for sustained development in asymbiotic culture range from $200\,mg\,l^{-1}$ to $2\,g\,l^{-1}$ (i.e. 0.02–0.2%). Mortality in seedlings of terrestrial species can be high when the concentration is as low as 100–400 mg, but above $3\,g\,l^{-1}$ there is an increasing proportion of unhealthy seedlings, described as swollen seedlings with a glasslike shoot tip, probably signs of vitrification (Fig. 9.5; Van Waes, 1984).

There is an inverse relationship between the concentration of mineral salts and the density of rhizoids on the young seedlings (Eiberg, 1970; Van Waes, 1984). When protocorms were grown on a medium without any macronutrients but with 20 g sucrose and micronutrients, they assumed a rounded outline and were covered in rhizoids. Asymbiotic seedlings of a range of European species could be kept alive on this substrate for up to 6 months.

The concentration of macronutrients also affects organogenesis in older seedlings; at concentrations above 60 mM the mineral salts promoted the formation of the shoot in seedlings of *Bletilla striata*,

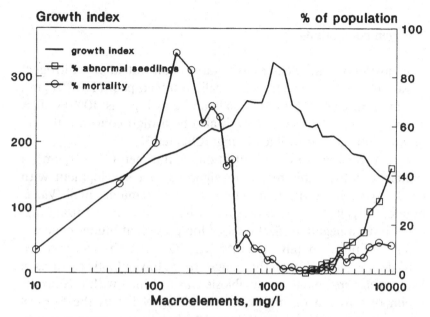

Fig. 9.5. Seedling growth index, mortality and frequency of abnormal seedlings in *Orchis morio* at varying macroelement concentrations. Growth index was defined as: (% of stage I) × 1 + (% of stage II) × 2 + (% of stage III) × 3, where stage I is the germinated seed, II the protocorm and III the protocorm with defined shoot tip. Abnormal seedlings are described as having a thickened base and vitrified shoot tip. Asymbiotic culture in darkness at 20–25 °C, basic medium BM. Data were recorded after 8 months in culture. Data from Van Waes (1984, table XXXIV).

while below 20 mM they tended to stimulate root development (Ichihashi, 1979). In particular, low concentrations of nitrate and phosphate appear to increase root growth in *Bletilla striata* (Ichihashi, 1978). This reaction is logical in ecophysiological terms since the plant is stimulated in its pursuit of soil minerals. As the concentration of accessible ions in the soil varies with the seasons, this could be a regulating factor in the development of new roots, and so perhaps indirectly determine the change-over between photosynthesis and mycotrophy. The same reaction could provide a means of managing organogenesis *in vitro*. This possibility evidently requires more investigation.

Nitrogen compounds

The most important element as regards plant nutrition is nitrogen. Most orchid soils are characteristically nitrogen poor (10–20 ppm), but the humus content of the soil can be as high as 30% (Möller, 1985a), in which case there will also be a high concentration of organic nitrogen accessible to fungi (Chapter 5).

When there are low concentrations of nitrogen (40–60 ppm) in symbiotic culture this results in vigorous root development with dense mycorrhizal infection, while the shoot remains small (*Vanda* hybrid: Burgeff, 1936). According to Dijk (1990) mycorrhizal fungi could have a negative effect on seedling growth at higher concentrations of nitrogen (above 3 mM NH_4NO_3), but the effect varied considerably according to the fungal strain. In symbiotic cultures of *Dactylorhiza* spp. balanced symbiosis was obtained with a concentration of nitrogen of approximately 10 mg N l^{-1} in the form of nitrate, ammonium or asparagine, but when the concentration was increased 10 times the fungus became virulent and all seedlings died (Penningsfeld, 1990b; Beyrle et al., 1991).

In asymbiotic culture the problem of virulence does not arise, but on the other hand the seedlings must take up their nitrogen direct from the agar which means that it must be available in a form that can be utilized and in adequate concentrations without reaching toxic levels. If nitrogen is supplied as NH_4^+ the uptake results in a release of H^+ which acidifies the substrate (Fitter & Hay, 1987) and may inhibit the growth of calciphilous plants. Conversely, the uptake of NO_3^- from the substrate results in a rise in pH. These effects tend to obscure the effects of the individual ions on seedling growth.

The fact that seedlings often grow better on ammonium than on nitrate, and in some cases only develop on substrates that exclude inorganic nitrogen and supply certain amino acids, led Raghavan & Torrey (1964) to suggest that the seedlings lack certain enzymes in the metabolic pathway leading from the uptake of nitrogen ions to the synthesis of amino acids and proteins, primarily nitrate reductase. Apparently some seedlings also lack nitrite reductase and glutamine synthetase, so that the nitrogen must be supplied as amino acids (see below). This deficiency diminishes as the seedlings grow older and its extent and duration probably varies with the species since the requirements are not the same in all species.

In many species both inorganic forms of nitrogen can be used and in many media recipes there is a mixture of nitrate and ammonium (Appendix A). For seedlings of *Cymbidium* and *Cattleya* ammonium proved to have a better effect on growth than nitrate (Curtis & Spoerl, 1948), but for *Bletilla striata* nitrate appeared to be superior and ranked as the most important anion. When seedlings of *B. striata* were grown *in vitro* the growth rate increased by the addition of nitrate up to extreme concentrations when it comprised 90% of the anions in the substrate (Ichihashi & Yamashita, 1977; Ichihashi, 1978), suggesting that nitrogen was the overriding limiting factor in the development of the seedlings.

Van Waes (1984) found that in the absence of nitrogen from the substrate the development of asymbiotic seedlings of European species ceased when they were about 3 times the size of the imbibed embryo. Most species develop better on organic than on inorganic nitrogen and in certain species, such as *Aceras anthropophorum* and *Epipactis* spp., seedling mortality was high when nitrogen was supplied only in inorganic form (Van Waes, 1984). Malmgren (1993) reported dramatic improvements in growth and in organ differentiation in *Cypripedium calceolus* and a number of other problematic species when inorganic nitrogen was entirely replaced by organic nitrogen (see Chapter 13). Several other investigators have also pointed to the benefits derived from organic nitrogen (e.g. Nakamura, 1982), particularly amino acids, although there is no agreement as to which of the amino acids are essential. Harvais (1972) and Harvais & Raitsakas (1975) recommended aspartic acid, glutamic acid, arginine and ornithine, whereas Van Waes (1984) obtained better results with glutamine than with glutamic acid. Besides increasing the growth rate amino acids also improve the development of rhizoids and are essential to asymbiotic seedlings even after they have developed a green shoot tip (Harvais, 1973). Ueda & Torikata (1972) found that arginine in the substrate increased the development of shoots and to a lesser extent roots in seedlings of *Cymbidium* spp.

Most culture media contain some amino acids either as specified components or in mixtures such as yeast extract, peptone or plant juices (Appendix A). Malmgren (1989*a*) cautioned against concentrations that are too high; peptone at about $3\,g\,l^{-1}$ caused browning of the tissues and led to high mortality in asymbiotic seedlings of *Cypripedium*, *Ophrys* and *Orchis*.

In the soil amino acids and other kinds of organic nitrogen could be assimilated by the fungi from organic debris and translocated to the plants, or they could be synthesized by the fungi (Harvais, 1973; Dijk, 1988*b*). According to Harvais & Raitsakas (1975) glutamic acid and aspartic acid were the dominating amino acids in the hyphae when the fungus was grown in pure culture on a substrate with inorganic nitrogen only.

Other mineral ions

Stuckey (1967) noted low concentrations of ions, in particular of potassium (K^+), in North American orchid localities, but in European orchid habitats Möller (1985*a*) found that the concentration of K^+ was about 10 times higher than the usual level. Fast (1985) reported an unusually high concentration of phosphorus and Möller (1985*a*) a remarkable amount of manganese and zinc in European orchid soils. Unfortunately, the effects of these ions on seedling and plant development have not been investigated to any great extent.

Ichihashi & Yamashita (1977) ranked the macroelements with respect to their effect on seedling growth: anions $NO_3^- > H_2PO_4^- > SO_4^{2-}$; cations $K^+ > Ca^{2+} > NH_4^+ > Mg^{2+}$. Reasonably satisfactory seedling development was obtained with most solutions except those in which the concentrations of ammonium or calcium were high (Ichihashi, 1978).

Harvais (1972) noted that the presence of iron (Fe^{3+}) in the substrate is of importance for root and rhizoid development, recommending concentrations up to $25\,mg\,l^{-1}$ ferric ammonium citrate. When Knudson (1946) introduced his C medium he emphasized its advantages for growing *Cattleya in vitro* with reference to the availability of iron and the relative proportion of iron and manganese (2.5:1).

Möller (1985*b*) reported improvement in the cultivation of adult plants of *Orchis purpurea* after applying a fertilizer containing Zn^{2+}.

Acidity

Most investigators agree that the distribution patterns of holarctic species are often limited by soil reaction (e.g. Sheviak, 1974). In surveys of North American species pH and moisture conditions of the soil were found to be the factors that correlated best with the geographical distribution (Wherry, 1918; Currah *et al.*, 1987*b*). The species have widely differing preferences; the pH of the soil in orchid localities in North America, for instance, ranges from 3.7 to 7.5, some species having a bimodal pH preference (Stuckey, 1967; Sheviak, 1983). A low pH tends to delay the breakdown of organic substances thus increasing the acidity of the soil by an accumulation of humus (Sundermann, 1961). Many species seem to thrive in organic debris, which has a low pH, on top of or mixed with a highly calcareous basic mineral soil, the resulting pH often being on the basic side, or up to between 8 and 9, according to Sundermann (1962*a–d*).

Substrates used for *in vitro* culture are often slightly acid when mixed, but after they have supported the growth of orchid seedlings for up to 1 year the pH can fall until it lies between 3.2 and 3.8, apparently without detrimental effects on cultures of epiphytic species (Ernst, 1967). Terrestrial species may differ in this respect, however. Fast (1980) considered pH values of 4.8–5.5 to be most suitable for protocorm development, indicating a lower limit of between pH 3 and 4 below which growth is seriously retarded, and an upper limit at pH 6.5. Eiberg (1970) noted an increase in the incidence of browning and mortality when the pH was below 5. In my own experience it is advisable to adjust pH to 7.0 for calciphilous European species.

Vitamins and plant growth regulators

In symbiotic culture the addition of biotin, thiamine or *p*-aminobenzoic acid to the substrate has no effect on protocorm growth, according to Beyrle *et al.*, (1991). It is possible that the fungi were able to produce these substances in adequate amounts, but the ground medium contained peptone and may have contained enough vitamins to support seedling development. Asymbiotic media with a high content of coconut milk, thiamine, pyridoxine or nicotinic acid have a tendency to induce excessive branching in the seedlings (Borris, 1969; Fast, 1982). Mariat (1952) noted that some species, for example *Bletilla striata*, differentiate whether or not they are provided with pyridoxine, nicotinic acid and biotin, whereas other species respond to these substances, as when growth is stimulated in seedlings of *Cattleya* sp. and *Thunia* sp. by the addition of nicotinic acid to the substrate (Noggle & Wynd, 1943; Henrikson, 1951).

Plant growth regulators (plant hormones) may affect germination and seedling development in different ways, so that it may be preferable to alter the hormone content of the substrate when the seedlings are transplanted the first time (Fast, 1980). Although benzyladenine (BA) increases the germination level in certain species (Chapter 4), browning and mortality in the seedlings increases, at a concentration of $2.2\,\mu$M BA in *Epipactis helleborine* and even at $0.44\,\mu$M in *Listera ovata* (Van Waes, 1984). Recommended doses of cytokinins range from 0.2 to 5 mg per litre of medium (cf. Appendix A), but at kinetin concentrations above *c.* $1\,\text{mg}\,l^{-1}$ there is a tendency to excessive branching and callus formation in seedlings (Borris, 1969). Species of *Dactylorhiza*, in particular, tend to respond to cytokinins with shoot proliferation (Fast, 1982).

In asymbiotic cultures cytokinins tend to stimulate shoot development and inhibit the growth of roots in the seedlings (Harvais, 1972). If the plants are still in the protocorm/mycorhizome stage, the cytokinins accelerate the transition to the leafy stage; for instance in *Cymbidium goeringii* kinetin at $10\,\text{mg}\,l^{-1}$ caused the differentiation of a leafy shoot from the elongated and branched mycorhizome (Ueda & Torikata, 1969*b*, 1970). The cytokinins zeatin and kinetin reduce the amount of rhizoids forming on the seedlings. The same effect is produced with potato extract, which contains appreciable amounts of cytokinins, if applied in concentrations above 10% (Harvais, 1973, 1974). Seedlings of a tropical hybrid of *Cymbidium* responded in almost the same manner but were much more sensitive, root and rhizoid development being affected by concentrations as low as 1 ppm; at 10 ppm there was a general inhibition of seedling development, including that of the shoots (Rücker, 1974).

Harvais (1972) advanced the hypothesis that the root–shoot nature of the protocorm is determined by the concentration of cytokinins. Protocorms in some species, particularly those belonging to *Cypripedium*, are mostly glabrous and stem-like and have high intrinsic cytokinin contents whereas protocorms in other species often are predominantly root-like and covered with rhizoids. This should explain why young seedlings of *Cypripedium* are insensitive to low concentrations of external cytokinins and seedling growth is negatively affected by high concentrations (see Section 13.1).

Van Waes (1984) noted that the survival rate of protocorms was reduced if the germination medium contained auxin, but naphthaleneacetic acid (NAA) is beneficial to rapid growth in seedlings of *Bletilla* sp. and a number of tropical species, causing earlier development of both roots and leaves ($0.1–1.8\,\mathrm{mg\,l^{-1}}$: Fonnesbech, 1972$a$; Strauss & Reisinger, 1976). An increase in concentration of NAA can influence the development of the shoot in relation to the root, in much the same way as the concentration of sucrose does (see above), i.e. the number and length of roots increase and the shoot is less well developed, as seen in seedlings of *Cymbidium floribundum*; higher concentrations of auxins also inhibited chlorophyll synthesis in seedlings of *Cymbidium* sp. (Ueda & Torikata, 1969a; Fonnesbech, 1972a).

Harbeck (1963) noted that shoot development could be stimulated by the addition of gibberellic acid in the substrate ($50\,\mu\mathrm{g\,l^{-1}}$), but this treatment resulted in underdeveloped protocorms without roots. Gibberellin (GA_4) applied as $5\,\mathrm{mg\,l^{-1}}$ can promote the differentiation of shoot and roots in *Cypripedium calceolus* (Borris, 1969). When applied in doses of $5\,\mu\mathrm{g}$ per seedling gibberellic acids are reported to break dormancy of the apical bud in seedlings of *Dactylorhiza*, in this way replacing cold treatment (Borris & Albrecht, 1969; Gruenschneder, 1973). At higher concentrations (above $0.57–1.4\,\mu\mathrm{M}$) GA_3 can cause browning in the seedlings (Van Waes, 1984).

Undefined organic substances

A number of organic additives have an appreciable effect on seedling growth and development; an astounding list of the substances used is given in Ramin (1983). However, there have been few attempts to identify the most active components in these complex additives. The difficulty is that fruit juices and similar additives have a variable composition, so that it is never possible to reproduce exactly a substrate in which they are included. The same problem arises when the formula of a substrate includes brands of commercial fertilizers, vitamin solutions, etc. Even if the product is widespread at the time the recipe is published, investigators working in other parts of the world or experimenting a number of years later can experience great difficulties in reproducing the substrate.

Coconut milk has for many years been popular as an additive to tissue culture media. In orchid cultures it is reported to stimulate the development of tubers and roots (McIntyre *et al.*, 1974). Van Waes (1984) obtained rapid development and moderate seedling losses when either peptone, casein hydrolysate or yeast extract was included in the substrate at concentrations of about $1\,\mathrm{g\,l^{-1}}$.

The effects of organic additives are less pronounced in symbiotic culture. It is usual to add yeast extract to the substrate, presumably for the benefit of the fungus.

Charcoal and anti-oxidants

The addition of activated charcaol to the agar substrate improves seedling survival, particularly in species that tend to exude phenolics, for instance in *Anacamptis pyramidalis*, *Dactylorhiza* spp., *Epipactis* spp., *Gymnadenia* spp. and *Listera ovata* (Van Waes, 1987). Gruen-schneder (1973) reported that charcoal both reduced the problem of browning and stimulated root development in *Dactylorhiza maculata* and it also improved the development in seedlings of *Paphiopedilum* sp. (Ernst, 1976). In tissue culture of *Cymbidium* charcoal in the substrate was beneficial in establishing polarity so that roots became positively geotropic (Werckmeister, 1971).

The application of anti-oxidants (1,4–dithio-L-threitol and 2–mercaptoethanol) in the medium also prevented accumulation of polyphenolics in seedlings of *Listera ovata* and *Orchis morio*, but survival was not as good as with charcoal (Van Waes, 1984).

10

Life history and phenology

Since the young seedling of terrestrial orchids lives underground, the early life history is largely unknown. Some investigators have painstakingly dup up seedlings and underground structures around the year and endeavoured to deduce the developmental process from this material (Fuchs & Ziegenspeck 1924*a,b*, 1926*a,b*, 1927*a–c*). This is usually the only source of information that we have as regards the life history since many species has never been grown from seed to the adult stage in culture or their life history has not been described when cultured plants were available.

Both excavation of plants in natural populations and observation of cultured specimens have their limitations as methods of investigating the life history. Unless the same individuals are repeatedly dug up and carefully replanted this method gives information only on the sequence of organogenesis and a rough seasonal timetable. Construction of an absolute timetable requires that there is a close correspondence between size and age, but this assumption is unrealistic, since cultivated seedlings of many species show considerable individual variation under uniform culture conditions. Estimates of the duration of seedling stages that have been made on the basis of excavations, particularly by Fuchs & Ziegenspeck (1922–7), have been justly criticized for being exaggerated, but it is mainly the most extreme estimates such as for the underground phase of *Orchis ustulata* that have been singled out (Summerhayes, 1951) and been met with scepticism. Most of the estimates made by Fuchs & Ziegenspeck do not fall wide of the mark if they are compared with those now emerging from population studies, as can be judged from Table 10.1.

The difficulty attached to drawing conclusions from dug up material is neatly displayed in an analysis of monthly soil samples taken from the site of a population of *Dactylorhiza fuchsii* (Fig. 10.1; Leeson *et al.*, 1991): The proportion of the seedlings that were protocorms decreases in the samples from October to September, suggesting that germination begins in October shortly after seed dispersal. However, no seasonal pattern can be seen in the relative number of larger seedlings, so it is impossible to discern the older cohorts of the previous years. Either germination is spread out over a long period of time, or individual growth rate varies considerably, or both. In natural populations of many other species there seem to be similar problems with age estimation of young plants.

Studies of cultivated plants have the advantage that the absolute age of the seedlings is known, and moreover growth conditions can be made to simulate seasonal changes in light and temperatures which can provide information on the way in which organogenesis is regulated in nature. However, this approach also has its shortcomings, since the cultivated seedlings are in a protected environment, so that the growth rate and organ development are probably much faster than they would be under natural conditions. Life history as described on the basis of cultivated material is usually shorter than the estimates predicted from studies of excavated material. Moreover, during the early life stages of cultivation *in vitro* the relative development of shoot and roots can be radically altered by the addition of growth regulating substances to the agar medium.

Only when observations derived by different methods are critically compared is it possible to construct a realistic, if not exact, picture of the life history of a species. Neither method reveals the life history on a statistical basis, accounting for the individual variation in fate. When the sequence of life stages is recorded in older plants, as in the study by Gregg (1991), it can be seen that many plants progress steadily from vegetative to reproductive adults, but Gregg's data also showed plants that experienced developmental setbacks or remained in vegetative stage. A timetable for development from protocorm to reproductive adult must either include individual variation to show

Table 10.1. *Life expectancy and duration of various life stages*

Species	A	A+B	B	C	D	E	References
Cypripedium calceolus				→192			Kull (1988)
Listera ovata			→20	→41			Inghe & Tamm (1988)
L. ovata					70	75%	Tamm (1991)
Spiranthes gracilis		3–5					Wherry (in Correll, 1950)
S. lucida		2					Case (1987)
S. spiralis		3–5					Wherry (in Correll, 1950)
S. spiralis			3				Willems (1989)
Goodyera pubescens		3					Case (1987)
Dactylorhiza incarnata				→25			Tamm (1972)
D. sambucina			→12	→43		67–76%	Inghe & Tamm (1988)
Orchis mascula			→8	→13		58%	Inghe & Tamm (1988)
O. simia	3		3–6				Willems (1982)
Amerorchis rotundifolia				Few seasons			Case (1983)
Himantoglossum hircinum	3		0	→10	2		Wooster (1935)[a]
Ophrys sphegodes							Hutchings (1987a)
Calypso bulbosa				Few seasons			Case (1983)
Corallorhiza odontorhiza		5–6					Wherry (in Correll, 1950)

Notes:
A, number of years from germination to appearance of seedling above ground. B, number of years from first appearance to flowering. C, maximum recorded lifetime after first appearance. D, half-life, i.e. the number of years elapsing from the first recording of a group of specimens until 50% have died. E, chance of flowering two consecutive years.

[a]See Section 13.23.

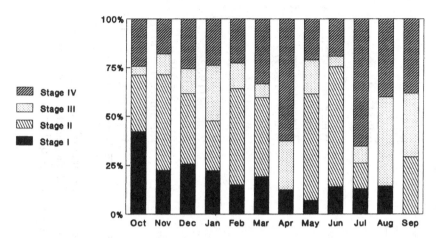

Fig. 10.1. Relative frequency of four seedling stages found in soil samples from a population of *Dactylorhiza fuchsii*. The smallest seedlings appear to decrease in number from October to September; there is no obvious seasonal pattern in the larger seedlings. Data from Leeson *et al.*, (1991).

the average life history or express the development in terms of least possible duration, the ideal life history.

10.1 Organ sequence from protocorm to mature plant

The mycotrophic protocorm grows monopodially into a short or long mycorhizome. *In vitro* there can be a proliferation of buds from the protocorm tip if cytokinins are present in the substrate (e.g. Fonnesbech, 1972*a*), but this should not be interpreted as something that would normally occur in nature. Barabé and colleagues (1993) put forward the theory that the main axis of the protocorm is almost immediately replaced by an axillary shoot, but this view is based on seedlings grown asymbiotically *in vitro* (St-Arnaud *et al.*, 1992), and is probably not a natural phenomenon. At least it is not corroborated by published reports of seedlings in the field or by my own observations.

As the mycorhizome develops there is considerable variation among the species as regards its length and orientation, its function with respect to mycotrophy and storage, and the point in time when the first axis is eventually replaced by a new shoot. Further variation is introduced with the development of storage roots and root-stem tubers in some of the groups (Chapter 7). In species where there is a well-developed rhizome the protocorm often persists for several years, and it is possible to find both mycorhizome and protocorm on seedlings that already have produced more than one leafy shoot, i.e. are at least a couple of years old. Conversely, in species where a tuber is formed the protocorm is soon discarded and the mycorhizome is usually both short and short-lived.

Life history group A: perennation by means of rhizome and roots

A.1 Rhizome→root→aerial shoot (Fig. 7.3a–f, p. 132)   In what is probably the simplest life history the rhizome continues to grow monopodially until the apical meristem produces the first aerial shoot (Fig. 7.3*a*). An axillary bud formed at the base of the aerial shoot then begins to grow so that the rhizome is continued as a sympodium. Before the first above-ground shoot appears some roots have usually developed and the underground structures have built up a store of nutrients derived from mycotrophy. This process probably goes on for several years before the leafy shoot emerges, but there is no exact information as to how long it typically takes for each species (see Chapter 13). The duration of the underground phase probably also varies somewhat according to local and individual factors such as the depth at which the seed germinates, the porosity of the soil, and the amount of humus building up on the surface, as well as year-to-year variation in climatic factors. In *Epipactis microphylla* the first aerial shoot is full-sized and flowers (Fuchs & Ziegenspeck, 1925), but most other species in this life history group produce leafy shoots of increasing vigour over a number of years, beginning with only one or a few small foliage leaves during the first season above ground.

In many genera the rhizome segments, i.e. annual increments, develop into swollen storage organs (Fig. 7.3*d*; Chapter 7), the first corm usually being formed at the base of the first leafy shoot.

However, in *Arethusa bulbosa* a swollen rhizome segment is produced from the protocorm before any foliage leaves are produced, at least when plants are grown asymbiotically *in vitro* (Yanetti, 1990). If this developmental pattern occurs in nature, the first corm must be filled entirely with mycotrophic products.

The heterotrophic species *Neottia nidus-avis* with its intensive production of mycotrophic roots presents a special variation in this life history group (Fig. 7.3f). The aerial shoot that eventually develops after an unknown number of years bears a flowering raceme that lacks foliage leaves and is virtually devoid of chlorophyll. The plant then withers, since there is no sympodial continuation of the rhizome, but the roots may survive to produce new plantlets (Section 7.2).

A.2 Rhizome→aerial shoot→root   It is unusual for the aerial shoot to appear before any roots have been formed, presumably because transpiration in the shoot makes great demands on the uptake of water. However, when a shoot develops on young seedlings of *Cymbidium goeringii* in spring, it appears about 1 month before the roots. In this species the rhizome consists of an elongated, branched and highly mycotrophic system both in nature and *in vitro* (Ueda & Torikata, 1970), so that there may be extensive hyphal contact with the soil to compensate for the initial lack of roots for absorbing water.

A.3 Rhizome→aerial shoot (Fig. 7.3g)   In *Corallorhiza* and *Epipogium* the mycorhizome also produces a branched rhizome system before any aerial organs develop, but roots are never formed. Some of the shoot tips turn upwards and elongate to appear above ground as inflorescences; there are no foliage leaves. A similar habit may exist in *Hexalectris*, to judge from illustrations in MacDougal (1899b) and Correll (1950), but I have not been able to examine any plant material.

Life history group B: perennation by means of a root-stem tuber

This group is characterized by a short-lived protocorm and the inconspicuous development of the mycorhizome and rhizome.

Leaves, rhizome and roots die each year, leaving only a tuber to survive the unfavourable season. As in group A roots often develop on the rhizome immediately before the leafy shoot emerges, but in some group B plants roots develop long before any leaves are produced and thus clearly have a primarily mycotrophic function.

In other of the life histories in this group a tuber develops one or more seasons before the leaves appear, indicating that the storage nutrients it contains must be products of the mycophagy taking place in the protocorm, the mycorhizome and the mycotrophic roots. The tuber enables the plant to survive physiological drought caused by either low precipitation or frost. In climates where severe drought is likely to occur soon after germination it may be advantageous for the plants to transfer nutrients from protocorm to tuber without appearing above ground; protocorms have a thin epidermis and a surface enlarged by rhizoids which make them more susceptible to desiccation.

It is remarkable that some species in this group produce a leafy shoot without having roots to supply it with water. This apparently occurs only in wet habitats, but nevertheless indicates that the amount of water absorbed through the mycelium under these conditions is adequate for the needs of the shoot. The same sequence of organ development seems to occur in some Australian tuberous species when cultivated *in vitro* (Clements, 1982c).

B.1 Root→aerial shoot→tuber (Figs. 10.2a and 10.3a) Plants with this life history first produce a root close to the meristem of the protocorm, which afterwards terminates in a leafy shoot. From an axillary bud at the base of this shoot the tuber emerges and begins to accumulate storage nutrients. When the leaves have died down the protocorm and the first root(s) disappear so that only the tuber survives.

After a period of arrested growth the bud on top of the tuber produces a short rhizome that develops one or more roots and is terminated by a leafy shoot. This sequence of events is common in *Coeloglossum, Gymnadenia, Dactylorhiza, Leucorchis, Orchis, Platanthera* and *Serapias,* among others, and is found in both summergreen (Fig. 10.2a) and wintergreen species (Fig. 10.3a) according to Beer (1863) and Fuchs & Ziegenspeck (1927b,c).

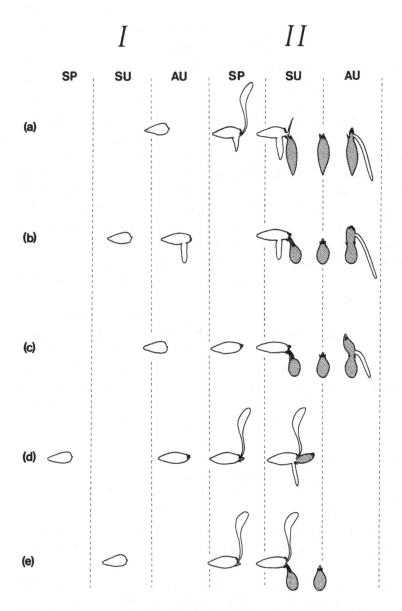

Fig. 10.2. Organ sequence from protocorm to adult in tuberous species with leaves in spring (SP), summer (SU) and autumn (AU), based mainly on analyses by Stojanow (1916) and Fuchs & Ziegenspeck (1927*b*, *c*); semi-diagrammatic. Four growth seasons are shown (I–IV) but the actual time frame is not known and probably varies individually. Shoot

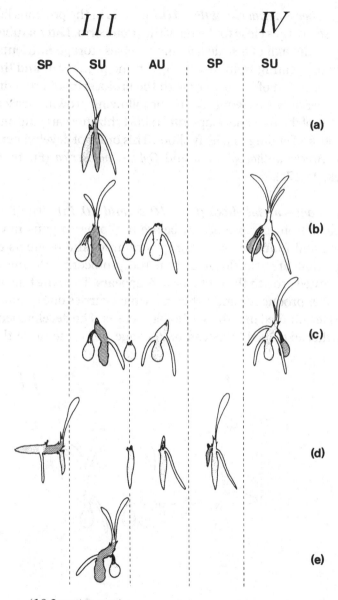

III   *IV*

SP   SU   AU   SP   SU

(a)

(b)

(c)

(d)

(e)

(*10.2 caption cont.*)
generations are shown with alternating density of shading. (*a*)
Root→leafy shoot→tuber. (*b*) Root→tuber→leafy shoot. (*c*)
Tuber→root→leafy shoot. (*d*) Leafy shoot→root→tuber. (*e*) Leafy
shoot→tuber→root.

B.2 Root→tuber→aerial shoot (Fig. 10.2b) As in the previous life history a root emerges close to the tip of the protocorm. Later a tuber develops from the axil of a scale leaf and receives storage nutrients, presumably originating from mycophagy in the protocorm and the root. During a period of arrested growth the protocorm and root die away, the tuber only surviving, and in the following growing season the bud on top of the tuber develops into a short rhizome carrying one or a few roots and ending in a leafy shoot. This order of development is noted in *Aceras anthopophorum* and *Ophrys holoserica* (Fuchs & Ziegenspeck, 1927c).

B.3 Tuber→root→aerial shoot (Figs. 10.2c and 10.3b) The first tuber develops from the protocorm before any other organs have been formed and functions as a storage organ for the products of mycophagy that are translocated from the protocorm. During a period of arrested growth the protocorm disappears. Then the bud on top of the tuber produces a short rhizome which carries one or more roots and is terminated in a shoot of foliage leaves. The development may be further extended if instead of a foliage shoot the tip of the

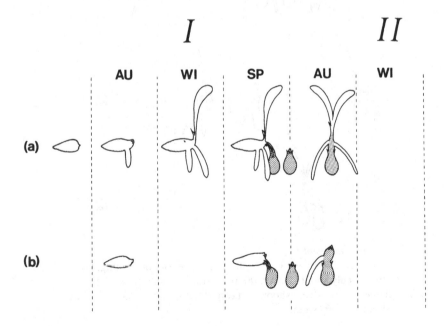

rhizome produces an arrested bud. In this case another cycle of tuber formation ensues, before the leafy shoot is eventually produced at the end of a rhizome segment. Examples of this organ sequence have been observed in *Anacamptis pyramidalis, Himantoglossum* spp., *Orchis mascula* and *O. militaris* and in summergreen (Fig. 10.2*c*) as well as wintergreen (Fig. 10.3*b*) populations (Fuchs & Ziegenspeck, 1927*c*).

B.4 Aerial shoot→root→tuber (Fig. 10.2d) In the European species of *Platanthera* the protocorm tip may begin by producing a leafy shoot from which basal roots later emerge. A tuber eventually forms from one of the leaf axils on the rhizome, but according to Fuchs & Ziegenspeck (1927*c*) not necessarily on the first segment that the rhizome produces. This sequence of organ development implies that the transpiration of the shoot must initially be provided for by the protocorm and its hyphal connections alone; perhaps characteristically this life history has only been observed under very humid conditions and in species normally developing along the B.1 organ sequence (root→shoot→tuber).

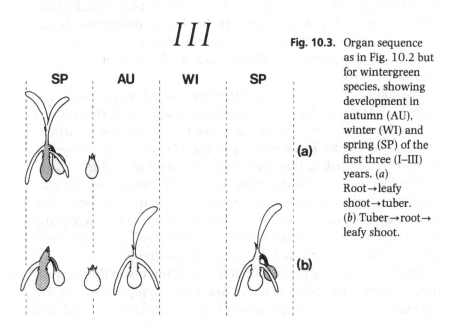

III

SP   AU   WI   SP

(a)

(b)

Fig. 10.3. Organ sequence as in Fig. 10.2 but for wintergreen species, showing development in autumn (AU), winter (WI) and spring (SP) of the first three (I–III) years. (*a*) Root→leafy shoot→tuber. (*b*) Tuber→root→ leafy shoot.

B.5 Aerial shoot→tuber→root (Fig. 10.2e)   In *Orchis pallens* Fuchs & Ziegenspeck (1927c) noted this unusual life history which, like B.4, requires that the leafy shoot functions for a considerable period of time without roots, while the tuber is accumulating nutrients. The flowering shoots of *O. pallens* appear above ground in very early spring, thus avoiding the problem of desiccation. In *Galearis spectabilis*, a species of damp deciduous forests, a leafy shoot is the first organ to be produced after the protocorm. The tuber begins to develops at the base of this shoot in spring and roots emerge immediately after (Fig. 7.5a; Rasmussen, unpublished data).

10.2 Phenology

Developmental events, such as the termination of the protocorm stage, the unfolding of the leafy shoot, the emergence of roots and the development of a tuber, largely coincide in all individuals, and at least some of them must be synchronized by external signals. The changes in mycorrhizal and organ development that take place in a recurring seasonal pattern could be induced by outside stimuli alone. However, in the regulation of events that lead to progression in the life history there must be an interaction between environmental stimuli and endogenous conditions such as chronological age, size and past history.

Phenology is still very incompletely described for terrestrial orchids and statistically based experimental inquiry into the effects of seasonal environmental factors on adult plants is lacking. Furthermore there is little knowledge as to the phenological changes in mycorrhizal activities. However, it may be useful to outline the kind of developmental regulation that could be in operation. Probable regulating factors in the environment can be identified by observing the seasonal changes that the plants are subjected to, mainly the fluctuations in temperature, precipitation and irradiation, and their presumed effects on mycotrophy and photosynthesis.

Since the phenology of the fungi is largely unknown there are no data regarding the point at which low winter temperatures or dry summer weather restrict the activity of the external mycelium, but it is known that established rhizoids collapse at high temperatures and at low humidity (Chapter 9), thus interrupting the lines of communication between the external mycelium and the plant. When the root system of *Ophrys holoserica* was observed *in situ* through a glass barrier it was seen that the mycelium surrounding the tuber and roots remained intact when the soil was wet, even at low temperatures down to nearly 0 °C, but that it was killed when the soil dried out and did not immediately regenerate after renewed watering (Vöth, 1980*b*).

The main season for mycophagy is probably in the wet autumn and early winter months when deciduous and herbaceous plants contribute a considerable amount of dead material to the organisms of the decomposition chain, among them the saprophytic fungi such as those living in symbiosis with orchids. Fuchs & Ziegenspeck (1924*b*) found that in species of *Dactylorhiza* the pelotons were alive and contained glycogen deposits during October and November. After that the lysis of hyphae began, followed by re-infection, and by December there were several dead pelotons in the each of the digestion cells. Although a few hyphae were still alive they contained hardly any glycogen. In *Bletilla striata* Masuhara and co-workers (1988) found that mycotrophic activity was greatest in spring and autumn, which coincides with the wettest seasons. Populations of *Goodyera repens* in Scotland were extensively colonized by fungi throughout the year, but the degree of infection was highest from December to May (Alexander & Alexander, 1984). These few observations suggest that in areas with high precipitation and mild winters orchid mycorrhiza is active most of the year, lack of moisture being the main restricting factor.

Photosynthesis requires light and humidity as well as moderate temperatures. In open habitats photosynthesis reaches its maximum in summer because of the high light intensities, except when limited by water deficiency. However, on the floor of a deciduous forest, as for instance in habitats of *Aplectrum hyemale*, light intensity in summer is reduced to 2–5% of incident sunlight which is close to the light compensation point for many species at summer temperatures (Auclair, 1972). In habitats of *Goodyera pubescens* the radiation measured in winter and early spring was higher than during the summer months (Stevenson, 1972) so that spring and autumn when the canopy is open are probably the most suitable periods for photosynthesis in woodland species; in mild climates there could be periods when conditions were favourable for photosynthesis even during the winter, at least when frost does not prevent the uptake of water.

The holarctic species display a diversity in leaf phenology. In the northern part of the range most species have leaves in summer, the leafy shoot appearing in spring with subsequent flowering from the same shoot. Members of the genus *Goodyera* are evergreen forest dwellers, often inhabiting evergreen coniferous woodland. *G. pubescens* produces new leaves continuously between June and October, the final rosette persisting all through the following winter (Rasmussen, unpublished data). Other species in shaded habitats, such as *Calypso bulbosa*, are almost evergreen, being leafless only for a few weeks in summer, while *Aplectrum* and *Tipularia* are wintergreen only, the leafless season lasting several months in summer (Fig. 7.3*e*, p. 132; MacDougal, 1899*a*).

In the southern areas that are dry during summer many species develop wintergreen leaves in the autumn, sometimes adding more leaves in spring and eventually flowering on the same shoot. All leaves of the wintergreen species die down some time in spring or early summer. A few of the wintergreen species are distributed relatively far north, for instance *Orchis morio* in Europe (Fuchs & Ziegenspeck, 1925; Mitchell, 1989) and *Aplectrum* and *Tipularia* in North America, but most of them are typical Mediterranean taxa such as the species of *Ophrys*, *Serapias*, *Orchis coriophora*, *O. sancta*, *O. laxiflora* and *O. palustris* (Fuchs & Ziegenspeck, 1925; Möller, 1987*a*; Mitchell, 1989). Some widespread genera such as *Orchis* and *Spiranthes* include both summergreen and wintergreen species.

A number of Mediterranean species such as *Aceras anthopophorum*, *Himantoglossum* sp., *Orchis simia*, *O. militaris* and *O. purpurea* are not wintergreen but sprout in spring (Fuchs & Ziegenspeck 1925; with respect to *Orchis* the different populations vary somewhat, see Section 13.17). Their life cycles resemble those of summergreen species in cooler climates but they have apparently adapted to a short photosynthetic season and it is possible that they are therefore more dependent on mycotrophy than their wintergreen relatives.

The seemingly dormant season of holarctic species, i.e. when they have no leaves, coincides with frost, with seasonal shade caused by the sprouting of a canopy in woodland habitats, or with a season of low precipitation and high temperatures. However, the leafless orchid is not necessarily physiologically inactive, as pointed out by Stoutamire (1974). The mycotrophic activity can probably continue unimpeded if the leafless season coincides with a period in deep shade or with low temperatures interrupted by milder spells.

Root phenology has not been fully investigated but there are several developmental strategies and they are apparently not correlated with leaf phenology. For example in field studies of *Liparis lilifolia* which is summergreen the roots develop only in spring, but in both the wintergreen *Tipularia discolor*, the summergreen *Galearis spectabilis* and the evergreen *Goodyera pubescens* the roots begin to appear in summer and their growth continues throughout most of the autumn (Rasmussen, unpublished data). All four species were found in the same habitat in Maryland, USA. Since the roots are generally long-lived in orchids the beginning of the mycotrophic season does not necessarily coincide with the emergence of new roots, and mycotrophic activity may end before the root ceases to function as an absorbing organ; this needs to be investigated further.

In most of the orchidoid species either all the new roots appear from late summer to autumn or the majority of them do so with a few more forming in spring when the leaves appear (e.g. Sharman, 1939). *Cypripedium calceolus* and *C. candidum* begin in late May to produce new roots that continue to grow until October; they may resume growth the following spring (Möller, 1968). In a sample of Canadian species (*Platanthera hyperborea, P. obtusata, P. orbiculata* and *Coeloglossum viride*) all new roots appear during the summer and are fully developed in autumn by the time the leaves and the old roots wither (Currah *et al.*, 1990). In *Spiranthes spiralis* the new roots develop in late autumn when the leaves unfold, and increase in size while they act as storage organs during winter and spring; when they reach full size the leaf rosette has begun to wither (Wells, 1981). In many other species the roots emerge in spring, as for instance in *Cleistes divaricata* (Gregg, 1989), *Liparis loeselii* (Mrkvicka, 1990*a*), and *Bletilla striata* which produces roots from May to September (Masuhara *et al.*, 1988).

In many summergreen species the leafy shoot dies down in autumn before roots begin to develop or shortly afterwards. The fact that roots continue to grow during the autumn suggests that the mycotrophic season extends the overall growth season into autumn in the summergreen species, as indicated diagrammatically in Fig. 10.4*a*. A considerable diversity in nutritional strategy can be imagined if the orchids make varying use of the two kinds of nutrition available to them in accordance with their habitat and climate. For Mediterranean species that produce spring leaves, the potential extension of the growth season could be substantial if conditions allow mycotrophy to continue all through the winter (Fig. 10.4*b*). In southern habitats summer drought is probably limiting to both photosynthesis and mycotrophy, which in wintergreen species may be active more or less simultaneously during the wet winter months (Fig. 10.4*c*). Conversely, in arctic-alpine habitats where physiological activity is limited by frost during winter the two kinds of nutrition probably overlap during the summer months. Mycotrophy may continue after the end of the phototrophic season when the leaves have died down in autumn, but temperatures may still be sufficiently high to make fungal activity possible on mild days (Fig. 10.4*d*).

Transition from the protocorm stage

The factor governing the transition from the protocorm stage probably varies according to the mode of growth subsequently adopted, i.e. the life history of the species in question, but few experimental data are available. For several of the summergreen species, low winter temperatures are responsible for the unfolding of the leafy shoot and accordingly determine when the protocorm stage comes to an end (Section 9.1). A combination of seedling size and duration of the cold period determines whether the dormancy in the bud is broken. This suggests that individual seedlings that have for some reason germinated later or been slower in development may require more than one winter period before they produce leaves.

Mitchell (1989) reported that *Orchis militaris* also required chilling before it could proceed to develop, but did not describe the effect on the subsequent organogenesis *in vitro*. According to the life history

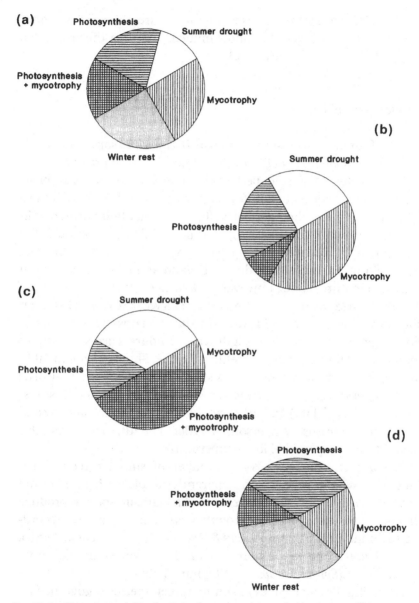

Fig. 10.4. Hypothetical distribution of mycotrophic and photoautotrophic nutrition throughout the year in species with varying leaf phenology and habitats, as discussed in the text. The upper part of the pie charts represents summer, the lower part winter. (a) Summergreen, north-temperate species. (b) Mediterranean species with spring foliage. (c) Mediterranean species with winter foliage. (d) Arctic-alpine species.

deducted from observation of seedlings in nature, this species cannot be expected to develop a leafy shoot immediately after the protocorm stage (life history B.3, see above).

Development of leaves

Leaves of summergreen species such as *Dactylorhiza* spp, *Gymnadenia conopsea* and *Platanthera bifolia* unfold after the large protocorm has been subjected to a cold period (Section 9.1). This takes place irrespective of length of photoperiods, and neither transfer to a fresh substrate nor long photoperiods suffices to break bud dormancy in seedlings *in vitro* (Borris, 1969). When chilling is omitted the protocorms of *Dactylorhiza* spp. still produce roots, but the leaf primordia in the apical bud do not expand and eventually growth ceases. In mature plants, however, Mrkvicka (1992) reported that if there was no exposure to low temperatures a weak response to longer photoperiods in spring could give rise to late and unsynchronized leaf elongation; the critical day length was 10 hours, corresponding to conditions in mid-February in central Europe. Shoot development in summergreen rhizomatous species such as *Cypripedium* spp. (Curtis 1943), *Epipactis gigantea*, *Goodyera repens* and *Listera ovata* (Rasmussen, unpublished data) and the cormous species *Arethusa bulbosa* and *Calopogon tuberosus* (Anderson, 1990a; Yanetti, 1990) is also brought on by a period of low temperatures.

In contrast, the leaves of a number of summergreen North American swamp species can apparently unfold with no external stimulus. These and wintergreen Mediterranean species produce shoots *in vitro* under constant growth conditions, apparently according to an internal rhythm (Myers & Ascher, 1982; Mitchell, 1988), but a decrease in temperature to 12–14°C can stimulate shoot development (*Spiranthes spiralis*: Stephan, 1988).

The dying down of leaves in summergreen species is governed by the decrease in night temperatures and the general ageing of the shoot (Mrkvicka, 1992). In *Orchis purpurea* and *Cypripedium calceolus* Möller (1985b) observed that continued watering could not delay the withering of the leaves, and there was a risk that the tubers would rot if more water was given during flowering. A decrease in temperature also brought the corm of *Arethusa bulbosa* into dormancy (Yanetti, 1990).

In wintergreen species such as *Spiranthes spiralis* (Stephan 1988) and *Orchis purpurea* (Möller 1990*a*) the leafy shoot is apparently induced to sprout in autumn by a decrease in temperatures, i.e. the same factor that brings about withering in the summergreen species. Decreasing photoperiods do not stimulate the development of shoot and roots. Lack of water can delay sprouting, which is resumed when the moisture content of the substrate rises, but watering of the plants cannot in itself induce sprouting without the temperature stimulus (Mrkvicka, 1992). However, the size of the leaf rosette increases if there is enough moisture when it is unfolding (Stahl, 1989). Similarly, in Australian orchids that are leafless in the dry season the subsequent sprouting of the shoot is elicited by a decrease in soil temperature; initially the shoot relies on the water and nutrients stored in the tuber or corm (Pate & Dixon, 1982), but these reserves can be a limiting factor in the development of the new leaves if the rainy season is delayed.

Some of the wintergreen species are induced by winter cold to produce additional leaves in spring. However, all leaves wither in late spring before the onset of dry weather. Water stress, rising temperatures and ageing of the shoot are all believed to influence this process, but there is no indication that day length is a factor of any importance (Mrkvicka, 1992).

Emergence of roots

In species with a summer resting tuber it is the decrease in temperature in autumn that acts as the stimulus for the production of roots in cultivated plants (Mrkvicka, 1992). The process can be delayed by keeping them dry, but cannot be induced by watering alone. The *in vitro* development of roots in *Cypripedium calceolus* can be stimulated by keeping the cultures at about 10 °C (Malmgren, 1989*a*), which suggests that a fall in temperature in autumn is responsible for the continued root development; the initiation of roots in the summer period is unexplained.

Light may also be a contributing factor in the initiation of roots.

Möller (1967) reported that in *Orchis mascula* the bud on the resting tuber attained a length of *c.* 30 mm in August, and roots began to develop. If the tuber was situated too deep in the soil no roots formed, which suggests that some signal produced when the bud reached the light was necessary for root initiation. He also observed that in species of *Ophrys* the roots develop simultaneously with the unfolding of the leaf rosette in autumn, which might also be connected with light-induced processes.

Spring roots could be induced by the same temperature stimulus that brings about the unfolding of the spring leaves, but experimental evidence is lacking.

Tuber development

The root-stem tuber is biennial or almost so, but for the first 6 months it remains within the sheath of the subtending leaf. When in *Orchis mascula* the slender roots emerge and the leafy shoot dies down in autumn the young tuber begins to develop as a bud which by January breaks through the leaf sheath. When the new leaf rosette sprouts in spring the young tuber has reached its full length but continues to increase in girth while the plant has leaves until flowering is over (Sharman, 1939). After the leaves have died down in summer, the tuber draws on the stored nutrients and provides sufficient nourishment for the new fibrous roots to develop and a new young tuber to be initiated in early autumn. By the following summer the tuber is finally exhausted (Sharman, 1939). In a similar manner, the young tuber in *Aceras anthropophorum* becomes visible in winter and attains full size in about 6 months, and then serves as a storage organ until the following summer (Wells, 1981). I have observed that the tuber in *Galearis spectabilis* appeared in May, at the same time as the leaves, and reached its full size in late autumn when the leaves and the old roots were disappearing (Fig. 7.5a, p. 138; Rasmussen, unpublished data).

In many species the tuber swells after the leafy shoot has appeared and presumably becomes filled with the products of photosynthesis. Those species in which shoot elongation is induced by chilling eventually produce a tuber, probably without further external stimulus. It is not known how tuber development is initiated in species with a life history where the first tuber is produced before the plant emerges above ground (Figs. 10.2b, c, and 10.3b).

In a study of *Orchis mascula* Möller (1967) observed that both the younger and the older tuber increased in size during spring; even after the unfolding of leaves and inflorescence there was no visible reduction in the old tuber. A possible explanation is that an increase in water content of the tuber could mask any transfer of biomass from tuber to shoot that might have taken place. The old tuber remained turgid until the end of the flowering season by which time the bud on the young tuber was about 5 mm long; then it began to wither. According to Sharman (1939) both tubers can survive if they are separated before this stage, and the old tuber has sufficient resources to support the development of a second young tuber (Section 11.2). It is possible that under natural conditions most of the nutrients that remain in the old tuber would be allocated to seed development. The shoot is still green during the fruiting stage, but the roots have already died (Möller, 1967).

Vöth (1980b) observed the growth of tubers in the soil through the glass side of a container and noticed that the new tuber stopped growing if the mycelium around roots and tubers had been damaged by drying out of the substrate. The undersized tuber entered into a premature resting phase and when the new shoot emerged it was correspondingly small, indicating that not only the water content but also the amount of stored nutrient was less than usual.

These observations imply that the development of the tuber is closely linked with the activities of the leafy shoot, but that hyphal connections are also required for successful development.

Transfer of infection between shoot generations

In many plants there arises a problem of transferring the infection from one generation of mycotrophic organs to the next if they are separated by non-infected tissue. For instance, cormous rhizomes are generally not infected, and since roots are often concentrated to certain parts of the rhizome segment new roots are separated in space from the old roots. It is often necessary to assume that the young roots receive their infection from the soil, but in some species there are arrangements to facilitate the retention of the original infection.

In *Aplectrum hyemale* the slender basal part of each rhizome segment is infected but the apical cormous part is not, the roots being concentrated in the zone between them, and probably becoming infected from the slender part. A new rhizome segment emerging from a bud on the cormous part is physically separated from the old infected tissue, but has at the base a zone of long rhizoids through which hyphae from the old roots can reach it with comparative ease (MacDougal, 1899a).

The tuberous species have a annual stage when only the tuber survives with no other organs present (Figs. 10.2 and 10.3). Burgeff (1909) suggested that the short rhizome that develops on top of the tuber might retain the infection for the next growing season. However, Fuchs & Ziegenspeck (1925) observed that this tissue becomes infected later than the new roots and therefore could not be the source of infection. Although there is little infection in tubers, the elongated tubers found in many species are colonized in the tapering parts and the ovoid tubers are probably slightly infected in the superficial cells layers (Section 7.4), which implies that they could retain the original infection. From observations of *Ophrys holoserica in situ* made through a glass window inserted in the soil Vöth (1980b) described how a mycelium extended from certain points on the tuber surface when the bud became active after dormancy. The new roots grew down through this mycelium and were probably infected through rhizoids. The younger tuber also became infected through rhizoids at the beginning of the flowering season.

In connection with vegetative reproduction there are often no mechanisms to secure a transfer of fungi from the mother plant, because the offspring typically is derived from uninfected storage tissue. *Gastrodia elata* produces lateral tubers that swell by means of nutrients transferred from the mother tuber, but they do not receive an infection from the mother plant. When they are released they die unless they are invaded by hyphae from the soil (Kusano, 1911). The same may be true of many protocorm-like propagules such as those produced on the margins of the succulent leaves of *Hammarbya paludosa* or on the stolons of *Hexalectris* and *Epipogium* (illustration in Correll, 1950, and in Irmisch, 1853).

11

Propagation

Practical cultivation of terrestrial orchids beyond the seedling stage is thoroughly described elsewhere (Fast, 1980; Cribb & Bailes, 1989), but since these plants are not subjected to large-scale production the present culture methods rely largely on a relatively small amount of accumulated experience. With more effective *in vitro* techniques for seedling production more plants will be available for culture experiments in the future. It would be desirable to have a more objective foundation on which to construct rational production schemes.

In this brief account I will focus only on the establishment in soil of the earliest life stages and on such aspects of cultivation as soil requirements and phenology that have a relation to the mycotrophic lifeform of the plants.

11.1 Weaning of seedlings from *in vitro* culture

Transfer of seedlings to the soil is perhaps the most challenging step in the cultivation of terrestrial orchids (Fast, 1982). Before they are removed from the protected environment of the culture vessel the seedlings should be as large and vigorous as possible, but there is usually an economic incentive to reduce the length of time they are kept *in vitro* and the number of transfers undertaken.

When seedlings are transferred to a fresh medium *in vitro* the growth rate always increases, particularly when the seeds have been sown densely (Fast, 1982). Wide spacing allows the seedlings to develop more rapidly (Tsutsui & Tomita, 1989; Rasmussen *et al.*, 1989). This advantage should be balanced against the greater number of culture vessels and amount of substrate necessary. Mitchell (1989) recommended that the protocorms be transferred for the first time when they are 3–5 mm long and a shoot tip has formed; even transfer to sterile compost may be possible at this stage.

Weaning of asymbiotic seedlings can involve a flask stage on inoculated agar or compost (Penningsfeld, 1990*a*) or explantation into clean pots containing sterilized soil which is subsequently inoculated. It will help the plants to adapt gradually to the parasitic microorganisms of the natural environment if they are infected by a compatible fungus while the substrate is still reasonably free from other microorganisms. When the mycorrhiza has been established, the seedlings appear to be healthier and survive the move to non-sterile conditions better than asymbiotic plants (Breddy & Black, 1954). This has been confirmed by Zettler & McInnis (1992) in a study of *Platanthera integrilabia* in which only the inoculated cultures produced plantlets that survived explantation into soil, but the rate of establishment was high – almost 20% of the seeds sown with the most suitable seed/fungus combination.

Apart from the risk of unwanted infection when the seedlings are removed from *in vitro* conditions there is also the danger of the plants drying out when exposed to less than saturated humidity. It is generally best to take the plants out when the root/shoot ratio is highest. Conditions *in vitro* can be adjusted to maximize this ratio, for instance by encouraging root growth by symbiotic infection and delaying the onset of photosynthesis. Myers & Ascher (1982) kept seedlings of North American swamp species *in vitro* in the dark as long as possible, i.e. until the leaves expanded, and weaned them when they had developed a relatively large root mass. A period of culture at cool temperatures (10 °C) can also stimulate root development (Malmgren, 1989*a*). In some species exposure to light or a reduction of the ion concentration in the medium could stimulate root development (Chapter 9). According to Mitchell (1989) the transfer of seedlings to sterile compost will also stimulate root development and is recommended if the seedlings begin to stain the agar dark so that the *in vitro* stage must be terminated.

Weaning during the summer is always risky, unless the plants have already formed tubers that are fairly resistant to desiccation. Weaning in late autumn and winter months is suitable for the wintergreen species, whereas spring weaning after a period of chilling is preferable for the summergreen species, if they are removed from *in vitro* conditions immediately after chilling (Mitchell, 1989). Spring explantation may be successful for summergreen species such as *Platanthera ciliaris* even if the plants are not subjected to a chilling period (Anderson, 1990b).

As regards moisture conditions the plantlets should be only gradually exposed to a lower relative humidity. Shading is required, and the plants are usually partially covered with glass, the humidity being kept high by means of indirect fogging (Myers & Ascher, 1982; Mitchell, 1989).

If the seedlings are produced out of season, a sojourn in pots in the greenhouse can help them to adapt to the seasonal rhythm outdoors so that they can be explanted immediately after the above-ground parts have died down (Mrkvicka, 1990b).

11.2 **Vegetative propagation**

Rhizomatous species can be divided by cutting the rhizome into sections that should represent not less than 3 years of growth. If the rhizome is partially severed by means of a shallow cut during the growing season before the intended division, this will encourage buds on the older parts of the rhizome to develop (Cribb & Bailes, 1989). The best time actually to divide them is at the end of the leafy season when the nutrient reserve in the rhizome is greatest and new roots are still short.

Many tuberous orchidoid species in genera such as *Anacamptis, Barlia, Dactylorhiza, Gymnadenia, Orchis, Ophrys* and *Serapias* tend to produce tubers from more than one of the basal buds, both in culture and under good conditions in nature (e.g. Ramin, 1973). The extra daughter tubers may form at some distance from the mother plant by means of various rhizomatous extensions (Chapter 7). This colonizing tendency can be encouraged by removing the young inflorescence (Sharman, 1939) and improving the growing conditions by organic mulching and by eliminating competition from neighbouring plants. If the daughter tubers are separated from the mother plant this serves to stimulate the proliferation further.

The species that usually produce only one replacement tuber can be propagated by division according to two methods depending on their leaf phenology. In summergreen species the two tubers can be separated during the flowering season; some authors advocate the beginning of the flowering season for this (Möller, 1987a; Mrkvicka, 1990b), others the time when flowering is coming to an end. Perhaps the optimum time varies between species. The young tuber is treated as though it is dormant and kept under dry conditions; Möller (1971) noted that this tuber will be just as vital as one that was allowed to remain on the old tuber. The leaf rosette remains on the old tuber and needs to be kept green for as long as possible so that a new young tuber can develop. Seed set should be prevented and the plant kept moist (Cribb & Bailes, 1989). Since a decrease in temperature is apparently the most important factor for causing the leaves to wither (Section 10.2) it should be possible to promote the growth of the new tuber by keeping the temperature at summer levels, but the critical range of temperatures is not known. If the plant is kept growing throughout the summer one or two new tubers will have developed by August or September and these may already be able to produce an inflorescence the following spring after a period of only 9 months of development (Möller, 1987a).

In the wintergreen species vegetative propagation can be carried out by separating the tubers in autumn when new leaves are appearing. Depending on species and locality the leaves unfold from October to around mid-December in European species (Ramin, 1976; Cribb & Bailes, 1989). The foliage leaves remain on the younger tuber that produces a replacement tuber and may flower as usual in the spring. The old tuber is terminated by a short segment of rhizome from the buds of which a new shoot and a new tuber develop (Ramin, 1976).

11.3 Growth conditions

Moisture is best supplied in the form of rainwater or demineralized water (Riley, 1983). For most species the general rule is that the soil should not dry out completely, but good drainage is essential (Riley, 1983; Pottinger *et al.*, 1988). For greenhouse cultivation of European species Bailes *et al.*, (1987) recommended a relative humidity of 50–60% in winter and 60–70% in summer. Species from boggy and swampy habitats may be cultivated in partially submerged pots provided that the compost is coarse and well aerated (Riley, 1983).

Species that are native to extremely summerdry Mediterranean regions should be fairly moist in winter, though in ventilated air, but when the leaves begin to die down in spring the watering must be reduced and the tubers kept dry. This period of drought is essential for the plants' survival, since too much moisture in cultivated plants during this period may cause the tubers to rot (Vöth, 1980*a*). Excessive drying, which could kill the living hyphae still present in the tuber, can be prevented by covering the pots with a mulch of leaf litter, but no water is given during summer (Ramin, 1976).

This treatment of tubers obviously does not apply to species growing in wet habitats or to those with a resting tuber during winter, which are best kept cool during the period of arrested growth.

Mediterranean species with a wintergreen rosette need additional light from November to February if they are grown indoors north of the natural range of distribution (Ramin, 1976).

For the north-temperate species a winter minimum temperature of 5 °C at night, rising to 10–12 °C during the day, is suitable. The spring temperatures should not fall below 10 °C, and the summer temperatures should not rise above 30 °C (Bailes *et al.*, 1987).

Mediterranean species prefer winter temperatures that lie between 12 and 20 °C during the day and a few degrees above 0 °C at night. Repeated freezing of wintergreen rosettes should be avoided. The protection of Mediterranean species and wintergreen species such as *Orchis morio* in a greenhouse in winter may also prevent damage by predators. The leaves often grow rapidly at fairly low temperatures, which is consistent with what happens in nature when the leaves sprout either in autumn or early spring. Burgeff (1954) found that leaf growth in *Orchis mascula* and *O. militaris* was better at 13–16 °C than at 20–25 °C.

It might be possible to extend the phototrophic season of summergreen species by inducing them to sprout early in the year by means of chilling, and to delay the dying down of the aerial shoot by several weeks by preventing a fall in temperature in autumn (Section 10.2).

Pot culture

When plants are to be transferred to pots the main concerns are to minimize damage to the roots, to preserve the symbiont and to prevent the plant from drying out.

It is best to transplant *Cypripedium calceolus* in March before the leaves have appeared and the youngest roots, that were initiated in the preceding year after flowering, have not yet attained full length. The plants can also be repotted when flowering is over, before new roots emerge, in which case mechanical damage to new roots is avoided (Möller, 1968).

Most tuberous species are fairly easy to repot after the leaves have died down since this is also the time when new roots emerge (Ramin, 1973). The plants must be planted with the bud directed upwards and at about the same depth as before, making sure that the new roots will not be exposed. The surface of the soil should be given a light organic mulch (Ramin, 1973). Although root damage is best avoided by repotting before new roots emerge the symbiont can be lost when there are no living roots on the plant (Bernard, 1909). For this reason some of the old soil should be retained (e.g. Ramin, 1973; Cribb & Bailes, 1989), although it is also possible that some living endophyte may remain in the surface layers of the tuber (Section 7.4).

If it is necessary to repot before the leaves die down, stress due to water loss and seed production can be reduced by removing the inflorescence. It is essential to provide plenty of moisture so that the vegetative season is not curtailed; watering should not cease until the leaves wither (Ramin, 1973).

All recommended potting mixtures have in common a high proportion of organic material and an open texture. Penningsfeld (1990a) suggested including 10% 'ventilation material' in the potting mixture, or as much as 30% in formulas used for *Ophrys* and other species of seasonally dry areas. The ventilation material varies from charcoal or chopped straw to perlite, vermiculite or sand (Burgeff, 1954; Harbeck, 1964; Riley, 1983). Good drainage can also be provided by stratifying the substrate in the pots; for the calciphilous species the basic layer could be made up of calcareous stones and gravel covered with a soil rich in loam and clay mixed with calcareous gravel, and an organic mulch (Möller, 1966).

For European orchids Bailes and co-workers (1987) recommended a mixture of 1 part unsterilized loam, 2 parts gritty sand, 2 parts fine pine bark, 1 part oak or beach leafmould. For calciphilous species some extra ground limestone is added, and extra peat is needed for rhizomatous species that have a higher requirement for moisture retention (Bailes *et al.*, 1987; Pottinger *et al.*, 1988). For *Dactylorhiza majalis* Mrkvicka (1990*b*) used a finely chopped moss substrate, consisting mainly of *Brachythecium rutabulum*, with a pH between 6 and 7.5. Pure *Sphagnum* is useful for *Listera cordata*; below the *Sphagnum* there should be a mixture of peat and calcareous sand, in the proportions of 3:1 (Mrkvicka, 1990*b*).

North American species of swampy habitats can also be grown in *Sphagnum* (Myers & Ascher, 1982), those from somewhat drier localities doing well in a mixture containing equal parts of *Sphagnum*, perlite, vermiculite and sand (Riley, 1983).

Burgeff (1954) experimented with germinating seeds of *Orchis mascula* and *O. militaris* directly in a sterilized organic compost that was inoculated with an appropriate fungus. The mixture consisted of chopped compost of *Polypodium* and leaf litter, the pH being adjusted by adding Ca CO_3. The best germination substrate was obtained from leaves of *Platanus* sp. that produce a basic mould with a pH of 7.9 (Burgeff, 1954).

Fertilizer

The use of artificial fertilizers is not generally recommended. Farmyard manure or a slow-release fertilizer are more suitable forms of fertilization. It is well known that orchids are weak in the competition with other plants when the soil is fertilized, but apparently the fertilizer may even harm the plants when they are grown singly in pots (Burgeff, 1954). The adverse effects of fertilizing may be related to a change in the relationship between the orchid and the fungus.

Most problems are apparently connected with nitrogen. High nitrogen levels tend to reduce the resistance of the plants to drought, frost and parasites (Möller, 1985c), and in symbiotic culture excess nitrogen may increase the virulence of the fungus, though a leaf fertilizer caused no problems (Penningsfeld, 1990b). Bailes and co-workers (1987) recommended light foliar feeds prior to flowering but these must be low in nitrogen. In contrast, special attention should be payed to supplying adequate micronutrients, such as manganese, boron and zinc (Möller, 1985c).

Growth rate in symbiotic cultures can be stimulated by the addition of a substance that is needed by the fungus, for instance cellulose (Penningsfeld, 1990b), and even after potting the seedlings can be watered with liquid fungus medium so that the symbiont is kept in a vigorous condition (Mitchell, 1989).

Parasitism and predators

When roots are damaged during transplantation it is often subsequent attacks by parasitic fungi that reduce the chance of establishment (Wherry, in Correll, 1950). There are not many reports on diseases in holarctic orchids, although Case (1983) mentioned *Botrytis* in connection with North American species of *Platanthera*. Invertebrates such as thrips, weevils, snails and slugs may sometimes be troublesome (Burgeff, 1954; Willems, 1982; Case, 1983).

The underground storage organs that contain large amounts of nutrients and occasionally are fragrant attract mice and other rodents; squirrels may cache tubers and corms for winter food (Burgeff, 1954; Wherry, in Correll, 1950; Case, 1983). Other vertebrates such as birds, rabbits, hares, deer and wild boars have also been observed feeding on leaves or below-ground parts of orchids (Burgeff, 1954; Stuckey, 1967; Gumprecht, 1980; Willems, 1982; Case, 1983; Whigham & O'Neill, 1988).

Grazing by deer may, on the other hand, be beneficial for species such as *Cypripedium* that require open vegetation (Case, 1964).

12

Effects of orchid mycorrhiza

Mycorrhiza is widespread in all groups of higher plants (Stahl, 1900; Allen, 1991). In general mycorrhizal relationships increase the plant's access to soil resources, i.e. water and mineral ions, thereby increasing its drought tolerance and ability to grow in poor and leached soils. Ectotrophic mycorrhiza is thus strongly associated with conditions of mineral nutrient stress (Harley, 1969).

Orchid mycorrhiza differs in important respects from most other types of mycorrhiza (e.g. Smith, 1974; Harley, 1984; Allen, 1991). The green plant functions as the producer of energy in ectomycorrhizal systems, VAM and ericaceous mycorrhiza, but in the mycorrhiza of orchids the fungus is the source of energy and thus provides the plants with a separate nutrition. This may alternate with, supplement or entirely replace the phototrophic nutrition. Although gametophytes of certain pteridophytes and a few specialized groups of higher plants such as *Monotropa* have established a similar parasitism on fungi, Orchidaceae is the only large plant group that makes consistent use of this kind of nutrition. It is necessary for seedlings of most plants to develop leaves before the nutrient reserves in the seed are exhausted, and for the plants to remain phototrophic throughout their life. In contrast, orchid seedlings have an option of living for extended periods as heterotrophic organisms; this opens habitats and evolutionary pathways for orchids that would not otherwise have been accessible.

12.1 Ecological consequences

Importance of light

Since the orchid mycorrhiza represents an alternative source of energy, it is easy to explain how some species of orchids thrive in deep shade where few other plants will grow. This is most evidently true of the species with chlorophyll deficiency.

Not all orchids are shade tolerant, however. Closing of the canopy can be a threat to species that inhabit grassland areas and open woods; in a mixed population of *Orchis purpurea* and *O. ustulata* Möller (1986*a*) observed a increase in the former but a decrease in the latter when the habitat became shaded. Plants of *Cypripedium candidum* and *C. reginae* first react to increasing shade by becoming etiolated, then they produce fewer flowers and smaller shoots, and after a few more years the plants disappear from above ground. Several other terrestrial orchids react similarly, for instance the 'fringed' species of *Platanthera* (Case, 1990). Many species characteristically occur in areas with unstable, continuously incomplete vegetation cover caused by various kinds of disturbance such as fluctuating water levels, grazing, cutting, clearing or burning (e.g. Stuckey, 1967; Case, 1983, 1990; Hutchings, 1987*a*). This is apparently connected with a requirement for light and unrelated to the mycotrophic nutrition of the plants. However, the tolerance to burning, grazing, etc., is clearly increased by the geophytic habit with a large part of the biomass below ground (e.g. Whigham, 1984), and possession of an underground source of energy.

The mycotrophic nutritional system is also responsible for the ability of orchids to remain below ground for one or more seasons. This phenomenon is well documented by long-term observations of individuals and occurs in both rhizomatous and tuberous species, though predominantly in the former (Table 8.2, p. 169). When monitoring some of the sparse populations of the extremely rare *Isotria medeoloides*, Gaddy (1983) found that the shoots did not appear above ground every year, and that about 40% of the plants could remain below ground in a typical year. This means that in theory a population could remain unnoticed in a locality where conditions no longer favour their photosynthesis and reappear, for instance, when forest is cleared. Fuchs & Ziegenspeck (1924b) cited an unpublished report on mass occurrence of flowering *Cypripedium calceolus* 2 years after a forest clearance in a locality where the species had not been recorded in recent times.

Natural substrate

The digestion of fungi requires oxygen and there is a need for a high content of organic debris in the soil on which the fungi can subsist. This is in agreement with the description of typical orchid soils as loose textured with high humus content (e.g. Sheviak, 1974).

Since the mycorrhizal fungi increase the efficiency of mineral uptake orchids and other mycotrophic plants have a competitive advantage in poor soils and soils with low accessibility of minerals, such as those with extreme pH. This can explain the frequent occurrence of orchids in acid bogs and on highly calcareous soil, and their success as epiphytes on the trunk and branches of trees, where the mineral supply is based only on dust and organic debris lodged on the bark of the phorophyte. Many orchid soils do in fact have a low content of minerals, in particular inorganic nitrogen (Section 9.4).

The generally low surface-to-volume ratio in the root system of orchids, and the existence of rootless forms, imply that the fungi assist in the uptake of water. Although the symbiosis requires some precipitation to function there is a considerable tolerance of seasonal drought, as can be inferred from the existence of a rich flora of tuberous orchids in Mediterranean climates.

Spatial distribution

Highly specific mycorrhizal relationships can only exist in a stable environment (Allen, 1991). Orchids can live in symbiosis with many different fungi, which implies that they can employ a succession of fungi during their lifetime when growing in a changeable habitat. If the specificity for fungi were high or the fungi local in distribution the great fecundity and dispersal potential of orchids would be illusory, but the fungi appear to be virtually omnipresent and able to exist without orchids.

On the other hand, a low specificity could ease interspecific competition between sympatric orchid species, if they could employ different species of fungi. There are some observations of mixed orchid populations in support of this idea (Dixon, 1991; Rasmussen, unpublished data), but further investigation is needed to show whether sympatric orchid species characteristically have complementary demands on the local fungi.

The successful spread of orchid seedlings requires a uniform availability of compatible fungi. The spatial distribution of orchid fungi is largely unknown; organic matter, which is the ultimate resource for the mycotrophic nutrition, is unevenly distributed in the soil. Competition between siblings for available mycelia could be an important factor in the spatial distribution of individuals as described in cultures *in vitro* (Alexander & Hadley, 1983; Rasmussen *et al.*, 1989; Tsutsui & Tomita, 1989). Some species grow in clusters, others tend to be widely spaced single individuals. Although this depends to a large extent on a differential tendency to produce vegetative offshoots, intraspecific competition among seedlings could also influence spacing patterns.

Phenology

Photosynthesis is a seasonally fluctuating but predictable source of nutrition whereas mycotrophy depends on the provision of organic matter that is unevenly distributed in the soil. In seasonal climates, saprophytic fungi generally have their optimal development in autumn with the fall of the leaves. Orchid species with mixed phototrophic and mycotrophic nutrition may have the strategic advantage of a longer growing season, or more than one annual growing season, but year-round studies of mycorrhizal dynamics and biomass acquisition should be undertaken to investigate this possibility. Some investigators have evaluated the mycorrhiza of the adult plant to be insignificant, except in the obviously chlorophyll-deficient species (Hadley & Pegg, 1989), and one study indicated that young plants of *Goodyera repens* relied to a decreasing extent on the symbiont as they matured (Alexander & Hadley, 1985). It is important to obtain information from a wider range of species, since undoubtedly the dependency on fungi varies, and a better understanding of the nutritional strategy might help to explain why some orchid species are relatively competitive and persistent whereas others are extremely rare and sporadic where they occur.

The year-round development of orchid fungi is largely unknown, but in seasonally dry habitats survival could be difficult, since *Rhizoctonia* rarely forms drought resistant spores. It has been suggested that accommodation within the plant tissue during unfavourable periods could be the only real benefit that the fungi have in the symbiosis with orchids (Burgeff, 1943). When symbiotic seedlings grown *in vitro* are transferred, hyphae invariably grow out from them and form a mycelium on the fresh substrate (Burgeff, 1936, and others), although presumably the external hyphae have all dried out during the transfer. In many species permanently infected tissues have been described, and in some species there are structural modifications that enable the infection to be passed from one shoot generation to the next (Section 10.2).

Seedling establishment

Fungal persistence around the adult plant might mean that seedlings have a greater chance of establishing in the vicinity, i.e. a kind of seedling nursing as suggested by Dixon (1991). In some habitats the substrate of the mother plant might still be suitable for seedling growth, but several observations suggest that the ideal germination site often differs from the site of the adult plant; in open spots with limited competition, such as newly exposed soil or sand, seedlings may occur in high numbers (Case, 1983). These opportunities for seedling establishment could be a main reason why many orchid species show a preference for somewhat disturbed areas. I almost never observed seedlings of *Tipularia discolor* close to flowering individuals, but they were abundant in a different habitat, i.e. decaying logs (Rasmussen, 1992a). The search for orchid seedlings is usually concentrated around adult plants, which could explain why they have been found so rarely in the field (e.g. Whigham & O'Neill, 1988).

Orchids are potentially long-lived but, by virtue of their massive seed production, opportunistic in their initial choice of site. Adult plants may occur in sites that would no longer be suitable for recruitment of seedlings, so that the population may need to move in order to maintain its overall size. This is due to the heterogeneous and unstable distribution of substrate for the fungi. Events such as windfalls or fire may deposit dead biomass for fungal decomposition in unpredictable locations and at unpredictable times. It is likely that the weedy and unstable nature of many orchid populations is partially a result of opportunistic use of the sudden food resources, although the prospering orchid population may not become visible until several years after the event. Typically the highest increase in population size occurs shortly after initial invasion; after a span of years the population may be declining because of low seedling recruitment (Sheviak, 1974; Case, 1983). Some of the populations that we observe today may in fact be senile rather than stable (Tamm, 1991).

Clones of *Cypripedium calceolus* can reach an age of about 192 years, if estimated on the basis of annual increment of the rhizome (Kull, 1988) – more than enough time for successional changes to occur in the habitat. Figures in about the same range were reached by Tamm (1991), who calculated the half-life in a declining population of *Listera ovata* to be 70 years. A lack of recruitment in populations of such plants will not be obvious for a long time. However, at the other extreme, *Ophrys sphegodes* has an almost weedy strategy, reaching the reproductive stage so quickly that about 84% of the plants that emerge above ground for the first time flower in the same year (Hutchings, 1989). In this species the rate of seedling recruitment has an almost immediate influence on the population size (Waite, 1989).

The contrast between the strategy of seed/seedling and that of the adult plant is most notable in the long-lived species. Although a successional habitat may be beneficial for seedling recruitment, a series of favourable years is usually required for the seedlings to reach flowering size. As pointed out by Curtis (1943) the long establishment phase does require a measure of stability in the environment because the seedlings are poor competitors with a minimum impact on their surroundings. In a population of seedlings Curtis (1943) noted a steep survival curve, i.e. among 100 seedlings only 4 were estimated to be more than 2 years old.

12.2 Evolutionary events and effects

Seedling mycotrophy

The lack of a radicular meristem and its tranformation into a mycotrophic tissue is considered to be a distinguishing feature of all members of the Orchidaceae (Vermeulen, 1966) and is probably a character unique to this family, making the orchid seedlings irreversibly modified towards mycotrophy. This modification must have been a decisive step in the early evolution of the orchids. Since germination in members of the Apostasioideae has never been observed it is not known whether they have mycotrophic seedlings, but it seems likely considering that the seeds are minute (Vermeulen, 1966). Adult plants in the Apostasioideae do form endotrophic mycorrhiza (Stern *et al.*, 1993).

The evolution towards increased dependency on mycotrophy entails a number of character losses and reductions, such as the loss of radicle, reduction in embryo differentiation and reduction or eventual loss of any endosperm formation and double fertilization. For this reason there is probably no option for evolutionary reversal, which would entail the reorganization of the radicle as an organ of absorption (water, mineral ions) for a photosynthetic seedling.

The physiological dependency on the fungi varies in degree, seedlings of some species becoming photosynthetic at an early stage while others have complex requirements that in asymbiotic culture must be met by components in the substrate. This diversity in seedling physiology reflects a progressive adaptation to mycotrophy, which if surveyed by means of minimalized substrates could help in tracing the course of evolution.

It is undoubtedly the mycotrophic nutrition of the seedling that has allowed reserve nutrients in the seed to be reduced to negligible amounts (Benzing, 1981). Since epiphytic species tend to have relatively smaller seeds than terrestrial species seed size may have been a decisive factor in the dispersal as epiphytes. Orchid mycotrophy can therefore be viewed as largely responsible not only for the great potential for dispersal that has enabled this group of plants to reach a world-wide distribution, but also for the evolution of epiphytic lifeforms. About 70% of all orchid species are epiphytes and they comprise more than two-thirds of the epiphytic plant species on earth, according to the estimates by Gentry & Dodson (1987). Orchid and ericoid mycorrhiza seem to be more suitable than other mycorrhizal systems for the epiphytic habitat (Lesica & Antibus, 1990). It is reasonable to assume that this is connected with the microspermy in both plant groups.

Germination

Test sowings *in vitro* and in the field show that some orchid seeds contain a fair amount of reserve nutrients that enable them to germinate in water and remain alive for some time. This lack of specialized germination requirements is probably a relatively ancestral condition in orchids, being comparable to conditions in other monocotyledons and, indeed, in angiosperms generally.

In contrast, other cultures *in vitro* and field observations indicate that some orchid species have acquired mechanisms that prevent the seeds from germinating in the absence of a compatible fungus. This germination strategy must have arisen after the mycotrophy of the seedling, and may have evolved more than once. A strong selection for this trait can be expected, since there is little advantage in germination without subsequent seedling development. The readily germinated species build up a short-term bank of seedlings that will exhaust their reserve nutrients unless a compatible fungus is close enough to reach them in time. In contrast, species with specific requirements can maintain a seed bank until conditions for seedling development are appropriate. Seed longevity in the soil may be in the order of months, even years, as judged by the first results of field sowing experiments (Rasmussen & Whigham, 1993). However, more research is needed to establish which orchid species depend on fungal presence during germination itself and the length of time the seeds remain viable. A dependency on fungal presence could have arisen independently in several evolutionary lines.

The role of the fungi in their relationship with seedlings and mature plants is no doubt primarily nutritional. In contrast, there is little information as to the nature of interactions that enable seeds to detect the presence of fungi and distinguish between different strains.

Mycotrophy in adult plants

Loss of chlorophyll and a reduction of foliage leaves in adult plants shows that some species have adopted a predominantly mycotrophic nutrition, and the structural diversity in these modifications indicates parallel evolution in different phylogenetic lines within the Orchidaceae (Brieger, 1976; Rasmussen, 1994*a*).

Since chlorophyll-deficient orchids survive as adults with an extreme dependency on mycotrophy there is no reason to ignore the possible benefits from mycotrophy in green orchids also. There are no actual measurements of the fungal contribution in any species, but a considerable body of indirect evidence (Chapters 7 and 8). Assuming some gains are achieved through mycotrophy the plants have two nutritional options; a differentiation in timing and relative importance of these could stimulate ecological diversity and speciation among related species. In the genera *Spiranthes*, *Epipactis*, *Cephalanthera* (including *C. austinae*) and *Orchis* there is a diversification among the species with respect to leaf phenology, including examples of chlorophyll-deficient forms.

Pollination biology

Mycotrophy as a mode of seedling establishment entails certain hazards, fungi being potentially either unavailable or incompatible with the seeds. The great potential for dispersal of the small seeds can only be effective if large numbers of seeds are produced. This, on the other hand, is possible only when the mother plant invests a relatively small amount of nutrients in each seed.

A large seed set results either from a high frequency of successful pollinations or from large numbers of seeds being produced from each pollination. Autogamous flowers do occur in some orchid species which produce numerous small fruits with – for orchids – comparatively few seeds; examples are *Corallorhiza odontorhiza* and *Listera cordata*. In most species, however, self-pollination is prevented by ecological or structural barriers (Dafni & Bernhardt, 1990) and at the same time the frequency of successful outcrosses is often limited by the availability of pollinators (e.g. Case, 1983, in *Arethusa bulbosa*). A large seed set is thus usually achieved by setting a few large fruits with numerous seeds.

This in turn requires the transfer of a large number of pollen grains to the stigma. The gynostemium, formed by adnation of filaments and the styles, the aggregated pollen (i.e. pollinia) and the shape of the stigma are orchid characteristics that ensure the deposition and pickup of ample pollen in a single pollination event (e.g. Benzing, 1981).

It is easy to see how the zygomorphic perianth and gynostemium is adapted to the need of regulating movements and behaviour of the pollinators to make them pick up and correctly position pollinia. Generalized attraction mechanisms aimed at a number of different pollinators may result in frequent visits but also in a risk of illegitimate depositions of pollinia and waste of pollen. This situation would favour the production of smaller pollinia with fewer pollen grains. Floral evolution in orchids has usually been towards increasingly intricate pollination syndromes and attraction mechanisms, aimed at a limited group of potential insect visitors whose behaviour is narrowly restricted.

Conclusion

Most of the traits that we consider to be unique or characteristic for orchids (gynostemium, pollinia, large seed production, minute seeds, etc.) are tied into a functional relationship with the orchid mycor-rhiza; they are either made possible by mycotrophy, such as the reduction in seed size, or necessitated by mycotrophy to ensure the success of this type of seedling establishment.

While orchid mycotrophy has not proved highly successful in terms of global biomass of its representatives, this group of plants is often singled out as particularly prolific in respect to speciation. The pollination syndromes typically encountered in orchids only work at a relatively short range, which is far exceeded by the dispersal of seeds. This sets a scenario in which small populations, perhaps consisting of only one individual, can easily become established at a long distance from the mother population, making further exchange of pollen unlikely. This in turn could stimulate speciation by founder effects, differential selection pressure or genetic drift. The orchid family is indeed characterized by a profusion of species, often with very limited geographical distribution. An intuitive recognition of such patterns may be expressed in comments such as: 'The orchid family is still in a state of evolutionary flux' (Garay, 1960), or 'Orchidaceae have all the earmarks of a group in active evolution; species, genera, tribes and subtribes are all difficult to delimit' (Dressler, 1993).

It is thus possible to infer that the evolution of mycotrophic seedlings was a founding and irreversible event in the history of orchids. Furthermore, the adaptations that are associated with the mycotrophy may have been a driving force in the speciation of the numerous orchids that we know today.

13

Descriptions of genera

13.1 Cypripedium

A circumboreal genus, *Cypripedium* occurs in bogs, swamps and woodland, especially in forest margins and clearings, usually not in deep shade. *C. calceolus* is virtually confined to calcareous soils, the topsoil being characteristically friable with a high humus content (Fuchs & Ziegenspeck, 1926a). *C. candidum* grows in black humus saturated with lime, the soil having a pH of 7.4 (Stoutamire, 1990). A stand of *C. reginae* in Canada was found growing in soil with about 20% more calcium than the average for the region (Harvais, 1980). Although most of the species thus prefer a neutral or alkaline substrate (Wherry, 1918; Curtis, 1943), *C. acaule* grows in acid soil (pH 3.5–5.4: Wherry, 1918; Curtis, 1943; Stoutamire, 1963).

In all species the adult plants have a horizontal rhizome and well-developed, dark green foliage leaves, and maintain this foliage from spring to autumn. They overwinter below ground. The vegetative period before flowering is assumed to be long: up to 10 years or more in American species (Curtis, 1943) and 6–10 years in *C. calceolus* (Kober, 1972; Fast, 1985).

In *C. arietinum*, *C. candidum* and *C. calceolus* var.*pubescens* the tips of 2 to 3-year-old roots may form buds as a means of vegetative propagation (Curtis, 1943).

Life history

C. *calceolus* presumably germinates in spring (Fuchs & Ziegenspeck, 1926*a*), although Irmisch (1853) observed very small protocorms of C. *calceolus* in December. C. *acaule* begins to germinate in spring and early summer (Curtis, 1943). The seeds usually lie fairly deep (2–5 cm), typically occurring below a layer of living mosses and partially decomposed plant debris on top of the humous mineral soil (Fuchs & Ziegenspeck, 1926*a*; Curtis, 1943).

The young protocorm of C. *calceolus* is greenish, though subterranean (Fuchs & Ziegenspeck, 1926*a*). In spring it produces long, infected rhizoids, the glabrous appearance observed by Irmisch (1853) perhaps being due to the withering of rhizoids during late summer and autumn (Fuchs & Ziegenspeck, 1926*a*). During the first autumn after germination a root emerges behind the shoot tip, and continues to grow throughout the winter and spring. During the following year the rhizome increases somewhat in length and produces a new root in the second autumn (Fuchs & Ziegenspeck, 1926*a*; Curtis, 1943). The previously positively geotropic rhizome now grows upwards and during the third summer produces a large apical bud, while on the lower internodes one or more additional roots begin to develop in the third autumn. According to Fuchs & Ziegenspeck (1926*a*) and Curtis (1943), the first foliage leaves typically do not emerge until the third (C. *reginae*) or fourth spring (C. *acaule*, C. *calceolus*). According to Irmisch (1853) seedlings of C. *calceolus* may produce one small leaf, as early as the spring immediately following germination. This is confirmed by field sowings conducted by Böckel (1972), who found that seedling leaves began to appear 24 months after sowing, which may have been roughly 1 year after germination.

The first root produced on the protocorm of C. *calceolus*, which is mycotrophic from the beginning, bears numerous rhizoids and contains little xylem and a considerable amount of phloem (Fuchs & Ziegenspeck, 1925). The cortex is densely infected with pelotons in various stages of lysis. The following roots are progressively less specialized for mycotrophy and from about the fifth root development is endogenous. Roots that develop when the rhizome is beginning to form the first foliage leaf are more typical slender roots for conducting water, and they usually lack pelotons. Thus the extent of mycorrhiza declines from year to year, while the above-ground organs increase in volume (Fuchs & Ziegenspeck, 1925).

In adult plants of *C. calceolus*, the bud containing foliage leaves and flowers for the next season differentiates during the summer. The new rhizome segment begins to elongate even before flowering ceases; the new roots emerge about 2 weeks later, and by the onset of winter they have attained their final length (Möller, 1968). The old leaves wither away in September and October (Möller, 1987a). In the other species the leaf phenology is probably similar. In *C. reginae* young roots emerge from the beginning of July and are fully developed, though still uninfected, by the end of August (Harvais, 1974).

Roots that are a couple of years old contain only patches of infection, mainly with dead pelotons which could represent the remains of temporary mycorrhizal infection during the underground season, although Harvais (1974) interpreted the infections in old roots of *C. reginae* as fortuitous and possibly pathogenic. The fact that new roots are initiated during late summer and autumn at a time when the leaves are about to die down and water requirements are decreasing to a minimum, suggests that even long and slender roots have an initially mycotrophic function. Old roots are able to resume growth, adding a new segment each year (Harvais, 1974; Möller, 1987a). The individual roots of *C. candidum* live for at least 3–4 years, spreading in the topsoil 2–5 cm below the surface (Stoutamire, 1990).

Although the roots reportedly display progressively less mycotrophic activity during the first years with leafy shoots, established plants have been observed to survive in culture in a leafless condition (Sheviak, 1974). In one natural population of *C. acaule*, some individuals reappeared after having been underground for 1–5 years (Gill, 1989).

Endophytes

During the first growing seasons the seedlings of *Cypripedium* are wholly dependent on a symbiotic fungus. Judging by the anatomy of the roots, the extent of mycotrophy decreases as the plants produce more and larger leaves, i.e. as the photosynthetic capacity increases. Living pelotons are notoriously rare in the roots of adult plants (Curtis, 1939; Harvais, 1974; Stoutamire, 1990). There may be small pockets of infection along the roots but the cortex usually contains mainly starch (Mitchell, 1989; Stoutamire, 1990). According to the phenology outlined above, late autumn should be the best time for observing active mycorrhiza, since in August the roots of the current season have not yet become infected and the old roots contain only dead pelotons (Harvais, 1974).

Both in *C. calceolus* var. *parviflorus* and var. *pubescens* MacDougal (1899*a*) observed that the hyaline hyphae were concentrated to the epidermis and exodermis; one or two outer layers of cortex contained digested pelotons only and the rest of the cortex was filled with starch. Stoutamire (1991) indicated that in *C. candidum* the densest infection is located about 15–20 cm from the rhizome in long, mature roots; segments of the root that are less than 1 year old are rarely infected. However, the roots of established plants are endogenous and develop from the uninfected part of the rhizome, so that infection must come from the soil. Isolates from mature plants could therefore differ entirely from the seedling fungus.

This explains the series of unsuccessful attempts to germinate seeds with isolates from adult plants. Germination and seedling development may be stimulated by the isolated strains but no mycorrhiza is formed, and so far no vigorous symbiotic seedlings have been produced *in vitro*. Isolates from adult plant roots identified as *Rhizoctonia repens* and *R. monilioides* (however, see Chapter 5 concerning the identity of these taxa) did stimulate germination in seeds of *C. reginae* that failed to germinate in asymbiotic controls (Curtis, 1939). Seeds of *C. calceolus* var. *parviflora* germinated with a fungus called *R. lanuginosa*, but several other species of *Cypripedium* germinated with none of the numerous isolates from *Cypripedium* that were tested by Curtis (1939). Germination was initiated by unspecified soil fungi found on or in roots of *C. calceolus* and *Cephalanthera rubra* (Weinert, 1990); these fungi did not form pelotons in the seedlings, but germination percentages were higher than in aseptic controls. Light (1994) succeeded in improving the development of asymbiotic seedlings by inoculation with two isolates from roots of adult plants, both of them Ascomycetes; they did not form endomycorrhiza with the seedlings, and there were incidents of parasitism. Although some fungi have been isolated from roots of *C. calceolus* at Kew none of the strains have had convincing effects on the seeds and seedlings (J. Stewart, personal communication). The same applies to fungi isolated from *C. candidum*, which neither affected the growth of the seedlings nor established seedling mycorrhiza (Stoutamire, 1990).

Before foliage leaves begin to develop the seedlings presumably maintain the original fungus in the rhizome, the roots being infected from within. Such young seedlings should be looked for and examined so that the identity and characteristics of the fungi that are active in the germination of *Cypripedium* species can be established. However, this alone may not solve the problem. I have extracted numerous apparently living pelotons from small seedlings without obtaining a growing mycelium on agar. It is possible either that these endophytes become irreversibly altered as soon as they enter the plant cells, or that they have specialized nutritional requirements.

Seed storage and survival

Seeds of *C. reginae*, that are dried and kept at − 5 or − 20 °C remain viable for more than 2 years (Harvais, 1980). Stoutamire (1990) found that seeds of *C. reginae* could be made to germinate after 3 years in dry, refrigerated storage, and those of *C. acaule* after 10 years.

Several observations suggest that mature seeds retain their viability for many months, even under moist conditions. Most species of *Cypripedium* have durable infructescences that are apparently adapted to the dispersal of the seeds over a long period of time (Stoutamire, 1974; Lucke, 1982). Furthermore, seeds may retain viability when incubated on agar even though they do not germinate (Ballard, 1987, see below). Seeds of *C. reginae* and *C. acaule* that had inadvertently been subjected to warm and moist conditions during transport and had become slightly fermented before they were sown, germinated extraordinarily well (Fast, 1982).

Germination in culture

The generally low germination percentages obtained with mature seeds of this genus can be explained either by persisting dormancy or by poor seed quality perhaps related to pollination efficiency or genetic or climatic factors. Both selfing and cross-pollination result in very variable germination percentages in *C. reginae*, irrespective of the age of the flower, but the quality of the seed can be improved by loading the stigma with a surplus of pollen (Ballard, 1987). Seed quality depends on provenance, and within populations of *C. calceolus* varies from year to year (Fast, 1974; Riether, 1990).

The measures taken to improve the germination response in *Cypripdium* spp. can be divided into five categories: (*a*) scarification of the testa, (b) lengthy soaking and cold stratification, (c) control of the ambient atmosphere, (d) adjustment of the light regime and (e) chemical factors in the substrate.

(a) Scarification   According to Van Waes & Debergh (1986*b*) the optimum duration of surface treatment for *C. calceolus* is 4 hours in 5% $Ca(OCl)_2$, which is rather short in comparison with their recommendations for other species. The resulting germination is about 40% as compared with virtually zero if sterilization is minimal, i.e. no more than is required to decontaminate the seeds (Van Waes & Debergh, 1986*a*). In *C. calceolus* and *C. acaule*, Riether (1990) also found marked differences in germination percentage according to sterilization time. In general, lengthy treatment in a weak concentration of hypochlorite has proved to be better than brief treatment in a strong solution. Frosch (1982) found that 100 minutes in a 0.3% NaOCl solution for *C. calceolus* and 40 minutes in 2% NaOCl for *C.*

calceolus var. *parviflorum* were suitable. Vacuum infiltration with water prior to sterilization further raises the germination percentage; this was used by Anderson (1990) followed by a treatment with 10% Chlorox for 6 hours (*C. acaule*), 2 hours (*C. candidum*) or 1 hour (*C. calceolus*). An initial rinsing of the seeds of *C. calceolus* in 0.2–5.0% H_2SO_4 before surface sterilization in NaOCl has proved beneficial for germination, but the optimum concentration and time is not yet known (Malmgren, 1993).

(b) Lengthy soaking and cold stratification   The results of soaking and cold stratification vary with the species to such an extent that they must be considered separately.

C. *acaule*: Cold stratification is not required for germination, which is initiated after about 8 months at room temperature with or without chilling, but germination takes place over a long period, of up to 2–3 years (Stoutamire, 1974; Ballard, 1990; Ling *et al.*, 1990). In contrast, simultaneous germination of *c.* 70% of the seeds was obtained if immediately after sowing they were kept for 3–5 months at 5°C before incubation at 25°C (Coke, 1990).

C. *calceolus*: Burgeff (1954) imbibed the seeds in water for 2–3 months, but did not give data for comparing the results with those obtained without soaking. Butcher & Marlow (1989) advocated removing the testa. Other attempts to break dormancy in fresh, mature seeds include shorter periods of submergence (1–8 days) in water, in a 0.5% solution of sucrose, or in 0.05% solution of KCl, none of which gave positive results (Fast, 1974). The effect of high (30°C) and low (−5 to 5°C) temperatures was tested on seeds from closed capsules and on dry and moist mature seeds for 1–8 weeks, but all tests proved equally unsuccessful (Fast, 1974). Some workers have tried cold stratification without positive effects (Lucke, 1981; Riether, 1990), but Ballard (1990) obtained 16% germination after chilling for 4 months compared with 8% after 3 months of chilling and 1% without chilling, and Coke (1990) obtained 50% germination after 3–5 months at 5°C.

C. *candidum*: Eight weeks at 5°C did not promote germination that was poor on a range of substrates (Stoutamire, 1990); discolouration and death occurred in most seedlings when they were *c.* 1 mm in diameter.

C. reginae: This seems to be easier to work with than most species and responds well to cold stratification (5 °C) for at least 2 months (Stoutamire, 1974; Ballard, 1987). Ballard sowed the seeds on agar in culture vessels and stored them for several months before subjecting them to chilling; he obtained a germination above 90%. This shows that there was no loss of seed viability due to storage under moist conditions at room temperature (Ballard, 1987). Reyburn (1978) could not raise the level of germination by cold treatment, but contamination seems to have been a general problem in his investigation. Lengthy soaking of the seed (45 days) in sterile water before sowing has been tested in several species, but apart from *C. reginae* and *C. californicum* the resulting germination percentages were extremely low (Oliva & Arditti, 1984). In experiments where a liquid substrate was used throughout the germination percentage of *C. reginae* was low, and the seedlings turned brown and gradually died (Harvais, 1982).

(c) Atmosphere It has been reported that more seeds germinate when the culture vessels are sealed (*C. acaule*: Kano, 1968) or the seeds are covered with culture medium (*Cypripedium* spp.: Curtis, 1943, *C. calceolus* var. *parviflorum*: Weinert, 1990). Ballard (1987), who obtained very high germination percentages, sealed the corked vessels with paraffin.

(d) Light All information suggests that the seeds are light sensitive and should be germinated in the dark (Curtis, 1943; Withner, 1953; Burgeff, 1954; Harvais, 1973; Stoutamire, 1974; Light, 1989). Ballard (1987) found that in *C. reginae* germination was inhibited by light, even after cold stratification. In a survey of species by Arditti and colleagues (1985), germination was found to be inhibited by light in seeds of *C. californicum* and *C. acaule*, but in those of *C. calceolus* var. *pubescens* × *C. reginae* there was no reaction. However, most germination percentages were so low that the results were inconclusive.

(e) Medium *C. reginae* is the only species for which the chemical requirements have been investigated by reducing the number of ingredients in the substrate (Table 4.3, p. 62; Harvais, 1973). This

species did not germinate on water agar or on a mineral salt substrate. If 1% glucose was added the seeds germinated, showing that they require sugar for germination. A supplement of kinetin had a pronounced effect, raising the germination percentage from 5% in controls to near 100% with 0.5–1 mg kinetin per litre substrate (Harvais, 1980). Auxins inhibited germination and the protocorms developed poorly. On Knudson C medium 50–75% of the seeds of *C. reginae* germinated, the best results being obtained when it included a buffer solution of KH_2PO_4 and K_2HPO_4. Other media tested, such as Chang, ZAK, Fast FN, Burgeff Eg 1 and Harvais have been found to be inferior or useless for *C. reginae* (Ballard, 1987). However, Anderson (1990) found that *C. reginae* germinated more rapidly on improved Harvais medium (Harvais, 1982) than on Knudson C, namely 8 weeks versus 4–6 months. Seeds of *C. reginae* have also been germinated on Thomale GD (Haas, 1973), Burgeff N_3f (Reyburn, 1978), modified Curtis (Oliva & Arditti, 1984), Norstog (Henrich *et al.*, 1981) and Norstog modified with $0.2 \, mg \, l^{-1}$ benzyladenine (BA) and reduced content of mineral salts (Pauw & Remphrey, 1993).

Some seeds of *C. calceolus* germinated on Fast FN, but almost none on other media tested, such as Knudson C and ZAK (Fast, 1974). Lindén (1980), however, obtained good germination results on ZAK as well as on Fast and Eg1, but all media were supplemented with microelements, vitamins, kinetin and auxins. Smreciu & Currah (1989*a*) also found Fast medium in Lindén's modification useful. Curtis and Norstog media were used by Butcher & Marlow (1989).

C. acaule germinated on Burgeff N_3f supplemented with peptone (Withner, 1953), on a modified Knudson C medium (Ballard, 1990), or on improved Harvais (Anderson, 1990*a*). Improved Harvais was also found to be suitable for *C. candidum* and *C. calceolus* (Anderson, 1990*a*). *C. debile* and *C. japonicum* germinated on Knudson C and on Karasawa medium (Sawa *et al.*, 1979, according to Nishimura, 1982). About 25% of seeds of *C. calceolus* var. *pubescens* and *C. californicum* germinated on Norstog (Henrich *et al.*, 1981) and germination also occurred on Hyponex medium (Oliva & Arditti, 1984).

The addition of amino acids, particularly glutamine or glutamic acid, increased germination in *C. reginae* (Haas, 1973; Harvais,

1974, 1982). These supplements were added in doses of $50 \, \text{mg} \, l^{-1}$, during germination and early development only. The addition of vitamin B stimulated germination in *C. calceolus* (Malmgren, 1989*b*), as did the addition of coconut milk ($100 \, \text{g} \, l^{-1}$: Muir, 1989; Anderson, 1990*a*), kinetin or BA ($0.5 \, \text{mg} \, l^{-1}$: Borris & Albrecht, 1969). Germination in *C. calceolus* could be raised from 1.2% to 13.2% by the addition of organic nitrogen to the basal medium and to 38.3% with the further addition of $0.2 \, \text{mg} \, l^{-1}$ BA (Van Waes & Debergh, 1986*b*). Fast (1974) treated seeds for 24 hours in 0.001% gibberellic acid and observed a slight stimulatory effect which could not, however, be repeated.

Fast (1974), Malmgren (1989*a*) and others advised the use of a dilute macroelement concentration in substrates for *Cypripedium*.

Immature seeds have probably given rise to most of the seedlings of *Cypripedium* spp. at present in culture. Most workers recommend excising and incubating seeds either 6 weeks after pollination (*C. reginae*: Frosch, 1985, 1986; *C. calceolus*: Lucke, 1982; *C. candidum*: Pauw & Remphrey, 1993), 7 weeks (*C. calceolus*: Malmgren, 1989*b*; Light, 1989), 7.5 weeks (*C. acaule*: Withner, 1953) or even 8 weeks after pollination (*C. calceolus* var. *parviflorum* and *C. reginae*: Pauw & Remphrey, 1993). Environmental and genotypic variation may account for the differences in these reports. The optimum excision time is always after the 5 weeks required from pollination to the beginning of embryo formation (Hildebrand, 1863; Table 2.1, p. 18). Allenberg (1976) reported that germination in immature seeds of *C. calceolus* was more rapid than in mature seeds, beginning 3 weeks after sowing, the seedlings being about 1 mm long after 2 months. Stoutamire (1964) has, however, noted poor growth in seedlings from immature seeds.

Culture of seedlings

Seeds may take many months to germinate and their further development *in vitro* varies considerably (Fast, 1974; Ballard, 1987; Butcher & Marlow, 1989). The protocorms benefit from transplantation at 2–month intervals (Frosch, 1985; Malmgren, 1989*b*). When the mycorhizome has elongated and given rise to three or four adventitious roots the leaves expand and turn green when exposed to light (*C. calceolus, C. reginae*: Harvais, 1973; Fast, 1974; Stoutamire, 1983; Malmgren, 1989*b*). The development of roots also seems to be stimulated by light (Fast, 1974). Asymbiotic seedlings of *C. candidum* that survive the first critical stage of development form two or three roots within a year as well as a bud that will be ready to unfold after 18 months (Stoutamire, 1990).

A high content of iron (Fe^{3+}) in concentrations of up to $5\,mg\,l^{-1}$ has been found to promote protocorm development in *C. reginae*, as did high nitrogen concentrations (up to $1400\,mg\,l^{-1}$) in the form of ammonium nitrate (Harvais, 1982). However, inorganic nitrogen in the substrate seems to increase mortality, apparently because some seedlings cannot make use of it. A dramatic improvement in organ development can be seen if nitrogen is supplied entirely in an organic form mainly as amino acids (Malmgren, 1993). On the other hand, Anderson (1990) found that the addition of kinetin, NAA or 2% coconut milk to the substrate inhibited seedling growth in several species. In *C. reginae* seedling development is affected by undefined additives: negatively by yeast extract and casein hydrolysate, and positively by potato extract, though this gave rise to some seedling mortality (Harvais, 1973, 1982).

When an apical bud is well developed a period of chilling seems to be necessary to produce organogenesis in both North American (Curtis, 1943) and European species (Malmgren, 1989*b*). Malmgren found that a protracted period of chilling was beneficial, first for up to 2 months at about 10 °C after the development of a shoot tip to stimulate root development, then for 3 more months at 2 °C. After chilling the seedlings can be transferred to soil in pots, preferably in early spring (*C. reginae*: Frosch, 1985; *C. calceolus*: Malmgren, 1989*b*). Functional leaves develop after chilling, rarely otherwise (*C. calceolus*: Borris 1970; Malmgren, 1989*b*).

According to Harvais (1982) the addition of kinetin has the same effect as vernalization, and when seedlings of *C. reginae* are grown at a constant high temperatures they develop a requirement for kinetin which cannot be replaced by potato extract. He also found that when auxin (NAA or IAA) was added to kinetin media this further improved seedling health in *C. reginae*. Some vitamins, in particular pantothenic acid but also thiamine and pyridoxine, can cause a broadening of the seedling leaves (concentrations of up to $0.5\,mg\,l^{-1}$: Harvais, 1973).

Problems with organogenesis are not uncommon, and often arise when cytokinins and other growth regulators are present in the growth medium. For instance, a high content of BA in the substrate makes the seedlings proliferate into callus and roots from which it is difficult to induce the formation of shoots (Borris, 1970).

A complete production programme for *C. acaule* has been described by Ling and colleagues (1990). The seedlings should be grown at room temperature until they become dormant, which apparently happens without any environmental signals when the rhizome is 5–15 mm long and a long tapering winter bud has been formed. Although the roots may continue to grow shoot growth ceases and the seedlings will die at this stage if not subjected to chilling. A period of 3–4 months at 3–5 °C is required; 2 months or less results in continued dormancy when the plants are returned to room temperature. After vernalization the plants can be transferred to a new substrate and may grow at room temperature for 4–8 months before dormancy again sets in at which time a new period of vernalization is required.

Ling *et al.*, (1990) transferred seedlings of *C. acaule* to soil after the second period of chilling but mortality was high on all soil mixtures. Seedlings of *C. reginae* can be transferred to soil after 7–8 months *in vitro* (Frosch, 1985). A propagation scheme that involves less handling was attained by non-sterile sowing in 'mycelium-enriched humus' (Muick, 1978), the underground seedlings producing 2–3 cm long rhizomes after 18 months. In May, 31 months after sowing, foliage leaves appeared above ground (*Cypripedium* spp.: Muick, 1978; see, however, Möller (1977) for critical comments on this method). Further plant development is clearly seasonal, requiring an annual cold period of dormancy (Whitlow, 1983). Flowers did not appear until the third year at the earliest (*C. reginae*: Frosch, 1985).

For pot culture, Muick (1977) recommended an acid compost over calcareous sand for plants of *C. calceolus* that were presumably symbiotic, whereas Frosch (1986) found that a coarse mineral soil was successful for asymbiotic plants. Vöth (1980a), in turn, used a mixture of leaf mould, calcareous gravel and loam for *C. calceolus*. Good drainage in combination with a constant soil moisture content seems to be essential, as well as protection from frost (Muick, 1977, 1978; Frosch, 1985, 1986). In apparent contrast to field reports, most American species in culture can be cultivated in a slightly acid humus-rich soil without lime (Vöth, 1980a), only *C. reginae* preferring a sandy texture. This species seems to be tolerant of shade and of competition for space from neighbouring plants but not of too little moisture (Harvais, 1980). If the pots kept under culture are vernalized indoors in a refrigerator they should be protected from drying out (Ling *et al.*, 1990).

According to Olver (1981), flowering can be induced in cultivated plants by simulating the seasonal rhythm in temperature and daylength regimes.

Vegetative propagation by division of the rhizome can be carried out in late autumn (Whitlow, 1983) or March–April (Muick, 1978). One rhizome segment ('one growth') can suffice as a propagule (Holman, 1976).

Status and prospects

In spite of considerable efforts these highly attractive orchids have not yet been brought into mass cultivation. While our understanding of seedling organogenesis *in vitro* steadily increases, considerable difficulties remain as regards the germination phase. With extensive sowings, growers may be content with low germination percentages, in some cases supplemented with vegetative *in vitro* propagation (Fast, 1974). However, these practical measures and the circumvention of seed dormancy by using immature seeds do not provide any clue to the factors governing the establishment of *Cypripedium* seedlings in nature. Although chilling has in some cases proved effective, and has indeed yielded some of the best germination results ever obtained, cold stratification does not seem to be the only requirement for breaking dormancy in *Cypripedium*. The unsuccessful attempts to germinate *C. arietinum* (Ballard, 1990) with many types of temperature regimes and substrates support this supposition. Further progress would be made if it were possible to trace and identify appropriate germination fungi.

13.2 **Epipactis**

Epipactis is mainly Eurasian genus with a south-central distribution. One species occurs naturally in North America, another is naturalized there. They are all rhizomatous, summergreen plants with habitats ranging from bogs to dry forest. The soil is usually alkaline (Buttler, 1986), but *E. helleborine* has a wide pH tolerance (Sundermann, 1962*d*). Nutrients are stored in the rhizomes and roots during winter (Fuchs & Ziegenspeck, 1925). Several albino forms and partially chlorophyll-deficient forms have been described (Renner, 1938; Salmia, 1986, 1989*b*, with references). Autotrophic individuals of *E. helleborine* may live up to 3 years in a heterotrophic state underground (Light & MacConaill, 1991).

The roots are perennial, living for at least 3 years, those of *E. microphylla* being reported to continue to grow for 8–9 years; they are highly mycotrophic when young and later function as storage organs (Fuchs & Ziegenspeck, 1925). *E. palustris* produces both horizontal roots that penetrate the humus layers and vertical roots that often grow deep down into the mineral soil. The xylem is mostly well developed and there is no evident anatomical specialization for mycotrophy in the roots. Infection may be lacking or sporadic; according to Fuchs & Ziegenspeck (1925) plants of *E. palustris* generally are more extensively infected in habitats that are poor in accessible nitrogen. The extent of mycorrhiza varies with the species but apparently not so that species confined to shaded habitats are most dependent on mycotrophy: *E. microphylla* is always infected, but *E. purpurata*, also a shade-tolerant species, is reported as having roots that are free from fungus. *E. atrorubens*, which requires open situations, is only irregularly infected (Fuchs & Ziegenspeck, 1925; Summerhayes, 1951).

In a mixed population with green and white forms of *E. helleborine*, the white individuals declined in numbers during a succession of dry summers, suggesting that albinos are less robust (Salmia, 1986).

Life history

Seedlings of *E. atrorubens* have been found on one occasion in great numbers as small resting protocorms, presumably originating from a germination that had occurred in spring. These seedlings were covered with rhizoids and densely infected (Fuchs & Ziegenspeck, 1924*a*, 1926*a*). Ziegenspeck (1936) also found germinated seeds of *E. helleborine* in spring but generally seedlings of *Epipactis* are difficult to find which suggests that reproduction is often predominantly vegetative. Fuchs & Ziegenspeck (1926*a*) mentioned that they searched for a long time but never succeeded in finding seedlings of *E. purpurata* and *E. palustris*. This has been my own experience with *E. palustris* (Rasmussen, 1992*b*).

The protocorm grows into a mycorhizome from which roots develop. The first of these roots emerge in the autumn and become densely infected, but when the photosynthetic stage is reached some roots develop in spring immediately before the leafy shoot sprouts, and infected tissue gradually decreases in extent. After a few seasons the rhizome is no longer mycotrophic, and only the roots contain pelotons. In most species the first aerial shoot is vegetative only, but in *E. microphylla* the first leafy shoot also bears flowers. The original protocorm was still preserved at the posterior end of the rhizome on some flowering specimens of *E. helleborine* and *E. microphylla* which were estimated to be 11 and 6–7 years old respectively (Fuchs & Ziegenspeck, 1924*a*, 1926*a*).

The first two or three roots are developed from the cortical layers (exogonous). Anatomically they are typically mycotrophic (Chapter 7) and can be infected from the mycorhizome. Larger plants, presumably in their second year but still underground, have been observed with no living infection remaining in the rhizome but with new endogenous roots that were less specialized for mycotrophy and apparently infected from the soil. Seedlings producing leafy shoots in spring were estimated to be 3 years old. However, the duration of the purely mycotrophic life stage can probably vary with depth of germination and nutritional conditions. Leafy plants had sporadic infections in the roots (Fuchs & Ziegenspeck, 1924*a*, 1926*a*).

Endophytes

An endophyte with brown and fairly coarse hyphae was observed by Fuchs & Ziegenspeck (1926*a*) in an old protocorm found in Germany on a flowering specimen of *E. helleborine*, but no attempt was made to isolate a living culture. A similar fungus has since been isolated from populations of *E. helleborine* in two different localities in Finland, and from green as well as albino plants (Salmia, 1988). It is a slow-growing mycelium, turning brown when mature and with thick-walled, verrucose hyphae. The main hyphae are straight, and there are intercalary monilioid cells. It is not known whether it can establish mycorrhiza in seedlings. Though it has not yet been identified, it does not resemble a typical *Rhizoctonia*, according to Warcup (in a personal communication cited by Salmia).

However, a strain that my co-workers isolated from *E. palustris* (D412–1) is a fairly typical *Rhizoctonia* with hyaline hyphae and small monilioid cells. It formed well-developed mycorrhiza with seedlings when germination had been induced by chilling (Rasmussen, 1992*b*). In controls without a fungus germination was poor.

It would be reasonable to look for fungi in species that are reported always to be infected as adults and in those that preserve the original protocorm and oldest roots when fully developed, since roots formed later may contain fortuitous infections with a variety of soil fungi. *E. atrorubens* seems more prone to reproduction by seed than most other species and isolation from spontaneous seedlings should be possible.

Seed storage and survival

Stainability with triphenyltetrazolium chloride (TTC) declined rapidly in stored seeds (Van Waes, 1984). It was reduced to zero when seeds of *E. helleborine* were stored for about 4 weeks at room temperature or 12 weeks in a refrigerator; when seeds of *E. atrorubens* were stored at room temperature their ability to stain disappeared after only 2 weeks.

Germination problems such as those reported by Veyret (1969) with *E. palustris* and *E. helleborine*, Arditti *et al.*, (1981) with *E. atrorubens* and in the review by Fast (1978) could be attributed to the rapid loss of viability during storage. However, reports on germination in *E. palustris* after long incubation periods at low temperatures (see below) show that imbibed seeds of at least this species preserve viability for several months. The correspondence between stainability and germination capacity has been found to be good in asymbiotic *in vitro* experiments (Van Waes, 1984; Van Waes & Debergh, 1986*b*), but in germination experiments with soil fungi Weinert (1990) obtained a germination level of 20–30% in seeds in which stainability was much lower.

Germination in culture

Lengthy sterilization improves germination in species of *Epipactis*. For instance, Van Waes & Debergh (1986*b*) sterilized *E. helleborine* and *E. atrorubens* for 4 hours in 5% Ca(OCl)$_2$, and Van Waes (1984) found that up to 8, 10 and 12 hours gave the most satisfactory result with *E. atrorubens*, *E. palustris* and *E. helleborine* respectively. I found that 2 hours in saturated Ca(OCl)$_2$ was the optimum for symbiotic germination in *E. palustris* (Rasmussen, 1992*b*), but the subsequent growth of the seedlings was better after a short period of sterilization (0.5 hours).

Newly harvested seeds of *E. gigantea* germinated (rupture of testa) in water with 0.05% Tween 80 but did not form rhizoids (Eiberg, 1970). Riether (1990) germinated *E. gigantea* and *E. royleana* on Eg1, ZAK and HPZ media. The ZAK medium gave a higher germination percentage with *E. gigantea* than modified Eg1 (Borris, 1970). Seeds of *E. gigantea* also germinated on Curtis medium, Norstog, Pfeffer medium with glucose and potato extract, and on Mead & Bulard medium (Arditti *et al.*, 1981; Henrich *et al.*, 1981; Hadley, 1982*b*; Rasmussen, unpublished data). Of these substrates, ZAK, HPZ,

Norstog and modified Curtis are devoid of inorganic nitrogen; the others contain both inorganic and organic nitrogen.

The European species seem to be considerably more demanding with respect to substrate. It may be necessary to exclude all inorganic nitrogen from the substrate for seeds of *E. helleborine* and *E. atrorubens* to germinate (Van Waes, 1984; Van Waes & Debergh, 1986*b*). Arditti and co-workers (1982*a*) succeeded in germinating immature seeds of *E. helleborine* on Curtis and modified Curtis media, somewhat more rapidly than mature seeds, in spite of the fact that the Curtis media contain considerable amounts of inorganic nitrogen. On the other hand they found that mature seeds of *E. atrorubens* germinated only on a modification of the Norstog medium that contained no inorganic nitrogen, though the germination was less than 1% (Arditti *et al.*, 1981, 1982*a*)

Van Waes & Debergh (1986*b*) found that the addition of benzyladenine (BA, 0.88 μM) raised germination in *E. atrorubens* to 72% as compared with 3.5% in controls, and in *E. helleborine* to 27% as compared with 2% in controls.

Germinating seeds of *Epipactis* do not seem to be so intolerant of light as those of terrestrial orchids in general. Van Waes (1984) obtained approximately the same germination percentages in dim light (100 lux in a 14 hour photoperiod) for 4 weeks followed by 4 weeks in darkness, as in 8 weeks of continuous darkness. With stronger light intensities, however, germination was reduced.

Cold stratification, about 2–4 °C for several months, is a crucial factor in the germination of *E. palustris* according to Borris & Albrecht (1969), but no effect was observed by Lindén (1980) when seeds were given various cold treatments (3 and − 10 °C for up to 3 months), or by Lucke (1981) who stratified seeds at − 1 to 10 °C for 8 weeks. Nor did chilling improve germination in *E. helleborine*, according to Van Waes (1984) who found that the optimal temperature regime for germination was a constant 20–25 °C. However, evidence from my own experiments with *E. palustris* confirms the value of cold stratification in germination (Rasmussen, 1992*b*) and indicates that chilling may be unsuccessful if applied immediately after sowing and for less than 12 weeks. Incubation with a compatible fungus at room temperature before chilling proved to be an important factor, although asymbiotic cultures were also to some extent stimulated by cold stratification. Seeds of *E. atrorubens* that were incubated with the mycorrhizal fungus from *E. palustris* and then subjected to cold stratification failed to germinate, however (Rasmussen, unpublished data).

Weinert (1990) made use of various soil fungi to stimulate germination in *E. palustris* with fairly satisfactory results (20–30%), but these fungi did not form pelotons and could therefore not support further growth. They were apparently not orchid endophytes.

Culture of seedlings

Van Waes (1984) succeeded in producing asymbiotic seedlings of *E. atrorubens* and *E. helleborine* from seeds that had been surface sterilized for a long period. They were grown on a substrate that contained BA but no inorganic nitrogen (medium BM_4), but nevertheless the mortality was very high, amounting to 25–75% within the first 4 months. The seedlings produced by Arditti *et al.*, (1982*a*) were transferred to light after germination, but developed only slowly.

Recent observations on *E. palustris* indicate that seedlings from unchilled seeds grow only poorly and when raised asymbiotically display atypical organogenesis. Lengthy sterilization also seems to decrease seedling growth rate. In contrast, symbiotic seedlings developing from seeds that had been cold stratified developed rapidly and could attain a length of 2–4 mm within 4 months at 20 °C. Some had by then developed a distinct leaf tip and two opposite roots (Rasmussen, 1992*b*).

Asymbiotic cultures of *E. gigantea* have been brought to maturity on Norstog and Mead & Bulard media (Myers & Ascher, 1982; Rasmussen, unpublished data). In my laboratory we subjected asymbiotic seedlings to chilling (5 °C) for 3 months when they had grown for some months. By this treatment the development of the leafy shoot was stimulated and the plants attained a height of 5–10 cm. They continued to be grown *in vitro* for another 6 months at room temperature in 12 hour photoperiods (*c.* 7 W m^{-2}). The seasonal resting period in older seedlings can apparently set in without any environmental signals (Myers & Ascher, 1982; Rasmussen, unpublished data). If cultivated under constant conditions of light, temperature and day length, the leafy shoots wither and after some time new shoots appear. We usually chilled the plants during the resting period, after which they could be transferred to soil, but we have made no systematic observations that could show whether low temperatures were necessary for any part of the later development.

Adult plants in cultivation should be protected, by mulching and shading, both from heat and from the drying out of the soil (Vöth, 1980a). The growth substrate can be a mixture of leaf mould, calcareous gravel and loam. Vegetative propagation is best performed in autumn immediately before the aerial shoots wither.

Status and prospects

Among the species tested *E. gigantea* seems to be the only one that germinates readily under asymbiotic conditions and without cold treatment. It may be significant that this species differs from most other species of *Epipactis* in flowering early throughout most of its geographical range, beginning in March (Williams & Williams, 1983). Most other species of *Epipactis* flower from mid- to late summer, releasing their seeds much later in the year than *E. gigantea*. It seems possible that seeds of *E. gigantea* germinate in autumn in which case fewer problems connected with dormancy would be a natural consequence.

13.3 Cephalanthera

Cephalanthera is a minor Eurasian genus (*c.* 12 species) with a number of species in central and southern Europe and the Middle East and several in Japan. A single heterotrophic species in western North America occurs in moist coniferous forests (*C. austinae*: Correll, 1950). Typical habitats are shady forests, open woodland and macchia, on basic soils (pH 8–9: Sundermann, 1962*a*), often on lime (Summerhayes, 1951; Buttler, 1986). The aerial flowering and leafy shoots emerge in spring from a rhizome, and wither in late summer or autumn as the seeds ripen. Most species are green and autotrophic, but albino forms of the normally green species occasionally occur. An analysis carried out in June showed that there was more symbiotic infection in the roots in the albino form of *C. damasonium* than in the green form (Renner, 1938).

C. austinae is completely heterotrophic; MacDougal (1899*b*) could detect no trace of chlorophyll in extracts from the aerial stem. The 5–40 cm long rhizome of this species penetrates deep into the thick humus, interlacing with the roots of neighbouring trees. There are two kinds of roots, viz. fibrous roots, and storage roots that are about twice as thick, have patchy infections and may descend 50–60 cm below the surface. These roots develop short side roots that are densely colonized by the endophyte.

C. rubra develops foliage leaves, but it also has persistently infected roots and the aerial shoots wither early in autumn. The roots are heteromorphic in roughly the same manner as in *C. austinae* (Fuchs & Ziegenspeck, 1925, 1926*a*). *C. longifolia* develops only one type of root, which is long and sparsely branched with a few short side roots. Infection tends to be concentrated to the distal parts and the side roots (Fuchs & Ziegenspeck, 1925).

Life history

Germinating seeds have never been reported. However, the original protocorm is retained for several years and may be present as the oldest segment of the rhizome on leafy plants (*C. damasonium*: Fuchs & Ziegenspeck, 1926*a*). According to Irmisch (1853) the first foliage leaves may already appear in the spring immediately after germination, but this observation could be based on an underestimation of the age of the rhizome.

The first roots to develop on the mycorhizome of *C. damasonium* grow superficially in the humus and are mycotrophic (Fuchs & Ziegenspeck, 1926*a*). A change in structure and function towards a greater water-conducting capacity and less mycotrophy occurs in the roots at about the time when the sixth root forms, fully developed plants often being completely devoid of infection (Fuchs & Ziegens- peck, 1925). These later roots, that are up to c. 0.5 m long, grow down towards the mineral soil (Fuchs & Ziegenspeck, 1926a).

Endophytes

No endophytes have been isolated or identified from *Cephalanthera*. Presumed living pelotons that were excised from roots of *C. rubra* failed to grow on PDA and Czapek agar with $4 g \, l^{-1}$ yeast extract (Filipello Marchisio *et al.*, 1985). In a study by Weinert (1990) fungi were isolated from roots of *C. rubra*, but not from pelotons, and though these fungi did have a slight effect on the seeds of *C. rubra* the few seedlings produced did not form mycorrhiza and soon ceased to grow.

Germination in culture

The European species of *Cephalanthera* have not responded well to artificial germination. Fast (1978) reported four attempts at germi- nation which were unsuccessful or yielded only poor results. When the seeds of *C. rubra* were sterilized for long periods (14–16 hours in 5% $Ca(OCl)_2$) about 60% germinated, but only on a substrate containing benzyladenine (BA) and the seedlings did not develop far (Van Waes, 1984; Van Waes & Debergh, 1986b). A few seedlings of *C. rubra* were obtained in co-culture with soil fungi (see above, Weinert, 1990) after pretreatment for 3 months at 5° C. Riether (1990) succeeded in germinating a few seeds of *C. longifolia* with a fungus isolated from *Listera ovata* and some seeds of *C. rubra* on a fungus strain originating from *Dactylorhiza incarnata*, but the see- dlings soon died.

Immature seeds of *C. longifolia* that were excised and explanted *in vitro* with the placenta attached died within a few days (Lindén, 1980) on a selection of substrates (ZAK, Eg1, Fast).

Culture of seedlings

Because of germination problems information is naturally lacking on the cultural requirements of seedlings. Since most species are relatively uncommon and are protected in their area of distribution, there is little published data on the cultivation of transplanted wild specimens.

Adult plants are grown in a mixture of leaf mould, calcareous gravel and loam and are protected by mulching and shading. Hot and dry soil should be avoided (Vöth, 1980a). Vegetative propagation by cutting the rhizome into pieces should be performed in autumn immediately before the aerial shoots wither.

The obvious requirements for further progress are the isolation of endophytes and the development of better germination procedures.

13.4 Limodorum

Limodorum comprises 1 or 2 species of Mediterranean distribution in Europe, the Middle East and North Africa (Buttler, 1986). The habitats are typically dry or semi-dry calcareous soils, in open forest or scrub (Mossberg & Nilsson, 1980; Buttler, 1986). The only leafy appendages are scale leaves on the flowering scape, and the whole plant is pale violet due to the presence of anthocyan in the epidermis. In spite of its heterotrophic appearance a considerable number of chloroplasts are present in the above-ground parts of *L. abortivum*, but it is not known to what extent photosynthesis takes place (Blumenfeld, 1935). However, underground cleistogamous inflorescences have been observed (Bernard, 1902; Fuchs & Ziegenspeck, 1926a) which implies that the plant can complete its life cycle without any phototrophic contribution from the inflorescence.

Life history

Protocorms are unknown from the field (Fuchs & Ziegenspeck 1926a). The mycorhizome remains unbranched until the meristem produces the first inflorescence and then continues by sympodial growth. The almost glabrous rhizome has a well-defined stratum of host cells within the cortex which is always infected with living hyphae. Numerous thick and hairy roots develop below the axils of scale leaves. These roots have a thick cortex that at first is filled with starch but later becomes mycotrophic. The conducting tissue consists mainly of phloem (Fuchs & Ziegenspeck, 1925).

Endophytes

A diversity of fungi have been isolated from *Limodorum* but they have always originated from the roots of adult plants. It is not known whether these fungi can support seedling development.

The endophyte that Blumenfeld (1935) repeatedly isolated from *L. abortivum* had septate hyphae, *c.* 3–4 μm wide without clamp connections, which formed a rose-violet mycelium and developed typical pelotons in the roots. Fuchs & Ziegenspeck (1926*a*) observed coarse, non-coiling hyphae in the root cells, but since there were no pelotons it may have been a fortuitous parasite. All specimens of *L. abortivum* examined by Riess & Scrugli (1987), however, were simultaneously infected with two endophytes; one of these had dark hyphae that were 8 μm thick and the other had hyaline hyphae about 2 μm thick. The coarse type was also observed in certain species of *Ophrys* in the local flora. Nieuwdorp (1972) observed hyphae with clamp connections in the roots of *L. abortivum*. This fungus entered the root through rhizoids but the epidermis was soon replaced by a non-suberized exodermis. Beneath the exodermis there were a few layers of small cortical cells where the pelotons remained undigested, while lysis occurred in the larger cells of the inner cortex.

Germination in culture

Asymbiotic germination has failed on a range of growth media: Veyret medium (Veyret, 1969), BM and BM1 (Van Waes & Debergh, 1986*b*), Eg1, ZAK and HPZ (Riether, 1990). Only on BM2, with organic nitrogen and benzyladenine (BA), was there any response,

2.7% of the seeds germinating after 8 hours of surface sterilization in 5% $Ca(OCl)_2$ (Van Waes & Debergh, 1986b).

13.5 Neottia

Neottia is a small genus of heterotrophic plants widely distributed in Eurasia. *N. nidus-avis* typically occurs in the shade of hazel and oak in deep humus on calcareous soil (Summerhayes, 1951) or in coniferous forest (Fuchs & Ziegenspeck, 1925; Buttler, 1986). No foliage leaves are formed and there are almost no photosynthetic pigments in the inflorescence, only small amounts of chlorophyll *a* and carotenoids. Chlorophyll *b* seems to be entirely lacking and no CO_2 assimilation could be detected (Montfort & Küsters, 1940). When light is experimentally excluded, the inflorescences react with etiolation and loss of the characteristic brown pigmentation (Weber, 1920).

The plants are usually monocarpic (Bernard, 1902; Fuchs & Ziegenspeck, 1924a; Mossberg & Nilsson, 1977) and Bernard (1909) estimated their age to be about 10 years old when they reach flowering stage. Underground, cleistogamous inflorescences have been found (Bernard, 1899; Fuchs & Ziegenspeck, 1926b), but since no systematic search has been made it is impossible to tell whether this is an anomaly or a common occurrence, somewhat resembling the normal conditions in the Australian genus *Rhizanthella*. However, Bernard (1909) referred to the underground inflorescences in a manner that suggests he considered them to be fairly common.

Vegetative reproduction occurs when the tips of detached roots are directly transformed into shoot meristems by a process that is unique to orchids (Irmisch, 1853; Prillieux, 1856; Peklo, 1906, Champagnat, 1971; Rasmussen, 1986).

Life history

Bernard (1899) observed germinated seeds of *N. nidus-avis* in great numbers in a dry and partially buried infructescence from the previous year. In spring the seeds, still in their capsules and still partly covered with the ruptured testa, had formed protocorms about 5 mm in diameter. The fungi had reached the capsules by way of the scape and the pedicels of the infructescence. Fuchs & Ziegenspeck (1926b) confirmed that the axes of underground, cleistogamous inflorescences are commonly penetrated by hyphae, and that seeds germinate within the capsules. Bernard (1899) observed a typical infection pattern in the protocorms, with starch-filled parenchyma at the chalazal end and peloton-filled cells at the micropylar end, describing the young protocorms as being devoid of rhizoids as did Fuchs & Ziegenspeck (1926b).

The protocorms elongate during the following autumn. According to Fuchs & Ziegenspeck (1924a), when the infection has passed into the new part of the mycorhizome this section detaches itself from the old decaying part during the winter, the process being repeated so that the mycorhizome progressively moves through new soil each year. In the fourth spring some short, thick adventitious roots, which are typical mycotrophic roots, develop below the apical meristem (Fuchs & Ziegenspeck, 1925). By the fifth year the new rhizome segment has become somewhat longer and produces a number of typical *Neottia* roots, and in the following years new growth progressively produces more and longer roots. By about the ninth year the plant has reached the flowering stage and the mycorhizome, which is densely covered with short, thick roots, terminates in an inflorescence (Fuchs & Ziegenspeck, 1924a).

Since during the first years the previous part of the rhizome is believed to disappear and be replaced by a new but similar rhizome segment the initial stages are virtually impossible to distinguish from each other in soil samples, making any estimates of the duration of this life stage quite unreliable. However, it is necessary to assume that the mycorhizome progresses through the soil if the notion of 'Ferment-Mykorrhiza' (discussed below) is to be credible. Since there are several other reasons for doubting the existence of Ferment-Mykorrhiza it is likely that the rootless stage is stationary and perhaps only lasts for 1 year, so that *Neottia nidus-avis* may in fact reach the flowering stage within 3–5 years. However, in order to establish this it would of course be necessary to follow the fate of seedlings of known age.

Endophytes

The fungus observed by Bernard (1899, 1900) in seedlings of *N. nidus-avis*, and tentatively identified as *Fusarium* sp., had brown septate hyphae and clamp connections. It could be found in the imbibed seeds before the testa ruptured, but not in mature capsules nor in dry seeds. The fungi isolated by Wolff (1927) were brown on a glucose substrate and hyaline on a substrate containing tannin. The hyphae were septate with monilioid cells borne on short lateral hyphae and on a substrate with fructose they produced hyphal coils, i.e. were typical *Rhizoctonia*. Burgeff (1936) isolated a similar strain and mentioned that it was morphologically very distinct from other species of *Rhizoctonia* that he recognized. In contrast, Fuchs & Ziegenspeck (1924*b*) observed an endophyte that they could not distinguish morphologically from one they had seen in *Orchis militaris*. The endophyte observed by Barmicheva (1990) had dolipores and imperforate parenthesomes, suggesting that it was either a *Tulasnella* or a *Sebacina* (Table 5.1, p. 83). While the seedlings observed by Bernard (1899) indicate that germination is sometimes supported by a non-rhizoctoneous fungus, the fate of those seedlings is unknown. That the established mycorrhiza is usually based on an association with *Rhizoctonia* seems, on the other hand, very likely.

Living hyphae are presumably always present within the plant tissue and infections are continuous in all infected parts of the plant so that the original fungus will be maintained (Magnus, 1900; Bernard, 1909). Fuchs & Ziegenspeck (1924*a*) indicate that a pure culture can be kept on a 'gum arabic–salep agar'. In spite of this it is not considered easy to isolate endophytes of *Neottia* and they are apparently difficult to maintain in living culture for any length of time (Peklo, 1906; Burgeff, 1909; Fuchs & Ziegenspeck, 1924*b*). Wolff (1927) suggested that a fluent medium, even sterile water, was more suitable than agar for pure culture and used a substrate that supplied nitrogen in organic form and carbohydrate in the form of salts of organic acids.

Several workers have been puzzled by the mycorrhiza in *Neottia* because of the apparent paucity of hyphae connecting the pelotons within the orchid tissue with the soil. The protocorm, rhizome and roots all lack rhizoids (Bernard, 1899; Magnus, 1900; Fuchs & Ziegenspeck, 1924a; Niewieczerzalowna, 1933; Burgeff, 1936; Nieuwdorp, 1972; Barmicheva, 1990) but the histological pattern of infection is similar to that of other tolypophagous orchids. The roots, which become infected from the rhizome, contain pelotons in three or four cortical cell layers below the exodermis. The infected tissue consists of a central layer of host cells flanked by digestion cells (Magnus, 1900).

Burgeff (1936) presumed that nutrition was provided wholly through the original fungal connections, i.e. the hyphae that penetrate the suspensor at the time of germination. However, on older roots fungal connections with the soil were observed during winter (Burgeff, 1932). Nieuwdorp (1972) suggested that there were hyphal connections with the soil through the oldest parts of the rhizome. Another theory, not supported by Burgeff, was that of enzymatic mycorrhiza ('Ferment- Mykorrhiza'), in which the orchid absorbs, through the epidermis of the roots, nutrients that have been released through the enzymatic activity of external fungi. This presupposes a short life for each root and the continuous progression of the plant through the soil. This hypothesis, however, does not explain the role of the internal hyphae since without any connections to the outside mycelium these hyphae could neither add to nor subtract from the long-term nutrition of the plant. It is also difficult to see how the orchid could survive by simply absorbing nutrients, including carbohydrates, from the humus. The glabrous slow-growing roots of *Neottia* seem to be poorly adapted for competing with the hairy roots of other plants in the absorption of nutrients, let alone for competing with the multitudes of soil microorganisms.

It is far more likely that the endophyte in *Neottia* is connected with the soil through the surface of epidermal cells – an idea also expressed by Barmicheva (1990) – supposing that these connections are extremely sensitive to desiccation and therefore easily overlooked in material that has been dug up, rinsed and preserved. The penetration of the outer periclinal walls of the epidermis has been unequivocally demonstrated in the youngest protocorms of *Listera ovata* (Fig. 5.5d, p. 108), and hand sections made from preserved roots of *Neottia* show a similar infection pattern that suggests epidermal penetration (Rasmussen, unpublished data). There is no apparent reason for assuming that the mycorrhiza of *Neottia* differs in principle from the tolypophagic type as described by Burgeff (1943).

Germination in culture

Asymbiotic germination has failed repeatedly using substrates such as Veyret medium, Eg1, ZAK and HPZ, and water (Veyret, 1969; Eiberg, 1970; Riether, 1990). Peklo (1906) reported several fruitless attempts to repeat Bernard's observations from 1899 by planting the seeds with and without capsules in different types of soil. Nor have the seeds responded to symbiotic tests. A pure culture of the *Rhizoctonia* sp. isolated by Burgeff (1936) from an adult plant did not induce germination. Inoculation with strains of *R. solani*, unidentified strains extracted from *Platanthera bifolia* and a variety of other soil fungi proved to be equally ineffective (Downie, 1959a; Weinert, 1990). However, with strains identified as *Ceratobasidium cereale* (pathogen from grass) and *Rhizoctonia anaticula* (isolated from *Platanthera obtusata*) Smreciu & Currah (1989a) obtained a low germination percentage in *Neottia nidus-avis* and the seedlings grew to a length exceeding that of the testa.

13.6 Ponthieva

The only species to be included here is *P. racemosa*, distributed in southern North America in moist woodland and by springs with neutral soil (Correll, 1950). The rosette of foliage leaves arises from a short rhizome with hairy and fleshy roots. The plants have leaves in summer and flowering takes place from September to April on a leafless spike.

Germination in culture

Seeds from closed capsules have been germinated on a range of substrates by Baker *et al.*, (1987). When incubated at 25°C in 16 hour photoperiods 75–83% of the seeds germinated. Infiltration or soaking before sowing had adverse effects on germination, which was also inhibited by the addition of hydrogen peroxide to the substrate.

13.7 **Listera**

Listera is a widely distributed genus in boreal and temperate regions. The plants have a creeping rhizome and in the summer produce sterile or flowering shoots with a characteristic pair of opposite leaves. *L. ovata* is widely distributed in Eurasia, occupying a wide range of habitats (Summerhayes, 1951), whereas *L. cordata* and several of the small North American species are confined to humid coniferous forest and *Sphagnum* bogs, in deep humus layers. Chodat & Lendner (1896) found *L. cordata* growing on decomposing stumps. All species are shade tolerant (Summerhayes, 1951; Williams & Williams, 1983).

Life history

Seeds of *L. ovata* presumably germinate in spring and the young protocorm rests during the summer (Fuchs & Ziegenspeck, 1926*b*). In autumn a highly mycotrophic, short thick root forms below the apical bud. A similarly shaped root develops during the second autumn, while the previous root becomes partially filled with starch. One or two additional exogenous roots develop during the third autumn and are longer and more slender than the previous ones. They are mycotrophic but not infected throughout their whole length.

Still according to Fuchs & Ziegenspeck (1926*b*) the first leafy shoot develops during the fourth spring (rarely as early as the third), usually producing two small opposite foliage leaves. From this stage onwards the new growth of the rhizome and the new roots do not become infected, and the new roots do not emerge in autumn but in spring. The mycorhizome and the early, mycotrophic roots may function for about 14 years. Roots generally continue to grow and have a life span of about 10 years, but they never branch except in connection with the production of buds for new aerial shoots. This is the usual means of vegetative reproduction since the rhizome rarely branches.

It is believed that *L. ovata* begins to flower when the plants are about 15 years old (Fuchs & Ziegenspeck 1926*b*). However, Irmisch (1853) estimated that flowering begins 4–5 years after the appearance of leaves (which could be about equivalent to 7 or 8 years from germination).

Germination in *L. cordata* has not been observed in the field. Adult plants have a few long and hairy roots that are heavily infected (Chodat & Lendner, 1896). The rhizome is not infected and contains some starch. As in *Neottia nidus-avis* the tip of the mycotrophic roots of this species can be transformed into a shoot meristem (Brundin, 1895), disconnected roots serving as a means of vegetative reproduction.

Endophytes

Downie (1949*b*) isolated two symbiotic strains of fungi from adult plants of *L. ovata*, one of which stimulated germination. This fungus, identified only as *Rhizoctonia* sp., had very fine hyphae (Downie, 1959*a*). A number of strains of *Rhizoctonia solani* were also tested with seeds of *L. ovata* but none of them formed seedling mycorrhiza.

Since the rhizome of adult plants is without infection each root must obtain a mycorrhizal fungus from outside and could easily be invaded by a fortuitous infection that is not associated with germination. In my experience the roots of mature plants are most often devoid of infection. In general, we have found extremely few seedlings in the populations of *L. ovata* that we have examined; only one plant with very small leaves and a few short roots was possibly a third-year plant. The roots of this plant yielded the strain D131–1 with which seeds were germinated (Rasmussen *et al.*, 1991). This strain is binucleate with *Tulasnella*-type parenthesomes and can thus be referred to *Epulorhiza* sp. (Fig. 5.1*a*, p. 81; Andersen, 1990*b*).

The infection in the roots of *L. cordata* is persistent and widespread and should be easy to observe. A range of fungi were isolated from this species by immersing the intact roots in tap-water (Chodat & Lendner, 1896). Some of these strains were probably soil fungi unconnected with orchid mycorrhiza, but a few may have been endophytes referable to *Rhizoctonia* sp. (as illustrated in Chodat & Lendner, 1896, fig. 4:2–7).

Fuchs & Ziegenspeck (1925) saw only few rhizoids on the myco-rhizome of *L. ovata*, and none on the thick mycotrophic roots. They assumed that there are only a few insignificant connections between the endophyte and the external mycelium and therefore suspected this species of having an alternative type of mycorrhiza, enzymatic mycorrhiza ('Ferment-mykorrhiza', commented on in Section 13.5 above). However, in the symbiotic seedlings of *L. ovata* we obtained *in vitro* there were many hyphal connections with the agar, and both protocorms and the first roots were covered with numerous long rhizoids (Rasmussen *et al.*, 1991, fig. 2). Furthermore, hyphae frequently passed through the surface of the somewhat papillose epidermal cells of the protocorms (Fig. 5.5*d*, p. 108).

Seed storage and survival

In seeds of *L. ovata* the biochemical viability measured as staining in TTC approached zero after 4 weeks of storage at room temperature, and after 8 weeks in refrigerator (Van Waes, 1984). Downie (1949*b*) indicated that mature seeds of *L. ovata* lose their viability within a few weeks in dry storage but that they can be sown in water immediately after harvest. This would keep them viable but not enable them to germinate.

In contrast, we kept seeds of *Listera ovata* for 6 months at room temperature without taking any precautions to preserve viability and still obtained 7.5% germination in asymbiotic and 21.5% in symbiotic culture (Rasmussen *et al.*, 1991).

Germination in culture

L. ovata does not require lengthy sterilizing, 10 minutes in 0.3% NaOCl (Lucke, 1984) or 5 minutes in saturated $Ca(OCl)_2$ (Van Waes, 1984) being adequate. The seeds do not germinate in pure water (Downie, 1941; Eiberg, 1969), nor does it help to add Pfeffer's salts, trace elements and glucose, but the further addition of potato extract may initiate some sporadic germination (Downie, 1941, 1949*b*, Hadley, 1983). By omitting inorganic nitrogen and adding organic nitrogen compounds Van Waes & Debergh (1986*b*) succeeded in improving the asymbiotic germination considerably, raising the level to 33–38% compared with 3% in controls. The asymbiotic germination process can be extremely slow; in Lucke's (1984) experiments the young protocorms were barely discernible with the naked eye after 1–1.5 years.

The best results have been achieved with symbiotic germination. In the presence of a fungus, Downie (1949*b*) obtained 40% germination on water agar; with the addition of Pfeffer's salts and trace elements, however, no more than 10% germinated. With glucose in the medium the infection became uncontrollable, but the further addition of potato extract provided almost ideal conditions that resulted in 75% germination after 6 months of incubation. The assumption that undefined, organic additives are necessary even in symbiotic culture is supported by Hadley (1982*b*), who recommended yeast extract for germinating *L. ovata*. Our own isolate (D131–1) gave 21% germination within 6 weeks on an oat medium with sucrose, yeast extract and macroelements, compared with 7.5% on an asymbiotic substrate (Mead & Bulard) that contained considerable amounts of organic nitrogen but no yeast extract (Rasmussen *et al.*, 1991). These data suggest that the seeds require the presence not only of a fungus to stimulate germination, but also of unknown elements provided by organic additives such as yeast or potato extract, the combined effect increasing the germination percentage considerably.

Lucke (1984) found that cold stratification did not improve germination in asymbiotic experiments on Thomale GD, but the low response could have been due to the lack of organic nitrogen in this substrate. Germination is strongly impeded by light. Van Waes (1984) obtained a reduction from 19% in darkness to less than 2% in 14 hour photoperiods with an intensity of 100 lux, but pretreatment in light had no effect on the germination percentage when the seeds were subsequently transferred to darkness and germinated (Van Waes, 1984).

Downie (1941) tried to germinate L. cordata asymbiotically without success, and Stoutamire (1964) reported only poor germination results.

According to Lucke (1971b) seeds of L. ovata mature about 28 days after pollination. Immature seeds from capsules that have attained about two-thirds of their final length may be used for germination. Ovules that are still attached to the placenta can be successfully excised and inoculated with the symbiotic fungus and will subsequently germinate and establish a mycorrhiza in the same way as mature embryos (Rasmussen et al., 1991).

Culture of seedlings

Little is yet known about the later development of L. ovata seedlings in culture. A reasonable number of asymbiotic seedlings obtained by Van Waes (1984) survived, 10–25% being lost within 4 months on a substrate containing organic nitrogen, whereas the mortality was above 75% on substrates in which the nitrogen was inorganic and on those containing growth regulators (0.2 mg l^{-1} benzyladenine (BA). Symbiotic seedlings developed more rapidly than asymbiotic seedlings and the protocorms formed thick roots (Rasmussen et al., 1991) that probably correspond to the early mycotrophic roots observed in the field by Fuchs & Ziegenspeck (1926b). Within 5 months several seedlings had formed more than one such root. In many seedlings the apical bud sprouted after cold treatment for 20 weeks at 5 °C. The leafy shoots carried two foliage leaves which turned green when the seedlings were transferred to photoperiods but did not unfold completely (Rasmussen, unpublished data).

The rapid development that takes place in symbiotic culture

suggests that the interpretations of the field material may be inaccurate, development in nature being probably of shorter duration than was assumed by Fuchs & Ziegenspeck (1926*b*). If seeds of *L. ovata*, which germinate reasonably well *in vitro* without cold stratification, emerge as seedlings immediately after they are dispersed they might have time to emit one or two mycotrophic roots during the autumn and after the winter would then be able to produce leaves, in less than a year from germination.

13.8 Galeola

Galeola septentrionalis and *G. altissima* occur in evergreen forests of temperate Japan, although this is mainly a tropical Asiatic and Australian genus of heterotrophic plants (Ohwi, 1965). While in the tropical relatives the vegetative body withers after the seeds mature, *G. septentrionalis* has a persistent horizontal rhizome 0.5 m below the soil surface. The perennial root system runs *c.* 5–15 cm below the surface and consists of about 20 long glabrous main roots with secondary roots of determinate growth. Those main roots that are about 1 m long are still fairly young; they are usually unbranched and are not infected. When mature they are up to 5 m long, have short secondary roots and are temporarily infected in patches. Subsequently a permanent mycorrhiza is established, concentrated mainly to some of the secondary roots.

The aerial flowering shoots that appear in spring, are pale and lack foliage leaves (Hamada, 1939). In a spectrophotometric investigation carried out by Nakamura (1964) no chlorophyll was detected in the shoots, this species thus being completely heterotrophic.

Endophytes

Plants may be infected by single hyphae, or by rhizomorphs, but only the latter type of association is considered to be permanent (Hamada, 1939). The type of infection that takes place by single hyphae begins with the formation of an infection cushion on the root surface;

hyphae are emitted from the cushion, penetrate epidermis cells and infect the cortex. The hyphae may remain alive in the outer cortex but in the inner cortical layers they are eventually lysed by tolypophagy.

The main type of mycorrhiza in this species is connected with rhizomorphs in the soil. After penetrating the epidermis the hyphae form internal rhizomorphs that extend along the root within the velamen, single hyphae growing from them through the exodermis and into the cortex where they form pelotons and eventually become lysed.

Both *Armillaria mellea* and *A. tabescens* have been isolated from *G. septentrionalis* (Terashita, 1985; Terashita & Chuman, 1987). The association with *A. mellea* was confirmed by inoculating asymbiotic seedlings (Terashita, 1985).

Seed storage and survival

The fruit is a berry and the seeds can be separated from the pulp by chopping up the fruits in water, filtering the macerate through a sieve with a pore size of 0.75 mm, and finally drying the filtrate.

When stored at 30 °C the seeds lose weight and their germination capacity is reduced (Nakamura, 1964) whereas dry seeds, stored at 4–5 °C in an airtight container with silica gel, will retain full germinability for at least 5 years (Nakamura *et al.*, 1975).

Germination in culture

Surface sterilization in KOCl (1.8% active chlorine) for 15 minutes is adequate, and the seeds will germinate on an agar substrate containing 1% sucrose, 0.75% KCl and 1% yeast extract (Nakamura *et al.*, 1975). They require external nitrogen (supplied as yeast extract) to germinate, whereas the need for an external carbon source (sucrose) does not arise until the seedlings begin to develop (Nakamura, 1962).

The percentage and rate of germination both increased when the seeds were pretreated by rinsing in a solution of 0.1% KCl. Germination is temperature dependent, the optimum temperature being about 30 °C, and hermetically sealing the culture vessels also raised the germination percentage (Nakamura, 1962). The optimum values for various atmospheric components during germination were found to be 5% O_2, 8% CO_2 and 4 ppm ethylene (Nakamura *et al.*, 1975).

Culture of seedlings

Terashita (1985) obtained seedlings in symbiotic culture on a substrate of cherry wood chips inoculated with *Armillaria mellea*. The seedlings developed a typical infection pattern with pelotons of *A. mellea* in the cortex of their roots.

13.9 **Spiranthes**

Spiranthes is a large and widespread genus with numerous species in North America and a few in Europe and Asia. All have a basal leaf rosette and thick fleshy tuberous roots. The rhizome is condensed and more or less erect. In *S. spiralis* the leaves are produced in autumn and function during winter. In spring they are replaced by the flowering spike that develops during summer and carries spirally arranged flowers at the same time as new leaves appear. This species is confined to short turf and other low vegetation and occurs characteristically in sunny well-drained places, usually on basic soils (pH 8–9, Sundermann, 1962a; Summerhayes, 1951). The withering away of the shoot in late spring is presumably an adaptation to the periodically dry habitat (Fuchs & Ziegenspeck, 1925).

In *S. aestivalis* the new shoot is produced in autumn but the yellowish-green leaves do not unfold until spring and are retained during the flowering season in summer. This species prefers boggy and marshy, basic to slightly acid soil with low vegetation (Summerhayes, 1951; Sundermann, 1962b). In the north of Ireland the populations of *S. romanzoffiana* have the leaf phenology of *S.*

aestivalis, but in North America the plants are wintergreen (Godfery, 1924). This species has longer and more slender roots than those of S. aestivalis, particularly when it occurs in humid, fairly acid habitats where the roots are spread horizontally in the humus. In drier habitats the roots grow down more or less vertically and tend to be thicker.

The North American species may similarly be divided into those that bear leaves at flowering time and those that are leafless. The latter seem to be confined to dry or mountainous habitats but the flowering season, and thus the vegetative season, varies considerably (data from Williams & Williams, 1983). Many North American species occur in fairly acid, peaty or sandy substrates, in a variety of habitats such as roadbanks, open grassland or heathland, moist open pinewoods and temporarily inundated sites. They are often reported from somewhat disturbed habitats such as recently burnt or grazed areas (Wherry, 1918; Catling, 1990). S. parasitica has no foliage leaves and the flowering scape is pale and apparently devoid of chlorophyll (Correll, 1950).

The species may reproduce vegetatively by means of budding root tips (Catling, 1990) or side shoots (Fuchs & Ziegenspeck, 1927a) to produce dense colonies. In S. aestivalis and S. sinensis bulbils are produced at the base of the leafy shoot (Burgeff, 1936; Mrkvicka, 1991). S. spiralis is able to subsist for at least 1 year below ground and still flower the following year (Wells, 1967), which suggests that the holomycotrophic interlude can be a period of invigoration. S. cernua and S. magnicamporum are also reported to live subterraneously from time to time (Sheviak, 1974).

Life history

Ames (1921a) found seedlings of S. cernua with small leaves in September. The leafy shoot developed within the first year; the tuberous roots were not formed before the next autumn, when the seedlings were about 2 years old. Flowering begins before the plants have attained their full size (Ames, 1921a).

Fuchs & Ziegenspeck (1925) assumed that S. aestivalis survives as a protocorm the first summer, and that the first foliage leaves are produced during the second spring a year after germination. The first root, which is formed the following autumn, develops endogenously from the mycorhizome that becomes distorted when it breaks through the surface. Mrkvicka (1991) observed a somewhat faster development, starting with protocorms in spring, the first root beginning to form in July, and the tiny, first leafy shoot appearing above ground in late summer. In addition to the tuberous roots a few fibrous roots are formed in this species, appearing on older plants from about the third year. Mrkvicka (1991) indicates that they appear in late autumn, but his illustrations suggest that they emerge in spring and function throughout the summer season when the plant bears leaves.

According to Fuchs & Ziegenspeck (1924a), *S. spiralis* has a much longer subterranean phase than *S. aestivalis*, estimated as 8 years before the first root is formed. In the ninth year foliage leaves appear. However, later workers (Burgeff, 1936; Möller, 1986b) are sceptical of that estimate.

The roots of *S. spiralis* begin to develop in autumn, those of *S. aestivalis* in spring, often more than one appearing each season. The roots and the leaf rosette, which develop at about the same time, become progressively larger from year to year. The longest and most nearly vertical roots are always the youngest (Fuchs & Ziegenspeck, 1927a; Summerhayes, 1951; Mrkvicka, 1991). Tuberous roots from the preceding year become exhausted, shrink and are displaced towards a horizontal position which may give the impression that the plant has fibrous, horizontal roots in addition to the fleshy, vertical tuberous roots (Fuchs & Ziegenspeck, 1925). The root dimorphism reported in *S. aestivalis* and *S. romanzoffiana* (Summerhayes, 1951; Mrkvicka, 1991) should perhaps be explained in this way. The tuberous roots are contractile, and in combination with the slightly ascending condensed rhizome serve to adjust the depth of the plants (Fuchs & Ziegenspeck, 1927a).

Tuberous roots have essentially the same anatomy in *S. aestivalis* and *S. spiralis* in spite of the great difference in habitats. The exodermis, which contains many short cells, is covered with a velamen with a densely hairy surface. There is no strongly developed stele, the xylem apparently having a low conducting capacity, but the thick cortex can store considerable amounts of water and starch. During autumn and winter the fungus invades the root of *S. spiralis* through the short cells, and colonizes the outer cortex where the starch disappears, the central part of the cortex that surrounds the stele remaining uninfected. Later the exodermal short cells become impermeable and the fungal coils disintegrate (Fuchs & Ziegenspeck, 1925).

Endophytes

The protocorm is strongly infected except in the uppermost part. By the time the roots have appeared the rhizome is free from fungus and each root must accordingly become infected from outside. However, in *S. spiralis* the old decaying roots form a moisture-retentive environment around the new ones (Fuchs & Ziegenspeck, 1927a) so that the original infection is likely to be passed on.

Infection in the roots of *Spiranthes* is periodical (Mitchell, 1989), the mycotrophic season being the autumn and winter (Fuchs & Ziegenspeck, 1925). The best time for isolating the endophyte is probably during the autumn before extensive digestion takes place. In *S. cernua* the plants are uninfected at the time of flowering, except for the uppermost part of the root adjacent to the rhizome (Ames, 1921a). In *S. spiralis*, most pelotons have been digested by flowering time, but the outermost cortex may contain living hyphae. New roots become infected when they have attained their maximum size (Beau, 1913).

Beau (1920a) isolated fungi that apparently were identical from adult plants of *S. spiralis* and *S. aestivalis*, but these strains did not influence germination of the seeds as much as an isolate from a species of *Ophrys*. Anderson (1991) isolated a strain tentatively identified as *Epulorhiza repens* from the roots of mature plants of *S. magnicamporum*. This fungus did not stimulate germination (25% germination compared with 99% in asymbiotic controls on water agar) but it was very effective in supporting seedling development.

Seed storage and survival

Seeds of North American species can remain viable for 3 years stored in a refrigerator (Curtis, 1936). However, species such as *S. cernua* and *S. magnicamporum* with a tendency towards polyembryony often have split testa, which possibly make the seeds particularly sensitive to desiccation, so that it is better to keep seeds of these species within mature but still closed capsules. Surface sterilized capsules may be stored for considerable periods in a cool, damp place but the best results are generally obtained when the seeds are sown almost immediately after harvest (Anderson, 1990*a*).

Germination in culture

The importance of surface sterilization is indicated by experiments performed by Lucke (1984) and Van Waes (1984) on European species. Unfortunately, Lucke did not make directly comparable tests within a homogeneous seed batch of *S. spiralis*, but the data indicate that surface sterilization in 0.3% NaOCl for 1 hour, as compared with 10 minutes, improved the germination result, and that vacuum infiltration increased the effect of lengthy sterilization. Van Waes & Debergh (1986*a*) found that 10 minutes in 5% Ca(OCl)$_2$ is enough to decontaminate the seeds, but the germination percentage rose from 4% to 60% when the treatment was prolonged to 30 minutes.

For American species such as *S. cernua* and *S. magnicamporum* 3 minutes in 5% NaOCl appears to be sufficient (L. Zettler, personal communication) and there may be a danger of killing the seeds by longer exposure to hypochlorites, which is perhaps related to the fact that the embryos of these seeds are green and apparently continue to grow up to the time when the capsule is about to open (Anderson, 1990*b*; W. Stoutamire, personal communication). Other species, i.e. *S. ochroleuca* and *S. ovalis* var. *erostellata*, were sterilized for 20 minutes in 7.5% Ca(OCl)$_2$ without apparent difficulty (Anderson, 1990*b*). The Asiatic species *S. sinensis* required only rinsing for 2 minutes in 0.5% available chlorine (NaOCl) to prepare the seeds for sowing (Tsutsui & Tomita, 1989).

In general the species of *Spiranthes* appear to be tolerant of the substrates used for germination. *S. cernua*, *S. sinensis* and *S. romanzoffiana* can germinate in water (Stoutamire, 1964, 1974). *S. magnicamporum* also germinates readily on water agar, the germination percentage being higher than that achieved on other substrates (Anderson, 1991). *S. cernua* and *S. magnicamporum* germinate within 10–14 days on a range of asymbiotic media (Stoutamire, 1974), Knudson C in several cases proving the most successful (*S. magnicamporum* and *S. ovalis* var. *erostellata*: Anderson, 1990a). This medium can also be used with *S. romanzoffiana*, which has been tested on a range of substrates: Knudson C (Anderson, 1990a), Norstog (Henrich *et al.*, 1981), Curtis solution 5 and full-strength Curtis, which was also used for *S. gracilis* (Oliva & Arditti, 1984).

Van Waes & Debergh (1986b) succeeded in doubling the germination percentage of *S. spiralis* when inorganic nitrogen was omitted from the substrate and replaced by organic nitrogen. Stephan (1988) used a substrate with both peptone and yeast extract for this species, and Wells & Kretz (1987) used Thompson medium that contains nitrogen in the form of urea and ammonium.

In the presence of various fungal strains, seeds of *S. sinensis* germinated within 3–4 weeks (Tsutsui & Tomita, 1989; Masuhara & Katsuya, 1989). In the latter study several media (Knudson C, Hyponex and oat medium), were tested without fungi, but no seeds germinated. Tsutsui & Tomita (1986) found that neither macerated mycelium nor fungal extract could replace a living symbiotic fungus in stimulating germination.

Cold stratification (-1 to $10\,°C$ for 8 weeks) did not affect the germination of *S. spiralis* (Lucke, 1981). *S. spiralis* germinates in darkness (Stephan, 1988) and in 16 hour photoperiods (Wells & Kretz, 1987), but with markedly reduced percentages if the light intensity is 100 lux or more (c. $0.3\,\mathrm{W\,m^{-2}}$: Van Waes, 1984). When Van Waes (1984) subjected the seeds to strong light before placing them in darkness fewer of them germinated than when seeds were incubated in darkness from the beginning. *S. gracilis* requires darkness for germination (Oliva & Arditti, 1984). According to Oliva & Arditti (1984) *S. romanzoffiana* germinated only in light, but Anderson (1990) obtained germination in both light and darkness, though germination in light resulted in a high mortality among seedlings. *S. sinensis* also germinates in both light and darkness (Tsutsui & Tomita, 1986).

Culture of seedlings

When plants are 5 mm long they can be transferred to a fresh substrate and placed in light, after which a rosette of leaves and photosynthetic roots will develop (*S. cernua, S. romanzoffiana, S. magnicamporum*: Anderson, 1990*a*). Stephan (1988) transferred seedlings of *S. spiralis* to light after 13 weeks and when he reduced the night temperature to 12–14 °C shoot development was stimulated (Stephan, 1988). Uninterrupted chilling for longer periods does not, however, seem to be necessary for further organogenesis (Borris & Albrecht, 1969).

The addition of plant growth regulators to the substrate does not promote root development nor does it increase survival in seedlings of *S. spiralis* (Anderson, 1990*a*; Van Waes, 1984). A substrate containing $1 \, mg \, l^{-1}$ benzyladenine (BA) and $0.1 \, mg \, l^{-1}$ NAA inhibited root development in *S. magnicamporum*; when seedlings were transferred to a substrate without these additions root growth was observed within 3 months (Anderson, 1990*b*).

Seedlings can be planted out when they have leaves and roots, those of *S. cernua* and *S. magnicamporum* within *c.* 4 months (Anderson, 1990*a*); subsequent chilling for 8 weeks stimulates further development. For *S. spiralis* it takes 7 months for the leaves and first roots to form *in vitro* (Wells & Kretz, 1987; Stephan, 1988) which adds to the uncertainty as to the duration of subterranean life in nature (see Life History above). According to Wells & Kretz (1987) 5 years elapsed from germination to the first flowers of *S. spiralis* but *S. cernua* and *S. magnicamporum* flowered after 19 months in culture (Anderson, 1990*a*) and *S. sinensis* after 29 months (Stoutamire, 1974).

Status and prospects

Spiranthes spiralis is characterized by the inability of the seeds to germinate in light and seems to rely on a fairly long mycotrophic seedling stage; unless cultivated with the fungi a long *in vitro* period may be required before the plants can be planted out. This possibly applies to other species from dry habitats as well.

Species of marshy habitats seem able to change over rapidly to phototrophic nutrition if this is appropriate. Some of them are able to germinate in light but they can probably be raised in both darkness and light, though there are indications that cultivation in darkness may be more successful (Anderson, 1990*a*).

Dependence on a fungus in the course of germination is expressed in several species by their reluctance to germinate asymbiotically when no organic substances are supplied in the substrate. Species with a short mycotrophic seedling stage would be expected to be rather less dependent on such additives than those with a long stage underground.

13.10 **Goodyera**

Goodyera is a widely distributed genus occurring in the holarctic region from the boreal to warm-temperate regions of North America and Japan extending through Eurasia, in the southern parts mostly in mountainous areas. The habitats are coniferous and deciduous forest, typically on acid soils with deep humus, conifer needles and moss (Wherry, 1918; Fuchs & Ziegenspeck, 1925; Williams & Williams, 1983). The pH of the soil in habitats of *G. repens* ranges between 4.8 and 7.5 (Sundermann, 1962*d*). *G. repens* is a characteristic inhabitant of old pine and spruce forest (Summerhayes, 1951). The plants form colonies that are either large individuals connected by a system of creeping rhizomes or colonies of seedlings recruited in the proximity of the mother plant (*G. pubescens*: Ames, 1921*b*).

The species of *Goodyera* are the only north-temperate orchids that are evergreen (Stoutamire, 1974), new leaves being added to the rosette during the summer season (*G. pubescens*: see Chapter 10). *G. repens* is an extreme ombrophyte, making effective use of the naturally available light and with little ability to exploit additional light when this is applied experimentally (Montfort & Küsters, 1940). This species showed no tendency to disappear below ground, i.e. to resort to pure mycotrophy (Alexander & Alexander, 1984).

Life history

The seeds are released in autumn though some seeds often stay in the dry capsules during the period of winter dormancy and remain fully viable in spring (Ames, 1921*b*, 1922; Curtis, 1936). The seeds of *G. pubescens* germinate in spring about 24 weeks after their release (Ames, 1922; Rasmussen & Whigham, 1993), and seedlings can occasionally be found in high numbers in the rhizosphere of adult plants. Even Beer (1863) may have observed protocorms of *G. repens*, but Fuchs & Ziegenspeck (1926*b*) presumed that what Beer saw could have been detached lateral roots functioning as propagules. During the summer the seedlings attain a length of *c.* 1 mm but develop no roots or leaves until the following spring and early summer (Ames, 1921*b*, 1922).

The embryonal axis continues as a creeping rhizome and may become partially epiterranean; it produces short densely hairy roots from the fourth growth season (*G. repens*, Fuchs & Ziegenspeck, 1925). These roots are mainly adapted for mycotrophy; the rhizoids are formed in clusters on scattered papillae, the xylem is insignificant and the cortical infection extensive (Fuchs & Ziegenspeck, 1926*b*). Each new root functions for 2–3 years and the infection within the rhizome is continuous with that of the roots (MacDougal, 1899*a*; Alexander & Alexander, 1984).

After the plants have flowered the whole of the flowering stem dies away, leaving a number of rooting axillary shoots as separate entities, thus forming a colony by vegetative reproduction. Summerhayes (1951) estimated that the age of the flowering stems of *G. repens* was about 8 years, but the survival rates found by Alexander & Alexander (1984) in natural populations indicate a much shorter life span, an average period of 2–4 years elapsing from the appearance of a rosette to its disappearance.

Endophytes

It is fairly easy to isolate the endophyte from G. *repens* and G. *pubescens*, presumably because living intracellular coils can be found throughout the year, and infection is extensive in both roots and rhizome (Mollison, 1943; Alexander & Alexander, 1984; Harvais, 1974; Mitchell, 1989). However, in G. *repens* the degree of infection in the rhizome fluctuates considerably, the infection being heaviest from December to May (Alexander & Alexander, 1984). According to these observations the roots were infected from the rhizome, not by re-infection from the soil. If the same strain tends to persist in the plant body throughout life, it would explain why isolates from the roots and rhizome of adult plants of *Goodyera* are so often effective in symbiotic propagation.

Fungi have several times been isolated from G. *repens*, first by Costantin & Dufour (1920), who always found one particular kind of mycelium, white to pale yellow, with 1–2 mm broad sclerotia, and which they described as *Rhizoctonia goodyerae-repentis*. Unfortunately, there is no extant nomenclatural type material, nor any extant original strains, and none of the characters mentioned in the original description make it possible to refer a sterile mycelium to this taxon with any certainty (Andersen & Stalpers, 1994; see Chapter 5). When isolates from G. *repens* have of tradition been referred to *Rhizoctonia goodyerae-repentis*, identification must have been based on their association with the orchid alone. The assumption of fungal specificity in G. *repens* is therefore founded on circular reasoning and is furthermore doubtful in view of the diversity of fungi attached to other orchid species.

Fungi isolated from adult plants in nature have in fact been found to differ in seedling compatibility; Mollison (1943), for instance, could easily synthesize mycorrhiza in asymbiotic seedlings with her isolates, but Purves & Hadley (1976) reported that many of their asymbiotic seedlings rejected the strains they had extracted from G.

repens. In addition to ten similar strains that Alexander & Hadley (1983) isolated and referred to *Rhizoctonia goodyerae-repentis*, and which stimulated seedling growth to varying degrees, they also isolated two different strains that clearly differed from the other group and which had no effect on seedling development. Unless irrelevant soil fungi were inadvertently isolated, these observations suggest that several fungi form mycorrhiza in the mature plants.

In my laboratory we have isolated three similar, binucleate strains from *G. repens* at different times, and on the basis of perforate parenthesomes referred them to *Ceratorhiza* (Andersen, 1990*b* and unpublished data). They all supported germination and seedling development in *G. repens*. Three isolates obtained from *G. pubescens* in Maryland, USA, also resembled one another but apparently differed considerably from our endophytes of the European relative. I germinated seeds and obtained mycorrhizal seedlings of *G. pubescens* with all of these isolates.

Of the seeds of *G. repens* sown by Alexander & Hadley (1983) up to 80% germinated, and mycorrhiza formed with a wide range of fungi, few of them, however, developing seedlings that grew better than asymbiotic controls. I have had a similar experience with *G. pubescens*, seeds of which germinated readily with a variety of orchid fungi from the area in which the species was growing, but the growth response of the seedlings with the two isolates originating from *G. pubescens* was very significantly better than with all the others (Rasmussen, 1994*b*).

The perfect state of isolates from *G. repens* has been identified as *Ceratobasidium cornigerum* (Warcup & Talbot, 1967). In a study by Andersen (1990*b*) that compared the sterile mycelia with perfect reference strains by means of restriction fragment length polymorphism (RFLP), two strains we had isolated from *G. repens* could both be tentatively identified as *C. cornigerum*. *G. repens* could not form lasting symbiotic associations with identified strains of *Rhizoctonia solani* (Downie, 1959*a*), and although Peterson & Currah (1990) successfully combined seedlings of *G. repens* with another pathogen, *C. cereale*, the histological pattern was not entirely normal since the hyphae proliferated in the epidermal cells and some parasitism was observed.

A strain of *Rhizoctonia* has been isolated from *G. oblongifolia* which stimulated germination in this species (Currah *et al.*, 1987*a*).

Seed storage and survival

The germination percentage in *G. repens* may decrease from over 80% to 50% from one year to the next in dry storage at 4°C (Alexander & Hadley, 1983). In dry storage at room temperature, our seeds completely lost their germinability within 30 months (Rasmussen, unpublished data).

Germination in culture

Van Waes (1984) recommended brief surface sterilization for seeds of *G. repens*, namely 5–45 minutes in 5% $Ca(OCl)_2$, and Arditti *et al.*, (1982*b*) used 10 minutes of saturated $Ca(OCl)_2$ for seeds of *G. oblongifolia* and *G. tessellata*. In contrast, I found after sterilization for 2 hours in saturated $Ca(OCl)_2$ that germination in *G. pubescens* was close to 100%, and very significantly better than with 1 hour of hypochlorite treatment (Rasmussen, unpublished data).

The species of *Goodyera* are generally easy to germinate asymbiotically; *G. pubescens* attained about 86% germination on water agar (Rasmussen, unpublished data) as well as on a variety of other substrates. *G. oblongifolia* germinated on all media tested except water agar (Harvais, 1974; Henrich *et al.*, 1981; Riether, 1990), and in experiments designed to identify the minimal medium, Downie (1940, 1941) found that *G. repens* germinated only in media that contained sugar, 1–2% dextrose (D-glucose) being identified as the most effective supplement, but not in water or solutions of inorganic salts even when they were supplemented with yeast extract or vitamins. The stimulatory effect of sugars could not be replaced by mannitol (Purves & Hadley, 1976).

A variety of complex media have been used successfully in a number of species. Germination amounted to 20–90% in seeds of *G. oblongifolia* and *G. tessellata* sown on half-strength Curtis medium

(Arditti *et al.*, 1981), and according to Nishimura (1982) Sawa *et al.* (1979) used Karasawa medium or Knudson C for germination of *G. macrantha, G. maximowicziana, G. schlechtendaliana* and *G. velutina.* Seeds of *G. pubescens* also germinate readily on Knudson B medium (Knudson, 1941) and on modified Knudson C (Coke, 1990).

On the other hand, it is possible to induce symbiotic germination in *G. repens* on almost any medium, including water agar (Downie, 1940). Riether (1990) also found that symbiotic germination occurred readily, but the same seeds did not germinate asymbiotically on Eg1 medium in spite of the 1% sucrose it contained.

Germination takes place in darkness (*G. repens*: Purves & Hadley, 1976; Stoutamire, 1983). For *G. oblongifolia*, Stoutamire (1983) recommended darkness, while Arditti *et al.*, (1982*b*) obtained germination in both light and darkness except with seeds of *G. tesselata* that germinated only in the light. *G. pubescens* was germinated at a low light intensity (Knudson, 1941).

Asymbiotic germination may take as long as 3.5 months (Arditti *et al.*, 1982*b*). Seeds of *G. pubescens* that have overwintered on the infructescences germinate more readily than those harvested in autumn (Curtis, 1936) which is consistent with the fact that cold stratification (5–10 °C) for several months also increased germination in this species (Stoutamire, 1974; Coke, 1990).

The addition of cytokinin or vitamins to Curtis medium improved the germination of *G. oblongifolia*, according to Arditti *et al.*, (1981).

Both rate of germination and subsequent seedling development are, however, greatly improved by symbiosis; in symbiotic cultures of *G. repens* the hyphae entered through the suspensor 5–6 days after incubation and pelotons formed as soon as the embryo cells enlarged (Mollison, 1943; Hadley, 1983). We found that 20–50% of our seeds of *G. repens* had germinated symbiotically within 28 days. Growth stimulation is apparent immediately after infection and some time before the first collapsed pelotons can be observed (Mollison, 1943).

Culture of seedlings

The stimulation of growth due to symbiotic infection is noticeable even when no carbohydrates are included in the substrate; optimum growth requires about 1% D-glucose, but a further addition of potato extract may result in a dominance of the fungus (Downie, 1940). The seedlings of *G. repens* are fairly tolerant to the acidity of the substrate within a range of pH between 4 and 7 (Rasmussen, unpublished data).

The asymbiotic protocorms of *G. oblongifolia* are initially glabrous (Harvais, 1974), the first rhizoid cushions forming on the lower part of the protocorm when the first leaf appears. The lack of rhizoids could be associated with asymbiotic culture, however. Development is similar in *G. repens* except that symbiotic seedlings are densely hairy. Both *G. oblongifolia* and *G. repens* are slow to produce roots *in vitro*, which is consistent with the growth pattern observed in the field (see above).

When the rhizome is 5–10 mm long it is no longer light sensitive and develops chlorophyll if subjected to light. Leaves are formed when the rhizome is more than 10 mm long (Stoutamire, 1983).

At this stage seedlings of *G. repens* benefit from chilling for three months (5 °C), which stimulates the development of foliage (Rasmussen, unpublished data). This is also a good time for transferring them from the culture vessels and planting them in soil after a period when the seedlings are gradually exposed to decreasing moisture levels. *G. pubescens* can be successfully grown in leaf mould (Coke, 1990) and we have cultivated *G. repens* in a mixture of living moss and pine needle mould.

Status and prospects

Although several observations indicate that spring is the natural germination season, which has in fact been observed in the field as regards *G. pubescens*, there is apparently little thermoregulation of seed dormancy. However, in spite of being evergreen *G. pubescens* has a seasonal vegetative growth rhythm in roots and leaves, and seedling development in *Goodyera* is affected by low temperatures. The thermoperiodic reactions need further investigation to be used rationally in propagation of the species.

G. repens was the first species to be brought to maturity in symbiotic culture. Considering the speed and relative ease of symbiotic culture there is no real need for asymbiotic propagation techniques in this species. There is little experience in raising the other species in symbiosis; symbiotic techniques could speed up the production but asymbiotic methods present few problems.

Since seedlings and adult plants of *G. repens* are almost permanently infected it would seem that dependency on the endophyte persists throughout life, but a better identification of the associated fungi is needed to ascertain whether the association of *G. repens* with *Ceratobasidium cornigerum* is specific.

13.11 **Herminium**

One widely distributed species of the genus, *Herminium monorchis*, occurs in temperate Europe and Asia, and a few more species are found in central and eastern Asia (Ohwi, 1965). *H. monorchis* grows in damp grassland, in short turf, mainly on calcareous soils (Summerhayes, 1951; Mossberg & Nilsson, 1977; Buttler, 1986). Two ovoid tubers are often formed each season, the larger one providing for leaves and flowering in the subsequent season, the smaller developing on a long stalk and becoming detached from the mother plant as a means of vegetative reproduction. The plant generally requires a period of some years before it attains flowering size (Fuchs & Ziegenspeck, 1925). The short and thin roots may be almost free from fungus, but the tip of the tuber usually contains some infection (Fuchs & Ziegenspeck, 1925).

Life history

It is probably because reproduction in *H. monorchis* is mainly vegetative that indisputable seedlings are rarely observed (Summerhayes, 1951) and its life history from seed in the field has not been described. On one occasion I found a leafless protocorm in a coastal marsh in southern Sweden in the month of July.

Seed storage and survival

After storage for 3 months at room temperature over silica gel germination was still as good as that observed in fresh seeds, which indicates that the seeds tolerate storage reasonably well.

Germination in culture

A sterilization time of 4 minutes in 5% NaOCl with Tween 80 was found to be adequate to decontaminate the seeds. In our experiments the germination percentages were generally low; for example they ranged from 3% to 13% in combination with a variety of strains of *Rhizoctonia* on oat medium with 3% of the seeds germinating in asymbiotic controls on Mead & Bulard medium (Rasmussen, unpublished data).

Culture of seedlings

Although germination was poor the symbiotic plants grew well and developed tubers *in vitro* without any environmental stimuli, after which the leaves died down and the resting tubers could be transferred to soil. Flowering plants were obtained within 2 years from sowing (Rasmussen, unpublished data).

13.12 **Platanthera**

Platanthera is a large genus, distributed throughout the whole of the holarctic region. The species occur on a variety of soils, often acid, and in a variety of habitats ranging from swamps and meadows to fairly dry woodland (Wherry, 1918; Correll, 1950; Summerhayes, 1951; Williams & Williams, 1983). The plants are summergreen, the foliage dying down in autumn, and they survive the winter by means of a tuber that can be rounded or pointed, rod-shaped, but only rarely divided (Correll, 1950). Normally one fresh tuber develops each year to replace the old tuber. According to Hesselman (1900) *P. obtusata*, an arctic to subarctic species, develops more than one tuber each season; such vegetative reproduction may be important in an area where a short flowering and fruiting season could make seed setting irregular.

Fuchs & Ziegenspeck (1925) indicate that the humidity of the habitat is reflected in the length of the tuber of *P. chlorantha* and *P. bifolia*; in dry soils it penetrates deep down and is hardly infected at all, whereas under moist conditions the tip of the tuber is short, even ascending, and is densely infected.

Life history

Seeds of *P. bifolia* and *P. chlorantha* presumably germinate in spring, usually not far below the surface of the soil. The seedlings exist during the summer as protocorms until the first autumn, when a large apical bud is formed which develops into a leafy shoot the following spring. After the shoot an exogenous root emerges from the upper portion of the mycorhizome and becomes infected. Since the upper end of the mycorhizome is not infected the root apparently becomes infected from outside. During the second winter the plant overwinters as a short rhizome with two segments, the younger being terminated by a bud that will subsequently produce the summer shoot and one or two roots. A more or less tuberous root usually develops during the third spring and summer, and the following year a tuber with roughly the shape and size that is typical of the species is formed (Fig. 10.2d, p. 204–5; Fuchs & Ziegenspeck, 1927c). In shady habitats the mycorhizome tends to persist for a longer time; the leafy shoot does not appear in the second spring, the terminal bud being used for monopodial elongation of the mycorhizome instead, while the seedling remains underground. Typically, the protocorm tissue decomposes during the same season that a true tuber develops (Fuchs & Ziegenspeck, 1927c).

In the seasonal cycle of the adult plant new roots appear in summer and become fully developed during autumn; the primordium of a new tuber will grow during the winter and spring months to attain full size late the following summer (Burgeff, 1909; Currah *et al.*, 1990).

Endophytes

The histological pattern of the mycorrhiza in *P. chlorantha* has been described in detail by Burgeff (1909). In adult plants the infections are located primarily in the roots that grow in the upper layers of the soil, and occasionally in the narrow tips of tubers (Fuchs & Ziegenspeck, 1925; Summerhayes, 1951). The main body of the tuber is usually not infected, except in *P. obtusata* where living pelotons occur in patches interspersed with starchy tissues throughout most of the tuber (Hesselman, 1900). In August, the roots of *P. obtusata* is densely infected but most pelotons have become lysed (Harvais, 1974).

Endophytes of *P. chlorantha* and *P. psycodes* were described as typical *Rhizoctonia* mycelia with monilioid cells (Burgeff, 1936). In my laboratory, a number of such mycelia have been isolated from *P. chlorantha*. Smreciu & Currah (1989a) tested two isolates derived from *P. obtusata* and identified as *R. anaticula* and *Sistotrema* sp. with seeds of *P. obtusata*, *P. bifolia*, *P. hyperborea* and *P. orbiculata* as well as a number of other terrestrial orchids. However, neither of these fungi stimulated germination of the seeds on a cellulose agar. A strain isolated from *P. orbiculata* was tentatively identified as *Sebacina vermifera* but its capacity to produce seedling mycorrhiza with *Platanthera* has not been tested (Currah *et al.*, 1990). The same applies to isolates identified as *Leptodontidium orchidicola* isolated from *P. orbiculata* and *P. hyperborea* (Currah *et al.*, 1987a, 1990).

Seed storage and survival

Seeds of *P. ciliaris* can be stored dry over $CaCl_2$ at 4 °C for up to 31 months without any noticeable reduction in germination (Gregg, 1990).

Germination in culture

Burgeff (1954) recommended soaking the seeds of the European species in water for 2–3 months, and Van Waes (1984) improved germination by lengthy sterilization in 5% $Ca(OCl)_2$: 8 hours for *P. chlorantha* and 10 hours for *P. bifolia*. However, other investigators have not found these pretreatments necessary. In symbiotic experiments in my laboratory, up to 80% germination was obtained in *P. chlorantha* after only 10 minutes in 5% NaOCl. In the North American species *P. ciliaris* the seeds were treated for 20 minutes in 7.5% $Ca(OCl)_2$ (Anderson, 1990b), which was sufficient for attaining 80% asymbiotic germination in 9 months.

Seeds of *P. bifolia* did not germinate in water, but some germinated on Pfeffer's salts + glucose, and still greater numbers on Pfeffer's salts + glucose + potato extract (Downie, 1941; Hadley, 1982b). Up to 32% of the seeds of *P. dilatata*, 21% of *P. obtusata* and 10% of *P. psycodes* germinated on a strictly inorganic medium, *P. obtusata* even germinating on water agar (Harvais, 1974). However, both *P. bifolia* and *P. chlorantha* responded with higher germination percentages when inorganic nitrogen in the substrate was replaced by organic nitrogen (Van Waes & Debergh, 1986b), and *P. hyperborea* also preferred a substrate to which amino acids had been added, 54% of the seeds germinating as compared with 23% on water agar (Harvais, 1974). Norstog medium, which has a high content of organic compounds, has been used with success for some species (for instance 25% germination was obtained in seeds of *P. dilatata* and 50% in *P. stricta*) although not all (Henrich *et al.*, 1981). Likewise, 80% of the seeds of *P. ciliaris* germinated on a highly organic substrate, i.e. Fast medium with the addition of 1 g peptone and vitamins (Andersen, 1990b), where no germination was obtained on a modified Pfeffer medium (Harvais, 1982). Fast medium was also useful for germination of *P. hyperborea* (Smreciu & Currah, 1989a).

Some of the negative germination tests that have been reported could be due to failure to record the very slow germination responses that occur in some of the species. While almost 100% of the seeds of *P. blephariglottis* germinated within 3 months on Knudson C, Fast and Harvais media germination in *P. ciliaris* and *P. leucophaea* was spread over a period of 12 months on the same substrates (Anderson, 1990*a*). Gregg (1990) found that *P. lacera* also germinated slowly, taking 7 months on a Lucke medium. In a number of species and hybrids Stoutamire (1964) noted that the only sign of germination for the first 4–6 months was a slight swelling of the embryo within the testa, after which rhizoids were produced, the embryos turned green and the growth rate increased rapidly. In *P. ciliaris, P. blephariglottis* and *P. sparsiflora* the final germination percentage was not reached until after 1 year (Knudson C with peptone: Stoutamire, 1974).

The slow reaction shows that viability is retained when seeds are stored in the imbibed state, and also suggests that under natural conditions germination would either be spread out through the whole year or have been synchronized by some environmental signal, for instance a period of low temperature (Gregg, 1990). A correlation between poor germination and low survival rate in the seedlings was noted by Arditti *et al.* (1985) and suggests that very low, or slow, germination represents erratic germination events in which the seed has not been suitably physiologically prepared for seedling development.

Germination in species of *Platanthera* is usually inhibited by light (Stoutamire, 1974, 1983; Hadley, 1982*c*; Anderson, 1990*a*). It is possible that the seeds of *P. stricta* are insensitive to light (Arditti *et al.*, 1985), the germination percentages being very low, however, in all treatments. In *P. integrilabia*, however, germination could be significantly increased if the seeds were subjected to a treatment in light before incubation in darkness (Zettler & MacInnis, 1994).

Hadley (1983) reported that the presence of symbiotic fungi did not affect the germination percentage in *P. bifolia* and *P. chlorantha* although the strains that were used were compatible and could establish a symbiosis with the seedlings. However, we tested a range of fungal strains on seeds of *P. chlorantha* against an asymbiotic control in which less than 20% of the seeds germinated; depending on the strain we obtained a more or less pronounced stimulation up to a germination of 80% (Fig. 5.2, p. 86). The strongest effects were obtained with strains of *Ceratobasidium* (*C. obscurum*, UAMH 5443, and *C. cornigerum*, CBS 132.82) and an unidentified strain isolated from *P. chlorantha* (D144–1). *C. cereale*, obtained as a pathogen from a grass, was fairly effective in the germination of *P. hyperborea* (Smreciu & Currah, 1989*a*).

Symbiotic germination was also achieved in seeds of *P. integrilabia* with an endophyte fungus (*Rhizoctonia* sp.) isolated from roots of mature plants (Zettler *et al.*, 1990).

Culture of seedlings

The best seedling growth responses are often achieved with fungi that exercise less than optimum stimulation on germination. A reference culture referred to *Tulasnella irregularis* (CBS 574.83) proved to be one of the best growth fungi for *P. chlorantha* (Fig. 5.3, p. 88). Seedling development was also rapid with imperfect strains of *Epulorhiza* sp. (D347–1, D47–7), tentatively referred to *Tulasnella calospora* and two other isolates (D93, D94), all taken from species of *Dactylorhiza*.

Burgeff (1954) obtained protocorms of *P. chlorantha* that were 3–4 mm thick with short, thick leaves and one root after 12 months of symbiotic development in darkness. With optimal fungi, the larger protocorms in our cultures of *P. chlorantha* attained a length of more than 10 mm within 5 months in darkness.

When Stoutamire (1983) subjected the asymbiotic seedlings of a number of species to light 6–12 months after germination, they responded by developing leaves and becoming photosynthetic. However, in *P. bifolia* and *P. chlorantha* the development of the leafy shoot is greatly stimulated by chilling at 2–5 °C (Borris, 1970; Hadley, 1970*a*), and at least 8 weeks of cold treatment is recommended for the American species *P. blephariglottis*, *P. ciliaris* and *P. leucophaea* before they are planted out (Anderson, 1990*a*). Seedlings of *P. dilatata* and *P. hyperborea* also require a period of chilling in order to promote further development (Smreciu & Currah, 1989*b*). Although shoots in some species develop fairly well without chilling, it is generally found that more seedlings survive if they are subjected to a cold treatment, and it is to advantage if the small plants are transferred to non-sterile conditions immediately after chilling.

Greenhouse plants usually reach the flowering stage within 3–4 years (Stoutamire, 1974; Fast, 1985) or sometimes less (32 months: Lindén, 1980).

Status and prospects

Platanthera contains several species with a considerable horticultural potential. However, there are problems with the propagation which mainly stem from insufficiently described seed dormancy mechanisms. The culture of seedlings and larger plants seems to offer few obstacles.

13.13 **Dactylorhiza**

Dactylorhiza constitutes a large group of species occurring mainly in northern and central Eurasia, and occupying a diversity of habitats such as bogs, dry meadows and woods (Summerhayes, 1951). The substrate ranges from basic to fairly acid (e.g. for *D. maculata* down to pH 5.5: Sundermann, 1962*b*).

Most species of *Dactylorhiza* are summergreen, the leafy shoots that appear in spring being preserved until the end of October in some places (Möller, 1987*a*). The foliage leaves are well developed, particularly in damp habitats (Fuchs & Ziegenspeck, 1925), and the overwintering tuber develops mainly during the leafy seasons of spring and summer. In December the bud on the new tuber already contains all leaf and flower primordia for the following season (*D. purpurella*, Harvais & Hadley, 1967*a*).

A wintergreen rosette is formed in a few species, i.e. the Mediterranean *D. romana* and some populations of its more northern relative, *D. sambucina* (Vöth, 1971; Lindén, 1980). Wintergreen populations of the latter occur as far north as the island of Öland (Sweden), plants retaining this habit when they are cultivated in other localities (Lindén, 1980).

The tubers are lobed or palmately divided, the finger-like extensions being short in some species, for instance *D. sambucina*. These extensions descend fairly deep down in dry habitats (e.g. *D. maculata*) but may be turned up towards the surface in waterlogged localities (e.g. *D. incarnata*: Fuchs & Ziegenspeck, 1925). Mycorrhizal infection is found in the slender roots and also occasionally in the extremities of the 'fingers' (Beau, 1913; Fuchs & Ziegenspeck, 1927c; Mitchell, 1989). Infection is reported to be heaviest in poor soils, particularly those deficient in accessible nitrogen (Fuchs & Ziegenspeck, 1927c).

In the population of *D. sambucina* studied by Tamm (1948) the plants were noted to be long-lived, but there was very little recruitment of new plants from seed.

Life history

Several investigators have observed protocorms in the field (Table 3.1, p. 30). In a study of several species Vermeulen (1947) observed young protocorms in summer, the protocorm stage lasting until the following spring when the leafy shoot appears. Möller (1990b) found seeds of *D. maculata* with ruptured testa in October. According to Fuchs & Ziegenspeck (1927c) *D. sambucina* germinates in spring. In this species, a rather extensive mycorhizome can be observed in autumn, the first root developing during the first winter. It is mycotrophic but contains a certain amount of xylem. During the spring, assumed to be 12 months after germination, the leafy shoot of *D. sambucina* unfolds, after which growth becomes sympodial and the rhizome is no longer infected. Only the endogenous roots and the short extensions on the tuber now contain mycotrophic tissue (Fuchs & Ziegenspeck, 1927b).

Development in *D. incarnata* and *D. majalis* follows the same lines as in *D. sambucina*, apart from evidence of a longer underground phase. According to Fuchs & Ziegenspeck (1927b) development from germination to shoot and tuber formation is unusually rapid in *D. sambucina* as compared with the period of about 4 years required in most species of *Dactylorhiza*. However, these estimates are based on the assumption that tuber girth increases from year to year so that the number of steles in the tuber indicates the age of the plant, and also that constrictions on the mycorhizome define limits between one season's growth and that of the next. These assumptions may not be valid (Chapter 7), but according to Ziegenspeck's estimates plants belonging to *D. majalis* and *D. incarnata* are about 16 years old when they reach flowering stage.

Compared with results obtained in culture these figures seem to be too high (Rasmussen & Rasmussen, 1991). Vermeulen (1947) found that temporary cessations in growth could produce narrowings on the mycorhizome which should not be interpreted as annual markings. He also observed that if the tuber that formed during the first summer was well developed an inflorescence could be produced in culture the following summer, i.e. only 2 years after germination. Schwabe (1953) observed leafy seedlings of *D. maculata* in a garden less than 2 years after the time when seed had been dispersed, the largest of them flowering 3 years later.

The tuber develops in the axil of one of the basal leaves (Chapter 7) and grows mainly during the phototrophic season. In its first year the tuber is elongated, usually rod-shaped with only one root-like extension which is the last part to form, emerging from the distal end of the tuber proper. The leafy shoot dies down in late summer leaving the tuber to survive a period of arrested growth, after which the apical bud grows into a short rhizome and develops one or a few new roots. Both the roots and the tip of the tuber remain mycotrophic until spring (Fuchs & Ziegenspeck, 1927c) when after a period of winter chilling the shoot begins to elongate and the leaves to unfold. In wintergreen genotypes the leaves unfold in autumn, dormancy in the apical bud on the tuber being broken soon after the summer drought, without the necessity of chilling, but possibly induced by a decrease in temperature.

Endophytes

Strains referred to *Ceratobasidium*, *Thanetephorus orchidicola* and *Tulasnella calospora* have all been isolated from the roots of adult plants of *Dactylorhiza* (Williamson & Hadley, 1970; Hadley, 1970b;

Filipello Marchisio *et al.*, 1985) and used in establishing seedling mycorrhiza (Williamson & Hadley, 1969, 1970; Hadley, 1970*b*). During such experiments a great diversity in effectivity was observed, some strains tending to become pathogenic, others being inactive or barely promoting seedling growth (Downie, 1959*a*; Williamson & Hadley, 1970). Strains of *Tulasnella calospora* are apparently almost universally compatible, whereas *Ceratobasidium* species and *Thanatephorus orchidicola* are dubious symbionts (Hadley, 1970*b*). In the case of *D. purpurella*, *Thanatephorus cucumeris* is also often compatible (Hadley, 1970*b*). Downie (1959*a*) identified this fungus as endophyte from four of eight populations of *D. purpurella* growing on moors and sand dunes.

The performance of the individual strain of fungus with respect to rate of germination is not closely correlated with its effect on the growth rate of seedlings (*Dactylorhiza maculata* and several others, Mitchell, 1988). When I tested a number of strains identified as *Tulasnella* spp. and *Ceratobasidium* spp. with seeds of *D. majalis* I could confirm this (Figs. 5.2 and 5.3, p. 86 and 88). Although some of our test strains that had been identified as *Ceratobasidium* induced a very high level of germination others did not, and none of them supported further seedling development. Other test strains referred to *Tulasnella* also gave rise to variable germination results, but some stimulated seedling growth considerably. One of these was a reference strain of *T. irregularis* (CBS 574.83), another was identified by restriction fragment length polymorphism (RFLP) patterns as *T. calospora* (D47–7) and the rest were tentatively referred to species of *Tulasnella* on the shape of their parenthesomes (Andersen, 1990*b* and unpublished data).

In contrast, Downie (1959*b*) found that strains referable to *R. repens* (which should correspond to *Tulasnella* sp., see Chapter 5) were less effective growth stimulators than other strains she isolated, and that varying growth rate in the fungi could not alone account for the differences observed in growth stimulating properties. Clements *et al.*, (1986) tentatively identified the fungi associated with *Dactylorhiza* as a species of *Ceratobasidium* and considered that it differed from the fungi effective in the mycorrhiza of *Orchis* and *Ophrys*.

The range of fungi that can be successfully used for germination and seedling development is greater when the carbon source is starch than when mono- and disaccharides are used (Downie, 1959*b*; Beyrle *et al.*, 1985).

Seed storage and survival

Seeds of *Dactylorhiza* are generally easy to store, their germination capacity being retained for up to several years (Van Waes, 1984). We used seeds of *D. majalis* that had been stored at room temperature for more than 2 years, after which the germination percentages became unacceptably low, but it is uncertain whether this was attributable to loss of viability or to secondary dormancy (Johansen & Rasmussen, 1992). Cold treatment for 3 months did not improve germination in old seeds, however (Rasmussen, unpublished data).

Germination in culture

According to Van Waes (1984), who used 5% $Ca(OCl)_2$ with 1% Tween 80, the optimal sterilization time varies from 6 to 10 hours depending on the species. Frosch (1982) sterilized the seeds for *c*. 45 minutes in 0.3% NaOCl. We have made it a practice to use 5% NaOCl for 6 minutes for seeds of *D. majalis* and its relatives.

A number of species germinate readily in water to which a little detergent has been added (0.05% Tween 80), i.e. *D. fuchsii*, *D. incarnata*, *D. maculata*, *D. majalis*, *D. purpurella*, *D. sambucina* and *D. traunsteineri* (Downie, 1941; Vermeulen, 1947; Eiberg, 1969; Veyret, 1969). Up to 70% of fresh seeds of these species germinated within 20 days. *Dactylorhiza foliosa* and *D. elata* gave poor results in water (Eiberg, 1969; Veyret, 1969), but considerable differences may be encountered in samples of the same species from different localities (Riether, 1990).

A Pfeffer–glucose medium is suitable for several species (Hadley, 1982*b*). Harbeck (1968) used Chang medium, but considered that those seed samples that germinated would do so on virtually any substrate on the condition that the right amount of moisture was provided. Substrates with a high molarity tended to give lower germination percentages, however. Additives such as various fruit and vegetable juices produced no positive effects, but ribose, nicotinic acid and adenine in small concentrations increased the germination percentage (Harbeck, 1968).

All species tested responded well to a substrate when organic nitrogen rather than inorganic nitrogen was used (Van Waes & Debergh, 1986b). *Dactylorhiza incarnata, D. maculata* and *D. majalis* germinate well on Ramin medium in which the nitrogen source is peptone (Galunder, 1986). Likewise, Vermeulen (1947) recommended a modification of Knudson B salts with a lower content of $Ca(NO_3)_2$ and found that nitrogen in the form of urea gave better results than ammonium. The addition of IAA inhibited germination in *D. purpurella*, whereas gibberellic acid (GA) and kinetin had no effect (Hadley & Harvais, 1968).

In symbiotic germination experiments Fuchs & Ziegenspeck (1922) obtained low germination percentages, but the results were improved by cold stratification. Borris (1970) greatly increased the germination in *D. majalis* by chilling for 3–4 months. However, chilling at 3 °C for 3 months did not raise the germination percentage in any of the species tested by Riether (1990); on the contrary, in some cases it was reduced.

Since seedling development is reported by Vermeulen (1947) to be promoted when protocorms were sunk in the agar substrate we wanted to test whether covering with agar exercised some influence on germination, but found a negative effect; seeds of *D. majalis* buried in agar seemed to be quite unlikely to germinate (Rasmussen, unpublished data). On the other hand, the atmosphere in the test dishes has some influence, since sealing and wrapping of the culture dishes raise the germination percentages considerably (Fig. 4.6, p. 56). An experiment with different concentrations of CO_2 had no obvious effect on germination, although an accumulation of carbon dioxide is to be expected in the tightly sealed culture containers. It has not been established whether ethylene is produced in the dishes either from endophytes or from seeds and seedlings, but the stimula-

tory effect of adding ethylene to germination dishes is quite significant (Fig. 4.7, p. 59; Johansen & Rasmussen, 1992).

During the first days of incubation germination is inhibited by
light, the more so as the intensity is increased (Fig. 4.5, p. 53; Van
Waes, 1984; Rasmussen *et al.*, 1990*b*; Rasmussen & Rasmussen,
1991). When seeds of *D. majalis* are incubated in light and subsequently moved to darkness the resulting germination is higher than
in controls that are incubated in darkness immediately (Rasmussen
et al., 1990*b*). This effect, although reproducible with other seeds of
the same batch, did not occur in all seed lots or in other species
(Rasmussen & Rasmussen, 1991).

Culture of seedlings

The development and survival of asymbiotic protocorms can be
promoted by the addition of GA (10 ppm) to the substrate. Kinetin
(10 ppm) also produced an increase in growth rate resulting in the
formation of large shoots, but also in a slight decrease in survival
(Hadley & Harvais, 1968). Borris (1970) found that seedling development was promoted when benzyladenine (BA) was added to the
substrate.

When the protocorms and medium darken as a sign of poor
seedling growth, this is the result of the exudation of phenolic
compounds. This problem is aggravated by exposing the culture to
light, high temperatures or oxidizing agents of the substrate such as
Fe^{3+} (Haas, 1977*a*), whereas browning can be reduced by placing
the cultures in dim light or darkness, and by transferring the cultures
frequently to fresh substrates which keeps the growth rate high. *D.
sambucina* is reported to be more than usually prone to browning
(Harbeck, 1968).

It seems that at optimal temperatures seedlings are able to develop
larger protocorms before the shoot tip develops than are seedlings
grown at temperatures that are low or too high (*D. majalis*: Rasmussen *et al.*, 1990*a*), which may affect the duration of their vegetative
state (Rasmussen & Rasmussen, 1991).

For a long time during their development protocorms are intolerant of illumination (Van Waes, 1984; Malmgren, 1988); when 20 weeks old, irradiation of 750 lux or more has deleterious effects on at least 70% of the seedlings. When applied earlier, light may be harmful to 90–100% of the seedlings (*D. maculata*, Van Waes, 1984). Seedlings of *D. majalis* do not seem to be as sensitive (Rasmussen, unpublished data), but light is in no way advantageous to seedling development.

When older seedlings are subjected to photoperiods this promotes the development of a shoot tip (*D. purpurella*: Hadley, 1970a), with the highest production of chlorophyll occurring at 15 °C or below. Except in the few wintergreen taxa the shoot elongates and the leaves expand after a lengthy cold treatment (Borris, 1970; Hadley, 1970a), such as when 12-weeks-old seedlings were subjected to 4 °C for 12 weeks (*D. maculata, D. majalis × purpurella*: Beyrle *et al.*, 1987; *D. elata*: Rasmussen, unpublished data); the required length of vernalization, however, depended on the size of the seedlings (Fig. 9.1, p. 178; see Chapter 9). Apparently a greenhouse with a minimum temperature of 9 °C in winter is not cold enough, colder though frost-free conditions being required (Harbeck, 1968).

After chilling, the seedlings can be potted out. Beyrle *et al.*, (1987) recommended the use of beech mould in the potting mixture and indicated that light manuring (0.1%) with a low content of nitrogen may be applied without harming the mycorrhiza. In Copenhagen, our survival rate was 50% for *D. majalis* in a soil mixture consisting of 2 parts coarse gravel to 1 part beech mould (Rasmussen, unpublished data).

It is also possible to transfer seedlings to soil during the summer without chilling, and subject the plants to naturally decreasing temperatures during autumn and winter. Survival rates with this method can be up to 90%, provided that when the seedlings are removed from the sterile containers they have roots that are at least 10 mm long (Malmgren, 1988).

With optimum culture technique, flowering specimens are obtained after 2–3 years (Vermeulen, 1947; Stoutamire, 1974) and in general after 4–5 years of culture (Malmgren, 1988).

Status and prospects

Dactylorhiza has probably presented fewer problems in both symbiotic and asymbiotic propagation than many other holarctic orchids and some workers have been able to raise species in considerable numbers (Harbeck, 1968; Malmgren, 1988). Poor germination in some species and some seed lots has usually been attributed to low seed quality, possibly as a result of self-pollination and partial self-incompatibility. However, the effects of cold stratification have not been studied in detail and it could be that the seeds of some species and some provenances require cold treatment before they can germinate.

Since many species lend themselves readily to cultivation, little effort has been made to study the few difficult species. The cultivation procedure that is satisfactory for many species involves vernalization to induce the sprouting of a leafy shoot. However, it is to be expected that the wintergreen species, such as *D. romana* and *D. sambucina*, have requirements that in this respect differ from those of the majority of species.

Symbiotic cultivation gives the most rapid development but asymbiotic germination and development is a good alternative, and has indeed been carried out on an amazing variety of substrates (e.g. Eilhardt, 1980).

The species of *Dactylorhiza* have been involved as test organisms in studies of several fundamental questions pertaining to orchid mycorrhiza. Some of these studies are treated in greater detail in previous Chapters. The ease of handling the species of *Dactylorhiza* will probably continue to make them favourable model species, for instance in attempts to solve problems such as those connected with seed survival and dormancy patterns.

13.14 Coeloglossum

Coeloglossum is circumboreal genus with one species, *C. viride*, occupying a wide range of habitats, mainly open places such as pastures, downs and sandhills, generally in short turf and mostly on chalky soils, both damp and dry (Summerhayes, 1951). Fuchs & Ziegenspeck (1925) also report it from mountainous habitats, both shady and open. The tuberous habit resembles that of *Platanthera*.

Life history

According to Fuchs & Ziegenspeck (1927c) germination occurs in spring in the lowlands, and somewhat later in the mountains. There is a resting period in summer, but this can be very short, particularly in shady localities (Ziegenspeck, 1936).

The protocorm persists and produces new exogenous roots each autumn until the third autumn when an endogenous root is formed. In the fourth spring a leafy shoot unfolds and during the following summer the first tuber develops at the base of the shoot. This tuber does not become infected, but the tubers formed in the fifth and subsequent years have slender extensions that are infected. The short rhizome that is produced from the apical bud of the tuber becomes more congested and exhibits progressively less infection from year to year, the mycotrophy becoming concentrated in the roots that appear in late summer and are fully developed in autumn when the leaves dies down (Fuchs & Ziegenspeck, 1924a, 1927c; Currah *et al.*, 1990). The rate of development depends on the habitat, the plants developing more rapidly in warm, open places.

The timetable set up by Fuchs & Ziegenspeck (1927c) is somewhat dubious because the authors apparently only visited the populations in the autumn, all illustrations being from this time of year, and only the size of the protocorms suggested that the seeds germinated in spring. If tubers are not formed until the fourth growing season, as suggested, the seedlings will be vulnerable to desiccation for several summers.

Endophytes

An isolate from *C. viride* which Downie (1959a) used for germination gave rise to successful symbiosis. This endophyte produced a creamy-white mycelium with encrusted sclerotial areas, and with a hyphal diameter of 4–5 μm. Another isolate from roots of *C. viride* was identified as *Thanatephorus orchidicola* (Hadley, 1970b). Symbiotic protocorms were produced with this fungus *in vitro*, but it caused no perceptible increase in the germination percentage (Hadley, 1983). Strains of *Rhizoctonia solani* from various sources, including parasitic strains, did not enter into lasting symbiosis, the seedlings either immediately being parasitized, forming only short-term symbiosis with the fungi, or else excluding them (Downie, 1959a).

Secondary infection is possible from the time when the first tuber begins to grown after a resting period. The risk of secondary infection may be increased in the first years since the tuber of a young plant has few parts where the fungus can be retained through a drought period, the tuber extensions not yet being fully developed. Since these areas are not continuous with the roots, each organ in older plants must become independently infected from the soil, which could lead to mixed colonization in each plant.

Germination in culture

For unknown reasons some workers have been unable to germinate seeds of *Coeloglossum* (Harvais & Hadley, 1967a; Riether, 1990), and even immature seeds have presented difficulties (Malmgren, 1989b). Other investigators have obtained germination in distilled water (Downie, 1941) or on various asymbiotic media such as Veyret, Knudson C with peptone, or Pfeffer medium with glucose (Veyret, 1969; Stoutamire, 1974; Hadley, 1982b).

The seeds germinate in darkness (Stoutamire, 1974).

Culture of seedlings

The asymbiotic growth of the seedlings can be promoted by the addition of potato extract (Downie, 1941). In older seedlings a period of chilling induces shoot development (Hadley, 1970a), as in *Dactylorhiza*.

Status and prospects

Although there is some evidence that germination takes place in spring, the seeds often germinate readily *in vitro*, suggesting that seeds can germinate a short time after dispersal. It is possible that the difficulties noted by some investigators are caused either by rapid loss of seed viability or by variable dormancy patterns in seed populations, but little has been done to investigate seed dormancy in this genus.

13.15 Gymnadenia and Leucorchis

Gymnadenia and *Leucorchis* constitute a small Eurasian group with a northern distribution, appearing as a mountain plant in the Mediterranean region (Summerhayes, 1951). *Gymnadenia albida* (= *Leucorchis albida*) extends to arctic North America. The species occur in open places on a variety of soils, mostly rich in lime, in habitats of varying degrees of moisture. *G. conopsea* occurs on basic soils (pH 8.0–8.5) and does not seem to be dependent on soil moisture content (Fuchs & Ziegenspeck, 1925; Sundermann, 1961; Sundermann, 1962b), but *G. albida* is to some extent a xerophytic species, growing in soil that temporarily dries out (Fuchs & Ziegenspeck, 1927c).

The species are summergreen, keeping the foliage leaves until the end of October (Möller, 1987a). During the phototrophic season nutrients are stored in a tuber which is slender, lobed or palmately divided.

Life history

Germination has not been observed in the field, but can be assumed to take place in spring since according to Fuchs & Ziegenspeck (1927c) seeds of both *G. conopsea* and *G. odoratissima* when sown in pots outdoors germinated after a period of frost and chilling. Ziegenspeck (1936) found small protocorms of *G. conopsea* in autumn, when they were presumably 6 months old. In the autumn Ziegenspeck also observed the appearance of the first one or two roots, on seedlings presumed to be 1.5 year old, and in the spring he saw the leafy shoots unfold on the rooted seedlings, apparently 2 years after germination. The first tuber was formed during the following phototrophic season (Fig. 10.2*a*, p. 204; Fuchs & Ziegenspeck, 1927*c*). In adult plants roots that are mycotrophic form in autumn and the root-stem tuber grows during the spring and summer. Beau (1913) found fungi in roots as well as in the tapering extensions of the tuber which extend quite deep down in dry soil and contain a fair amount of xylem. The mycotrophic roots grow in the upper layers of the soil.

G. albida is believed not to appear above ground before the fourth spring and to be the most decidedly mycotrophic species in this group (Fuchs & Ziegenspeck, 1925). It is facultatively autogamous and sets seed effectively, so that numerous seedlings are often observed in the natural populations (Summerhayes, 1951).

Endophytes

Downie (1959*a*) could not germinate *G. conopsea* and *G. albida* with a fungus isolated from adult plants, nor with any of her test strains of *Rhizoctonia solani*. One of the strains that we isolated from roots of *G. conopsea* var. *densiflora* was identified as *R. solani* on hyphal characters (Andersen, 1990*a*) but an ultrastructural study showed that it had ascomycetous septal pores (strain D227–3; Andersen, 1990*b*). There are no data on its compatibility with seedlings. Other strains we have isolated are all referred to *Epulorhiza*, presumably *Tulasnella* spp. One of these, from *G. odoratissima* (D228–3), produced effective mycorrhiza with *G. conopsea* as did our strain from *D. majalis* (D47–7), identified by restriction fragment length polymorphism (RFLP) analysis as *Tulasnella calospora* (Andersen, 1990*b*).

However, a reference strain of *Tulasnella calospora* (CBS 573.83, no host indicated) brought about only poor germination in *G. conopsea* and did not promote seedling growth, similar results being obtained with *Tulasnella irregularis* (CBS 574.83), *Ceratobasidium cornigerum* (CBS 132.83) and *C. obscurum* (UAMH 5443). Since the germination percentages were in general low, it was difficult to see whether these fungi differed in performance.

Seed storage and survival

The ability to stain in triphenyltetrazolium chloride (TTC) is preserved for 12 weeks when the seeds of *G. conopsea* are stored at room temperature, and for 24 weeks when stored in a refrigerator (Van Waes, 1984). In my laboratory we were able to germinate about 10% of the seeds after they had been kept for 19 weeks at room temperature.

Germination in culture

The optimal period of surface sterilization in 5% $Ca(OCl)_2$ is 2–4 hours for *G. conopsea* and *G. odoratissima*, according to Van Waes & Debergh (1986b). As a surface treatment Burgeff (1954) used soaking of the seeds for some weeks.

Several workers have had difficulty in germinating the seeds asymbiotically, obtaining little or no germination (Veyret, 1969; Eiberg, 1969; Malmgren, 1989b; Riether, 1990). Downie (1941) germinated seeds of *G. conopsea* in water and various fluid media, but the same treatment proved to be ineffective with seeds of *G. albida*. Conversely, Veyret (1969) found *G. albida* to be the easier of the two species to germinate and Harbeck (1963) obtained very few seedlings of *G. conopsea* while having almost complete success with *G. albida*, both on Chang medium.

A change in culture medium from inorganic to organic nitrogen raised the germination percentage considerably (Van Waes & Debergh, 1986b). Pritchard (1984) obtained about 45% germination in seeds of *G. conopsea* on a Norstog medium which contains amino acids. According to Hadley (1982b) both *G. conopsea* and *G. albida* require yeast extract (added to Pfeffer–glucose medium), but Galunder (1986) still obtained a fairly low germination level in *G. conopsea* (10–40%) on media containing yeast.

According to Riether (1990) cold stratification did not improve germination, but we have obtained a slightly higher germination percentage after chilling of hydrated seeds for 12 weeks.

When immature seeds of G. conopsea were excised from the capsules 4–5 weeks after pollination almost all of them germinated (Malmgren, 1989a). Embryos that were only 1 week older when excised failed to germinate.

Fuchs & Ziegenspeck (1927c) germinated seeds of G. conopsea symbiotically with a fungus isolated from Dactylorhiza majalis after subjecting the seeds to cold stratification. Hadley (1983) also obtained well-developed symbiotic seedlings of G. conopsea but although a number of fungi increased growth rate they did not seem to have any effect on the germination percentage. However, in one of my own experiments with fresh seeds of G. conopsea, germination ranged from 1% to 14% with a variety of fungi while not exceeding about 2% in asymbiotic controls on Mead & Bulard medium. In a later test with oat medium throughout, a number of identified strains gave results that were inferior to that of the asymbiotic control, whereas a selection of our own Epulorhiza strains and an identified strain of Ceratobasidium obscurum gave better results. These observations suggest that some fungi stimulated germination while others had an adverse effect, but the statistical basis is insufficient since the germination percentages never exceeded c. 10–15% and variation within treatments was high. Furthermore, we found little correspondence between the germination percentages and the subsequent growth of the symbiotic seedlings.

Culture of seedlings

Seedlings of G. conopsea grow rapidly, both asymbiotically and in symbiotic culture. A simple salt and glucose medium supports the seedlings, as do more complex substrates (Downie, 1941; Malmgren, 1989b). We obtained fully developed plants from symbioses with strains of Epulorhiza (D47–7 and D347–1), the former being identified as Tulasnella calospora. When a shoot tip is formed the seedlings require cold treatment at 2–5 °C for several months to induce the bud to elongate and the leaves to unfold (Borris, 1970; Lindén, 1980). However, Stoutamire (1974) obtained shoot elongation after cultivating the seedlings for 14 months on Knudson C + peptone medium, in darkness, without any changes in temperature. Further cultivation may take place along the same lines as for *Dactylorhiza* or *Platanthera*. The same cultivation procedure is effective for *G. odoratissima* (Vöth, 1980a).

Status and prospects

Since immature seeds of *G. conopsea* germinate so readily, and prolonged soaking and chilling are reported to increase germination, there is presumably dormancy in mature seeds. This enables seeds to remain dormant in the soil after dispersal and to germinate about half a year later in spring, as supported by a few field observations.

While the cultivation of seedlings seems to be relatively easy and is successful when carried out as described above, the usually low germination percentages and inconsistent germination results still present a problem. This situation could be improved by a detailed investigation of the dormancy patterns and the relations between seed age, storage and viability. Possibly a long period of cold stratification combined with symbiotic culture would constitute the laboratory conditions most closely resembling the natural environment.

13.16 **Nigritella**

Nigritella is a minor genus of Eurasian mountain plants. Although the plants inhabit damp localities, the shallow soils and drying winds may create conditions of comparative drought. The leaves appear in summer and are narrow and close to the ground. There is an extensive root system; the palmately divided tubers have numerous long extensions and a well-developed xylem system, but in the slender, superficial roots the xylem is not so strongly developed and these roots are heavily infected (*N. nigra* and *N. rubra*: Fuchs & Ziegenspeck, 1925).

Life history

Development from germination to phototrophic plants resembles that in *Gymnadenia* (Fuchs & Ziegenspeck, 1927c).

Germination in culture

The seeds take 2–3 months to germinate *in vitro* (*N. nigra*, *N. rubra*: Veyret, 1969; Haas, 1977b).

Immature seeds of *N. nigra* can be excised *c.* 1 month after pollination and almost 100% germination can be obtained within 2 weeks after incubation (Malmgren, 1989a). The result varied markedly with the provenance of the seeds, those of Swedish origin germinating as described, whereas those from central Europe were almost non-responsive.

Culture of seedlings

The addition of 2.5% pineapple juice to the substrate stimulates seedling development and 3–4 months after germination the seedlings can be large enough to have a leaf and a root (Malmgren, 1989a). When this stage is reached the seedlings can be transferred to soil. Since they are intolerant of both drought and excess moisture great care must be taken during the first few weeks after explantation.

13.17 **Orchis**

Orchis is a large Eurasian genus occurring in a variety of habitats. *Orchis palustris* prefers wet, seasonally inundated places but the majority of species are found in high grassland, open deciduous or evergreen forest and macchia. Typically the soil is shallow and alkaline, usually calcareous (Fuchs & Ziegenspeck, 1927c; Sundermann, 1962a; Buttler, 1986).

Some species, mainly in the southern part of the distributional range, produce a wintergreen rosette of leaves that withers during fruiting before the advent of the summer drought. Some northern species and populations adhere to the same seasonal pattern, but in most of them the leaves unfold in spring or, according to Rose (1948), as early as January or February and persist until fairly late in summer (*O. militaris*, *O. purpurea*, *O. simia*: Fuchs & Ziegenspeck, 1925; Summerhayes, 1951). In some species both the summergreen and wintergreen habits are reported (*O. purpurea*, *O. ustulata*: Rose, 1948; Fast, 1985; Möller, 1990a), which is perhaps an expression of different ecological races. Plants of *O. mascula* from wintergreen populations on Öland, Sweden, sprouted in autumn after they had been moved to the Botanical Gardens in Copenhagen, whereas the leaves of adjacent Danish plants did not unfold until spring (F. N. Rasmussen, unpublished data).

Another example of local variation in leaf phenology is *O. palustris*, one of the few species typical of wet habitats, which has practically no summer resting period. In late summer the terminal shoot on the new tuber immediately replaces the old aerial shoot and the rosette remains green during the winter, according to Möller (1987a), but in the same species Fuchs & Ziegenspeck (1925) observed a well-developed but not yet unfurled leafy shoot during winter, so that the plants assumed a summergreen habit.

During the summer resting period that is typical of the genus, the plant consists of a globose rootless tuber with a terminal bud that develops into a short rhizome. At first this rhizome is not infected, but living hyphae may be found in the two-layered epidermis of the tuber. In the autumn, when roots develop from the new rhizome, they become infected and the infection may spread to the rhizome, the extent of rhizome mycotrophy varying with the species (Fuchs & Ziegenspeck, 1927c; Sharman, 1939).

The summergreen species (or populations) may produce anatomically specialized and mainly mycotrophic autumn roots that differ markedly from the mainly water-conducting roots that develop in spring (Fuchs & Ziegenspeck, 1925). In wintergreen plants the function of the autumn roots is divided between mycotrophy and the transport of water to the leaves but, according to Fuchs & Ziegenspeck (1925), some wintergreen species such as *O. morio* also produce autumn roots that are mainly mycotrophic and that have little water-conducting tissue. They interpreted the water-storage tissues of the tuber and leaves of this species as resulting from the necessity for strict water economy during winter. With the rising demand for water at the time of flowering, however, long slender non-infected spring roots develop (Fuchs & Ziegenspeck, 1925).

In the early-flowering species *O. pallens*, Fuchs & Ziegenspeck (1927c) noted that leaves and flowers were often damaged by frost in the northern part of its range, so that plants had to subsist for a time on pure mycotrophy. Unfavourable conditions such as late frost or overgrowing vegetation may well promote mycotrophy, perhaps to the point where only subterranean organs develop. Young plants of *O. simia* occasionally disappear underground after they have produced their first foliage leaf, and reappear after remaining underground for 1–2 years, but a longer disappearance is usually an indication that they are dead (Willems, 1982). By long-term monitoring it can be shown that plants in a population of *O. mascula* also occasionally remain underground for one season (Inghe & Tamm, 1988).

Life history

All observations of germination in nature have been made in late summer or autumn (Table 3.1, p. 30). Rose (1948) observed mature seeds of *O. purpurea* germinating at 25 °C after being sown around the base of the parent plant in a greenhouse pot. Möller (1989) observed a large number of seedlings of *O. militaris* in the field, but noted that seedling mortality was high as the result of a period of drought in spring when they were about 6 months old. B. Johansen and T.F.

Andersen (unpublished data) observed that seedlings of *O. mascula* in shallow soil died after a very dry spring.

Development after germination appears to vary considerably according to the species. In an unusually well-documented case study of a recently established population of *O. simia*, Willems (1982) observed seedling leaves 4 years after the founder plant had flowered. The underground stage was evidently not longer than that, but it is not known how much of that time was spent as ungerminated seed. Other species are also believed to have a long initial life underground, for example *O. pallens* 4– 6 years and *O. ustulata* 10 years (Stojanow, 1916; Fuchs & Ziegenspeck, 1927c). There is, however, reason to be sceptical of the age estimates of these stages, especially with respect to *O. ustulata*. First, cultivated plants can sometimes reach the flowering stage in a much shorter time – in *O. ustulata* within 3 years (Möller, 1985a). Secondly, the age estimates of mycorhizomes are based on the number of constrictions, each of which is taken to represent one year's growth, but constriction lines may also arise from contraction of the rhizome, or from nodes. Cultivated seedlings are rarely constricted, perhaps because of more constant growth conditions. In contrast to *O. ustulata* rapid development has been reported in *O. mascula* and *O. tridentata*, both believed to produce their first leaf above ground in the second growth season (Beer, 1863).

Many species remain as protocorms until the first root and leafy shoot forms (root→aerial shoot→tuber, Chapter 10; Fig. 10.2a, p. 204 and Fig. 10.3a, p. 206); it is the initial stage as unbranched protocorm/mycorhizome that apparently differs in length, the extreme being *O. ustulata* whose mycorhizome may eventually attain a length of 20–30 mm, until roots and leaves are produced. The most rapid development is found in *O. morio* and *O. papilionacea* in which the protocorm/mycorhizome develops its first exogenous root some time during the autumn after germination, and even in the first winter, or the following spring (depending on the leaf phenology) may produce a small foliage leaf and more roots. A tuber later develops from the base of the leafy shoot and grows during the following leafy season. This tuber is the only part that survives the winter (Fuchs & Ziegenspeck, 1927*c*; Ziegenspeck, 1936; Möller, 1987*b*).

In another type of life history the first development involves a transition from the protocorm to a tuber, without any intervening leaf production (tuber→root→aerial shoot; Fig. 10.2*c*, p. 204 and Fig. 10.3*b*, p. 206). A succession from protocorm to tuber while the plant remains below ground was first described by Stojanow (1916) from *O. mascula*. The protocorm terminates its development in the first year, when a tuber is formed from the axil of one of the scale leaves at the tip of the protocorm, and the protocorm then withers. The following season the apical bud of the tuber produces a short rhizome on which roots, scale leaves and a new tuber develop. On the basis of the number of steles in the tuber and a general evaluation of the size distribution in the material, Stojanow judged the underground stage of *O. mascula* to last for 2 years. This is more than Beer's (1863) estimate (cf. above) and could mean that *O. mascula* in some places proceeds from protocorm to root and foliage formation instead of first producing a tuber. The life history is presumably affected both by properties of the habitat and by individual variation.

The life history of *O. militaris*, *O. purpurea* and *O. simia* is described as being basically similar to that of *O. mascula*, the plants existing below ground for more than 1 year, developing their first tuber during the second or third summer. This tuber grows into a short rhizome that sends out mycotrophic roots in autumn (cf. Fig. 7.1*c*, p. 124), and by the following spring the first leaf may emerge (Fuchs & Ziegenspeck, 1927*c*). With respect to *O. militaris*, this description is contradicted by Irmisch (1853) who observed the first leafy shoot during the first summer after germination, i.e. another observation that suggests that populations may differ in life history.

Möller (1987*b*) also observed seedlings of *O. mascula* and *O. purpurea* that were described as constricted in the middle and regarded these constrictions as the result of the accumulation of nutrients in the apical part of the protocorm, the 'pretuber' (Vorknolle), and not as the product of two growing seasons. The apical bud of the 'pretuber' developed into a small leaf during spring. Without anatomical and morphological details it is difficult to say whether this observation could represent a perennial protocorm, a protocorm in the process of producing a tuber, or another young life stage.

Endophytes

There is a chance of losing the original fungus already at the time when the protocorm dies away and the first tuber is left to survive the dry season. In order to identify the fungus that was active during germination, the young underground stages must be looked for in the autumn and winter months.

Orchis morio, O. laxiflora and *O. sancta* form seedling mycorrhiza with strains of *Ceratobasidium cornigerum* and with several unidentified strains, according to Muir (1987, 1989). We have grown seedlings of *O. militaris* for a short time with identified strains of *Tulasnella calospora*. Isolates from species of *Dactylorhiza* spp. are often compatible with species of *Orchis*.

Seed storage and survival

Seeds of *O. morio* remain stainable in triphenyltetrazolium chloride (TTC) for more than a year when refrigerated (Van Waes, 1984).

Germination in culture

The seeds of *Orchis* usually require little scarification. Frosch (1982) recommended surface sterilization for 45 minutes in NaOCl (0.3% active chlorine) for a number of species, and Mead & Bulard (1979) treated the seeds of *O. laxiflora* for no more than 15 minutes in 3% $Ca(OCl)_2$; Muir (1987) sterilized the same species for 10 minutes in 5% NaOCl. *Orchis morio* only needs treatment for 45 minutes in 5% $Ca(OCl)_2$ with 1% Tween 80 if maximum germination is to be attained, according to Van Waes & Debergh (1986b).

Apparently there is also a group of species that require a fairly intensive surface treatment. Van Waes & Debergh (1986*b*) found 4 hours to be the optimum treatment for *O. coriophora* and *O. mascula* with the same procedure as above. Before sowing seeds of *O. mascula* and *O. militaris* Burgeff (1954) soaked them for 2–3 months in water (Chapter 4). Rinsing in sulphuric acid before surface sterilization in hypochlorite raised the germination percentage in *O. militaris* and *O. purpurea* considerably; for *O. purpurea* the recommended treatment to obtain 20–30% germination was found to be 1% H_2SO_4 for 15 minutes, followed by 0.3% NaOCl for 45–60 minutes (Malmgren, 1993).

Fresh seeds of *O. militaris*, *O. morio*, *O. laxiflora* and *O. sancta* can germinate in water with the addition of detergent (0.05% Tween 80), but *O. mascula*, *O. palustris*, *O. purpurea* and *O. ustulata* failed to germinate with this treatment (Vermeulen, 1947; Eiberg, 1969, 1970). Readiness to germinate asymbiotically varies, however, even within a species (Fast, 1978).

A noticeably higher germination percentage was obtained in *O. coriophora* and *O. mascula* when organic nitrogen was added to the substrate instead of inorganic nitrogen, but the effect was less pronounced in *O. morio* (Van Waes & Debergh, 1986*b*). Vermeulen (1947) modified Knudson B by replacing ammonium with urea, which increased the level of germination; a further increase was achieved by reducing the concentration of $Ca(NO_3)_2$ to one-tenth of the prescribed amount. Addition of certain vitamins such as thiamine or riboflavin increased the germination percentage of *O. laxiflora*, but the effect on germination as such was much less apparent than the stimulatory effects on seedling development (Mead & Bulard, 1979). Fast medium with both peptone and yeast was effective for asymbiotic germination of *O. (Amerorchis) rotundifolia* (Smreciu & Currah, 1989*a*).

Riether (1990) achieved no positive effects of chilling on seeds of *Orchis* species, which is consistent with the assumption that they germinate in summer and autumn in the wild. However, with repeated freezing in liquid nitrogen followed by thawing Pritchard (1984) increased the germination in *O. morio* from 62% (controls) to 82% after four freezing and thawing cycles.

Muir (1987) obtained symbiotic germination in *O. laxiflora* with various strains of *Ceratobasidium cornigerum* and with unidentified strains originating from *O. morio* and *Dactylorhiza fuchsii*. We have germinated *O. papilionacea* with a fungus referred to *Epulorhiza* sp. (D347–1: Andersen, 1990*b*), but the subsequent growth of the seedlings was not satisfactory. We also established a mycorrhiza between *O. mascula* and a strain of *Epulorhiza* (D185–1) which resulted in a low germination percentage although the seedlings developed well (Rasmussen, unpublished data). In an experiment set up by Borris & Voigt (1986), seeds of *O. mascula* did not germinate asymbiotically whether they used fresh mature seeds, after-ripened seeds or immature embryos from green capsules; but when they introduced a fungus strain isolated from an adult plant the seeds germinated well. The illustration of this unidentified isolate suggests that it is a typical species of *Rhizoctonia*. Borris & Voigt (1986) tried to simulate the effect of the fungus chemically by adding various organic compounds to the substrate, but neither vitamin B (nicotinic acid) nor the plant growth regulators benzyladenine (BA), zeatin and gibberellic acid (GA_3) produced any effect on germination in *O. mascula*. Surprisingly, however, the addition of indoleacetic acid (IAA) generated an increase in germination (Chapter 4).

Immature seeds have been brought into culture in several investigations. Malmgren (1989*b*) noted that *O. morio* can be harvested about 4 weeks after pollination, but according to Frosch (1980) 6 weeks is the optimal time; *O. purpurea* and *O. mascula* can be germinated when about half ripe (Allenberg, 1976; Malmgren, 1989*b*).

Culture of seedlings

Several substrates with organic nitrogen compounds, for instance Ramin, Mead & Bulard and Norstog are recommended for the culture of *Orchis* species (Galunder, 1986; Muir, 1989). Malmgren (1993) used a substrate that contained no inorganic nitrogen and emphasized the importance of amino acids for the cultivation of *O. coriophora*, *O. palustris*, *O. purpurea*, *O. simia* and *O. ustulata*. BA and thiamine are reported to stimulate development in the seedlings (Borris, 1970; Mead & Bulard, 1979).

Cold treatment is not usually necessary for breaking bud dormancy in species of *Orchis* (Borris & Albrecht, 1969) but differences in chilling requirements could reflect the summergreen versus wintergreen habits of the species or populations in question. In *O. morio* and *O. spitzelii*, both of which are wintergreen, the leafy shoots unfold without previous cold treatment (Lindén, 1980). Nevertheless, it seems that the chance of survival is usually increased if seedlings are chilled before being subjected to non-sterile conditions, perhaps because of their subsequent rapid development. In the culture of *O. morio* it is advisable to transfer seedlings to fresh substrate every 6 weeks to maintain a high growth rate, and when root primordia appear the seedlings should be given a period at 10 °C before or after they are transplanted to soil (Frosch, 1983*b*). Plants can be brought to flower in the third spring after germination (Vermeulen, 1947).

Asymbiotic seedlings of *O. morio* do not tolerate illumination during the first many weeks; mortality may exceed 80% if seedlings are subjected to strong light within the first 24 weeks of growth (2500 lux, *c.* $7.5\,W\,m^{-2}$, Fig. 9.2, p. 180; Van Waes, 1984). Some fast-growing species such as *O. laxiflora* may become light tolerant at an earlier stage; Muir (1987), for instance, subjected her cultures to photoperiods only a few weeks after sowing, and described seedling development as being satisfactory.

Vegetative propagation can be induced by removing the developing tuber after the plant has flowered (*O. mascula*: Sharman, 1939), which causes another bud on the old stem to produce a new tuber. For the remainder of the growing season this second tuber continues to develop provided the leaves of the old tuber survive (Möller, 1987*b*). The excised tuber should immediately be treated as dormant (Cribb & Bailes, 1989; Chapter 11).

Status and prospects

Some of the inconsistent results with *Orchis* could be attributed to inappropriate conditions for storage and rapid decrease in seed viability.

In view of the horticultural possibilities of this genus it is surprising that more attempts have not been made to solve the dormancy problems. Cold stratification is probably not required when natural populations normally germinate in summer or autumn. The scarification technique apparently needs to be optimized for each species that is taken into culture and further improvement could be attained by paying attention to such possible requirements as an after-ripening period. Bearing in mind the effects a change in substrate composition or the introduction of a fungus can bring about, germination can probably also be induced by certain chemical stimuli.

Apparently the species that have so far responded best to *in vitro* cultivation are those described as having a relatively brief heterotrophic phase of seedling development in nature.

13.18 Aceras

Aceras is a monotypic genus with a Mediterranean distribution, preferring dry, open grassy places on lime-rich soils (Summerhayes, 1951; Sundermann, 1962a). Plants of *A. anthropophorum* are small with globular tubers and a basal rosette of leaves. Some populations are wintergreen, others not (Summerhayes, 1951). The roots die in summer, new and highly mycotrophic roots being produced in autumn (Fuchs & Ziegenspeck, 1925).

Life history

After germination a fairly thick mycorhizome is formed which produces an exogenous root in the first autumn. During the following summer a short rhizome segment develops, close to the tip forming a small tuber that survives the late summer drought. In the second autumn this tuber produces a rhizome segment and a lateral root, and in the following spring the leafy shoot sprouts, a new tuber forming at its base during summer (Fig. 10.2b, p. 204; Fuchs & Ziegenspeck, 1927c; this description obviously refers to summer-green populations).

The root formed in the first autumn may be wholly mycotrophic, whereas those formed in the second autumn and onwards also serve to conduct water to the shoot that subsequently develops.

Endophytes

During the first dry season, which the seedling survives as a small tuber, there is a chance of losing the original fungus. The reportedly heavy infection in the roots of adult plants could thus comprise a variety of different strains. We have isolated one typical *Rhizoctonia* mycelium from *A. anthropophorum* (D55–1) but its symbiotic capacities have not been tested. In our symbiotic germination experiments of *A. anthropophorum* (see below), other strains of *Epulorhiza* have been successfully used, but cultivation beyond early seedling stage has not been attempted.

Seed storage and survival

During a storage experiment set up by Van Waes (1984) seeds showed a rapid decline in stainability in triphenyltetrazolium chloride (TTC) from *c.* 30% at the onset of the experiment to zero within 4 weeks when the seeds were stored at room temperature, and within 12 weeks if they were kept in a refrigerator.

In my own experience of seeds stored in the dried state over silica gel at $-18\,^{\circ}\mathrm{C}$, germination was reduced from 40–80% (depending on which fungus was used) to 21% with the most effective fungus after 21 months of storage (Rasmussen, unpublished data).

Germination in culture

Van Waes & Debergh (1986b) recommended a fairly long period of surface sterilization for *Aceras*, viz. 2 hours in 5% $Ca(OCl)_2$ with 1% Tween 80. They found that the asymbiotic germination was improved by the addition of organic nitrogen to the substrate, but the percentage was still exceedingly low, being at best about 6%. Veyret (1969) and Riether (1990) failed to germinate seeds of *Aceras*.

Newly harvested seeds were, however, found to germinate in pure water with 0.05% Tween 80 (Eiberg, 1970), and our symbiotic germination on oat medium with various strains of *Epulorhiza* proved successful (40–80%), the best results being obtained with the strain of *Epulorhiza* sp. (D347–1).

13.19 **Traunsteinera**

Traunsteinera is a small genus, presumably closely related to *Orchis*, distributed in the Caucasus and mountains of Turkey, and reaching southern and south-eastern Europe. The plants are found typically in open woodland or damp grassland, often above the tree line (Mossberg & Nilsson, 1980; Buttler, 1986).

Traunsteinera globosa resembles *O. mascula* in habit but the leaves are less succulent. The roots are long and thick, whereas the tuber is fairly small (Fuchs & Ziegenspeck, 1925).

Germination in culture

Veyret (1969) succeeded in germinating seeds asymbiotically within 3 months on Veyret medium.

13.20 **Anacamptis**

Anacamptis is genus with a southern Eurasian distribution, preferring dry, open situations, taller grassland or open grassy woods on lime-rich soils (Fuchs & Ziegenspeck, 1925; Summerhayes, 1951; Sundermann, 1962a). *A. pyramidalis* has globose tubers and a few slender but fleshy roots. In autumn an overwintering leaf rosette is produced which elongates in spring to form a leafy flowering spike (Summerhayes, 1951; Vöth, 1980a). The leaves are generally short and narrow and wither early; the roots are infected. These features suggest that mycotrophy is of importance in the adult plant.

Life history

Protocorms of varying size may be found in March. The larger ones are believed to be 3 years old (Ziegenspeck, 1936), which would imply that the plants grow monopodially without forming roots for 3 years in succession. During the spring the largest protocorms form a terminal bud and a tuber which survives the summer drought. In the autumn the bud on the tuber grows into a short rhizome with one or two roots, and a leafy shoot may also develop. The following spring a new tuber appears and continues to grow until the onset of the summer drought (Fig. 10.3b, p. 206–7; Fuchs & Ziegenspeck, 1927c).

The rhizome and the first roots are highly mycotrophic. Those roots that develop immediately before or after the shoot appears are also water-conducting, and the amount of xylem is greater than in the first-formed roots (Fuchs & Ziegenspeck, 1925).

Endophytes

No endophytes have been identified from A. pyramidalis. The original fungus that is active during germination would be difficult to find, unless seedling stages prior to the development of the first tuber are used for isolation. After the late summer resting period, which only the tuber survives, the plant presumably becomes reinfected from outside.

Seed storage and survival

Van Waes (1984) reported that the seeds lose their ability to stain in triphenyltetrazolium chloride (TTC) within 8 weeks when stored at room temperature, and within 24 weeks when stored in a refrigerator.

Germination in culture

Van Waes & Debergh (1986*b*) recommended fairly intensive surface sterilization, namely 3 hours in 5% $Ca(OCl)_2$, and Galunder (1986) sterilized the seeds for up to 90 minutes in 1% NaOCl. However, when sterilization was preceded by rinsing in 0.5% H_2SO_4 for 3–4 minutes the required time in 0.3% NaOCl was only 5–6 minutes, the subsequent germination being 20–30% (Malmgren, 1993). This timing is critical, however, 7 minutes in NaOCl being destructive to the seeds.

Newly harvested seeds can germinate in water with the addition of 0.05% Tween 80, but no rhizoids develop (Eiberg, 1970). A substrate containing Pfeffer's salts, yeast extract and glucose can initiate asymbiotic germination (Hadley, 1982*b*). However, when inorganic nitrogen was replaced by organic nitrogen Van Waes & Debergh (1986*b*) obtained a considerable increase in germination percentage, and a another medium in which there were no inorganic salts (Ramin) was used successfully by Galunder (1986).

Germination is inhibited in light and exposure to light before the seeds are incubated does not stimulate germination (Van Waes, 1984).

Hadley (1983) obtained better results with symbiotic germination than with asymbiotic germination but did not specify the associated fungus. Riether (1990) incubated seeds with a strain of fungus from *Dactylorhiza majalis* but produced no lasting symbiosis.

In my laboratory we have excised embryos from immature capsules 21 and 28 days after pollination, and from mature capsules 35 days after pollination. Germination was sporadic, both asymbiotically on Mead & Bulard medium, and with a strain of *Tulasnella calospora* (D47–7) on oat medium. The reason for this failure could be lack of compatibility between the parent plants resulting in poor seed quality, since experimental pollination in successive years yielded a high proportion of empty testas. Seedlings originating from sporadic germination (of fresh seeds or immature seeds) did not establish.

Culture of seedlings

The survival rate of seedlings is much higher on BM3 than BM and BM4 media (Van Waes, 1984), indicating that seedlings thrive better on organic nitrogen than on inorganic nitrogen, and that they do not need benzyladenine (BA). Malmgren (1993) recommended a substrate in which all nitrogen is supplied in an organic form as amino acids.

The adult plants require fairly moist soil and high atmospheric humidity in winter, as well as protection from frost. The plants should be watered in spring until they reach the early flowering stage, and then kept dry but protected by means of mulching with pine needles or moss to prevent them from drying out completely (Vöth, 1980a). After resting, the plants sprout with no changes in culture regime, apparently following an endogenous rhythm.

Status and prospects

Seed dormancy is a major problem in the propagation of *A. pyramidalis*. If the limited field evidence can be taken to show that germination normally takes place in spring, it would be reasonable to test the effect that chilling may have on the seeds. Considering that the climatic conditions are warm and dry at the time of seed dispersal, a warm dry after-ripening treatment might also be required.

According to Fuchs & Ziegenspeck's interpretation (1927c) of the field evidence, the seedling exists during several years as a protocorm, i.e. oversummering without a tuber, in which case the seedlings must be vulnerable to desiccation. Under favourable growth conditions, however, seedlings may grow large enough to produce a tuber in the first year, which would make the life history less precarious and reduce the estimated time span from germination to leafy stage to 12–18 months.

13.21 **Amitostigma**

Amitostigma is an Asiatic genus with a few species in temperate East Asia and Japan (Ohwi, 1965). *A. keiskei* occurs in wet mossy habitats, often bogs in mountainous situations (Cribb & Bailes, 1989). The plants are summergreen with a single foliage leaf on the flowering stem and have tuberous roots, according to the figures in Masamune (1969).

Germination in culture

Both *A. gracile* and *A. keiskei* germinate on Karasawa medium and Knudson C (Sawa *et al.*, 1979, according to Nishimura, 1982). These authors recommend soaking the seeds of *A. gracile* in 0.1 M KOH for 5 minutes before sowing; the effect of this treatment could be due either to the potassium ions or to the basic reaction of the solution (Chapter 4).

13.22 **Barlia**

Barlia is a small genus (one or two species) in Mediterranean Europe, closely related to *Orchis*. Open woodland, macchia and abandoned fields are the typical habitats (Mossberg & Nilsson, 1980; Buttler, 1986).

A leaf rosette emerges in the autumn and the shoot elongates to produce an inflorescence in early spring. The aerial parts die down at the beginning of the summer drought and the plant survives as a rounded tuber.

Germination in culture

Asymbiotic germination can be achieved on a range of media such as Knudson C with peptone, Eg1, ZAK and HPZ (Stoutamire, 1974; Riether, 1990). In my laboratory we have found that germination is sporadic even when seeds are sown from green capsules, and very slow, taking up to 10 months on Mead & Bulard medium. The germination level proved slightly better in the presence of a fungus strain referred to *Tulasnella calospora* (D47–7), and a mycorrhiza was established.

13.23 **Himantoglossum**

Himantoglossum is a small Mediterranean genus confined to dry, calcareous soils. *H. hircinum* often grows in disturbed habitats such as railway banks, roadsides and chalk pits, as well as on downs and in dry meadows (Good, 1936; Sundermann, 1962*a*). The lower leaves may unfold in autumn (Summerhayes, 1951) or in the early spring of the succeeding year (Fuchs & Ziegenspeck, 1925). The upper leaves and the inflorescence appear during the spring, the leaves beginning to wither at flowering time. The short phototrophic season thus ends with the summer drought (Fuchs & Ziegenspeck, 1925), the ovoid tuber being the only organ to survive the dry season. With the onset of autumn rains mycotrophic roots develop (Fuchs & Ziegenspeck, 1925).

Life history

The development of *H. hircinum* from seed resembles that in *Orchis purpurea* and *O. simia* fairly closely (Figs. 10.2*c*, p. 204, and 10.3*b*, p. 206; Fuchs & Ziegenspeck, 1927*c*). Ziegenspeck (1936) first observed protocorms of *H. hircinum* in autumn and assumed that they had just germinated. The first tuber was formed during the first spring after germination from a bud in the axil of a scale leaf at the tip of the protocorm, the protocorm itself disappearing in summer. In the second autumn a short mycotrophic rhizome developed from the tip of the tuber, which produced a mycotrophic root and from the axil of a scale leaf the second tuber began to develop, reaching full size in early summer. The rest of the plant then withered, leaving the tuber as the only part to survive the summer drought. After a resting period, the apical bud on the tuber in autumn produced a short rhizome with a new mycotrophic root and eventually, when the seedlings were more than 2 years old, a leafy shoot.

Endophytes

Gäumann *et al.* (1961) isolated 15 fungi from the roots of *H. hircinum*, 14 strains being identified as belonging to Ascomycetes, and one as *Rhizoctonia versicolor*. They found that all these strains were able to induce defence reactions to fungi in samples of tuber tissue. One of the Ascomycetes, *Nectriopsis solani*, also occurred in the roots of other orchids in the area they investigated.

There are no data on fungi tested *in vitro* for compatibility during germination.

Germination in culture

Van Waes (1984) recommended sterilizing the seeds of *H. hircinum* seeds for as long as 10 hours in 5% $Ca(OCl)_2$. With NaOCl a 60-minute treatment in 1% is optimal, according to Frosch (1982); and 10 minutes in 0.3%, followed by washing in 6% H_2O_2 was sufficient to produce a certain amount of germination (Lucke, 1976).

The seeds are able to germinate in dim light, and germination can be observed within 3 months on TGZ medium (Lucke, 1976). Muir (1989) succeeded in germinating seeds on both Norstog and Mead & Bulard, and Riether (1990) obtained some germination on Eg1 medium, the result varying with the origin of the seed batches. Chilling for 3 months after sowing had adverse effects on the germination level.

Culture of seedlings

Asymbiotic protocorms can be transferred to a fresh substrate about 4 months after sowing, and develop rapidly on a substrate containing coconut milk (Lucke, 1976).

Mature plants that are brought into culture from the field often diminish in size and vigour after some years, or die after the first flowering season (Good, 1936). Wooster (1935) described a cultivated plant that was noted as flowering one year but then not reappearing in the following 2 years. In the third spring small plants were observed in the pot which might indicate that development from seed to aerial shoot can occur within 3 years, but it is also possible that the plantlets were vegetative offspring arising from accessory tubers.

Dry weather in summer is favourable to the maturation of the tuber; if the soil is wet during the summer resting period the tuber may rot. Thus the tendency for plants to die after flowering (monocarpy) could be promoted by high precipitation in the summer. A wet summer, on the other hand, is favourable for seedling establishment, according to Vöth (1980a).

13.24 **Serapias**

Serapias is a Mediterranean genus, occurring in open pine forest, garrigue and macchia on dry soils as well as in damp meadows and on both alkaline and slightly acid soils (Buttler, 1986). The pH of *Serapias* habitats ranges from 6.5 to 8.0 and the ion content of the soil is typically low (Vöth, 1976). In autumn a wintergreen rosette of leaves is produced which in spring elongates to form a terminal spike of flowers. A spherical tuber develops during the phototrophic season. After flowering the shoot dies down, and the plant survives the dry summer season as a tuber underground (Fuchs & Ziegenspeck, 1927c).

Endophytes

The species of *Serapias* are compatible with *Tulasnella*-like fungi during seedling development (Clements *et al.*, 1986).

Seed storage and survival

The seeds can be stored for up to 3 months at 3 °C without adverse effects (Vöth, 1976).

Germination in culture

Surface sterilization with 7% $Ca(OCl)_2$ or 3% H_2O_2 for 5–8 minutes is adequate (Vöth, 1976).

Asymbiotic germination occurs on a variety of substrates (Veyret, 1969; Fast, 1978; Pritchard, 1985a) being generally most rapid on substrates containing yeast extract (Vöth, 1976). A special substrate has been developed for *Serapias* (NB-1, Appendix A).

Temperatures around 17 °C and below 20 °C are the most suitable for germination, which takes place in darkness (Vöth, 1976).

Culture of seedlings

Seedling development is promoted by the addition of peptone to the substrates (Vöth, 1976), and organogenesis does not require cold treatment (Borris & Albrecht, 1969). In an experiment carried out by Vöth (1976) the tubers formed more readily if banana pulp was added to the substrate – possibly an effect of the plant growth regulators contained in the fruit. Unfortunately the result could not be repeated.

After they have been transferred to soil the plants should be watered from below so as to prevent the crown from rotting (Vöth, 1976). The summer resting period begins when the leaves wither in late spring, after which the pots are left to dry out until late summer. When the new shoot appears, possibly as a result of an endogenous developmental rhythm, watering is resumed. This is a good time to repot the plants, before any roots develop. During the winter the plants are kept at 5–10 °C (Vöth, 1976).

Two years after sowing, a few of the plants treated in this way may produce an inflorescence, and after 3 years about 60% of the plants flower (Vöth, 1976).

13.25 **Ophrys**

The large Eurasian genus of *Ophrys* has its main area of distribution in the Mediterranean region, a few species extending to northern Europe. They are almost exclusively inhabitants of dry soils with a pH of 8 or more (Sundermann, 1961, 1962*a*, *d*). The species usually have wintergreen leaf rosettes, the leaves dying away in spring during the flowering period. The plant survives the dry summer period as a tuber. In the northern part of its distribution range where the summers are less dry, *O. holoserica* flowers during the summer months so that the appearance of winter rosettes follows almost immediately after the old shoot has withered, with almost no summer resting period (Gumprecht, 1980).

The time when the leaves emerge appears to vary according to the weather in autumn: in a fairly dry autumn the leaves unfold later than if it is damp (Chapter 10). The primordium of a new tuber can be seen in November, but growth is slow until spring (March). Root initials are formed in September and the new roots grow through the leaf sheaths in November (*O. apifera*: Möller, 1987*a*).

O. sphegodes usually flowers in its first year above ground and compared with most other species has a short life span (Hutchings, 1987*a*). The half-life is about 2 years after emergence.

Life history

Fabre (1856) observed germination in *Ophrys apifera* around adult plants in March. All stages of germination occurred simultaneously, i.e. imbibed seeds with ruptured testa, protocorms with rhizoids, as well as plantlets with roots and a small leafy shoot. These seeds and seedlings could not be less than about 9 months old, originating from the previous release of mature seeds. Possibly the largest seedlings were some that had already germinated in the autumn, or they may have represented an earlier cohort. Plantlets with a tuber instead of a protocorm were considered to be at least 2 years old.

Fuchs & Ziegenspeck (1927*c*) also interpreted their field observations as evidence that the seeds germinate in spring. Beau (1920*a*) observed the germination of hybrid seeds (*O. bombyliflora × ciliata*) that had been sown in the pot round the mother plant; protocorms with rhizoids and pelotons were observed 2 months later. Although the time of sowing was not mentioned, it was probably in early summer when the seeds ripen, which implies that the seeds are able to germinate immediately after dispersal. This was, however, under cultural conditions.

According to Fuchs & Ziegenspeck (1927*c*) an exogenous mycotrophic root is produced from the protocorm in the first autumn. By the beginning of the dry season the protocorm has grown into a short mycorhizome which has produced a small, globose tuber from the axil of one of the scale leaves. Only that tuber survives the summer resting period. In the second autumn the apical bud on the tuber produces a short rhizome which is terminated by a leafy shoot and sends out highly mycotrophic new roots. Late in the mycotrophic season, in the early spring, water-conducting roots are produced to meet the demands of the leaves as the light increases and water becomes scarcer. The second tuber develops at the base of the leafy shoot during spring.

Endophytes

The young roots, which may be found in autumn and winter, are heavily infected (Mitchell, 1989). The fungi seem to survive the dry season in the outer cell layers of the tubers. The mycelium extends from the surface of tubers at the end of the dry summer period, and persists round tubers and roots until spring (*O. holoserica*: Vöth, 1980*b*). Drying out during this period may destroy the mycelium, thereby adversely affecting the growth of the plant.

Several fungus strains have been isolated from members of *Ophrys*. The fungi that Clements *et al.*, (1986) found to be effective in symbiosis with *Ophrys*, and that they tentatively referred to *Tulasnella* sp., were morphologically and physiologically distinct from the *Tulasnella* strains that were compatible with *Orchis* and *Serapias*. Our own isolates from species of *Ophrys* have all been fairly similar and have been referred to *Epulorhiza* spp. (Andersen, 1990*b*).

Germination in culture

Surface sterilization for 45 minutes with 0.3% NaOCl provided optimum conditions for germination in *O. holoserica* and *O. sphegodes* (Frosch, 1982). Germination may succeed with a shorter treatment, however: 10 minutes in 0.2% NaOCl, followed by 10 minutes in 3–5% H_2O_2, was used by Lucke (1971a). In contrast, Van Waes & Debergh (1986a) treated seeds of *O. apifera* and *O. sphegodes* for 4 hours with 5% $Ca(OCl)_2$ for optimum results.

Some species of *Ophrys* can germinate and produce rhizoids in water (Stoutamire, 1974). Hadley (1982b) used Pfeffer's medium with glucose for germinating *O. apifera*, but for seeds of *O. apifera* and *O. sphegodes* Van Waes & Debergh (1986b) found that a medium with organic nitrogen produced better results than one with inorganic nitrogen. *O. apifera*, *O. holoserica* and *O. sphegodes* also germinated well on other substrates with a high organic content such as Norstog, Ramin and Mead & Bulard (Pritchard, 1985a; Galunder, 1986; Muir, 1989). Full strength Curtis medium produced about 60% germination within 2 months in *O. fusca* and *O. lutea* (Barroso et al., 1990).

Other workers have found that germination is a slow process; for instance, Lucke (1971a) sowed seeds in August and did not observe the first protocorms until March. The germination percentages vary considerably from study to study (Fast, 1978), and some workers such as Veyret (1969) find that species of *Ophrys* are generally difficult to germinate. In my own experiments, fresh seeds of *Ophrys* germinated poorly on oat medium in association with a variety of fungi, as well as asymbiotically on Mead & Bulard medium, the only exception being *O. ciliata*.

When immature seeds of *O. insectifera* are excised from the capsules 5–7 weeks after pollination, about 90% of them germinate within a month (Borris & Albrecht, 1969; Malmgren, 1989b). In *O. sphegodes* the seeds germinate best if excised between 8 and 9 weeks after pollination (Lucke, 1971b).

The physical conditions used for germination are usually room temperature and darkness (Frosch, 1983a), but in some cases seeds also germinate in dim light or photoperiods (Lucke, 1971a; Clements *et al.*, 1986).

Culture of seedlings

The asymbiotic protocorms can be transferred to a fresh sterile substrate 3 months after sowing, and again at intervals of 4–5 weeks to keep them in vigorous growth (Malmgren, 1989b). Borris & Albrecht (1969) noted that chilling is not necessary for bringing about organogenesis. Leaves appear about 6 months after germination and the plants can then be transferred to continuous light or photoperiods. The roots form later than the leaves (Lucke, 1971a; Frosch, 1983a; Malmgren, 1989b). If plantlets with roots are transferred to soil in early autumn the root system will develop further and the leaf rosette will overwinter (Malmgren, 1989b). Similar conditions can be produced indoors by potting plants in coarse sand with a little fertilizer and placing them under light at low temperatures (2–8 °C), with an ambient humidity of *c.* 60% (Frosch, 1983a). During the leafy season the tuber is formed.

It is also possible to induce the formation of tubers while the seedlings remain *in vitro* (Lucke, 1971a). This was perhaps the result of adding activated charcoal to the substrate, which tends to improve root development in general. The observation that indoleacetic acid (IAA) and potassium ions (K^+) stimulate the production of 'minitubers' (Barroso *et al.*, 1990) could be helpful in developing a procedure for manipulating organ development *in vitro*.

The flowering stage is sometimes reached within 3 years from germination (*O. holoserica*: Möller, 1985a).

It may be difficult to introduce a mycorrhiza in asymbiotically produced seedlings. Although incubated with a number of orchid endophytes the seedlings of *O. apifera* did not form a mycorrhiza (Hadley, 1983). In my laboratory we have had the same experience with *O. ciliata*; the hyphae that entered the rhizoids of the asymbiotic seedlings disintegrated before they had entered into the protocorm body and formed pelotons.

It is interesting to note that the sequence of organ development in asymbiotic culture differs from that inferred from observations of field material. *In vitro* the leaves appear before there are any roots or tuber whereas in the field the roots develop first, followed by the tuber and then the leaves. It seems possible that premature leaf development and an early change-over to photosynthesis could be the result of asymbiotic culture.

Status and prospects

The cultivation of this genus is still far from easy, the main problems arising during the germination phase. More studies of dormancy patterns and the preservation of viability during storage would be of interest.

13.26 **Pogonia**

One species of *Pogonia, P. ophioglossoides,* occurs in eastern North America and eastern Asia in marshes, *Sphagnum* bogs and wet meadows on both highly acidic and highly alkaline soils (Wherry, 1918; Case, 1990). Another species, *P. japonica,* grows in damp places in open vegetation up to 2500 m (Ohwi, 1965; Cribb & Bailes, 1989).

The plants consist of a condensed rhizome with a number of slender roots, 3–8 cm long, and a solitary aerial flowering stem usually with a single foliage leaf (MacDougal, 1899a; Correll, 1950; Ohwi, 1965; Case, 1990). Vegetative reproduction occurs when the tip of a long horizontal root curves upwards and develops into a shoot tip (Holm, 1900; Carlson, 1938; Catling, 1990). At the base of this young shoot a secondary root emerges, seemingly in continuation of the old root. Both roots and rhizome are densely covered with long, unicellular hairs, and infection has been observed in the epidermis, hypodermis and the cortical layers (MacDougal, 1899a; Holm, 1900; Carlson, 1938).

Germination in culture

Seeds of *P. japonica* germinate on Knudson C and Karasawa media (Sawa *et al.*, 1979, according to Nishimura, 1982). Those of *P. ophioglossoides* germinate on Knudson C with the addition of peptone and on Curtis (1936) medium (Stoutamire, 1964; Arditti, 1982). The seeds can germinate in photoperiods. Elongated seedlings of *P. ophioglossoides* with small bulbous offsets rapidly turn green when subjected to natural light.

13.27 **Cleistes**

Cleistes divaricata, which is the only species of *Cleistes* included here, occurs in southeastern North America in highly acidic damp grassland (Correll, 1950; Gregg, 1989). It is a rhizomatous plant with slender roots developing in spring and persisting for more than a year; the solitary leaf is present during the summer and is borne halfway up the flowering stem (Gregg, 1989). Vegetative reproduction possibly occurs by means of budding root tips as in *Pogonia* (Holm, 1900; Catling, 1990).

Endophytes

Infection has been observed in the epidermis, hypodermis and the outer layers of cortex in both roots and rhizome (Holm, 1900). By midsummer the cortical cells of the roots are full of starch but hyphae are observed in the outer cortex of the older roots (Gregg, 1989).

Seed storage and survival

When the seeds were stored at 4 °C above $CaCl_2$ they remained viable for 3 years (Gregg, 1989).

Germination in culture

The seeds can be surface sterilized with 2% Chlorox for 25 minutes. Gregg (1989) found that about 5–6% of the seeds germinated within 5 months in darkness and at room temperature on Lucke medium supplemented with activated carbon. There was no apparent difference in the quality of seeds from spontaneous capsules and those from capsules produced by self-pollination (Gregg, 1989).

13.28 **Arethusa**

Arethusa is a small genus with one species in northern North America and another in Japan, the latter occasionally being referred to *Eleorchis*. The typical habitat of *Arethusa bulbosa* is extremely acid bogs with deep *Sphagnum* moss (Wherry, 1918; Correll, 1950). The aerial flowering stem arises from a short rhizome and bears a solitary foliage leaf during the summer. At the base of each stem a corm develops during the leafy season (Correll, 1950; Masamune, 1969; Williams & Williams, 1983).

Germination in culture

Yanetti (1990) used either immature seeds taken at least 54 days after pollination, or mature seeds. Various substrates, such as modified Knudson C and Lucke O_4, were found to be suitable, but the composition of the substrate does not seem to be of critical importance for germination. Mature seeds germinated after 6–12 months in darkness when, after sowing, the seeds were subjected to gradually decreasing temperature, kept at 4–6 °C for 3–4 months and gradually warmed again to room temperature.

Culture of seedlings

The protocorm produces a corm which enters a dormant phase when the temperature is lowered. The small plants can be transferred to a fresh substrate either before the period of vernalization or immediately after. When after a period of chilling the temperature is raised a shoot begins to grow which may or may not be terminated by foliage. If a leaf appears at the beginning of the growth season the plant should be transferred to the light. The base of the annual shoot then swells to form a new corm; when this has matured a new cold treatment is necessary. After three to six growth cycles, the diameter of the corm has increased to about 1–2 cm; the plants have now attained flowering size and can be planted out (Yanetti, 1990).

13.29 **Calopogon**

Calopogon is a small genus confined to southeastern North America, growing in marshy places such as grassy swamps, bogs or damp coniferous woodland (Correll, 1950; Williams & Williams, 1983). Most habitats are highly acidic but the species is also found in alkaline fens (Wherry, 1918; Case, 1990). The rhizome develops into a succession of corms. One or two narrow, upright leaves usually occur on the aerial stems, which may also be leafless (Correll, 1950).

Seed storage and survival

If the seeds of *C. tuberosus* are stored at about 5 °C in a desiccator some will still germinate after 5 years (Stoutamire, 1964, 1974).

Germination in culture

C. tuberosus germinates asymbiotically on almost any medium, e.g. Knudson B liquid medium, Burgeff N_3f, Knudson C, modified Curtis,

Norstog (Carlson, 1935; Henrich *et al.*, 1981; Stoutamire, 1983; Arditti *et al.*, 1985; Anderson, 1990*a*). Germination percentages are generally high and development rapid; Liddell (1952) found that the seeds germinate even more rapidly in a liquid medium than on solid substrates. The seeds germinate equally well in light and in darkness (Anderson, 1990*a*).

Culture of seedlings

The seedlings of *C. tuberosus* may turn green 2 weeks after the seeds are sown (Stoutamire, 1983) and develop two or three leaves within the first 2 months of culture (Carlson, 1935). Anderson (1990*a*), using Lucke medium, found that the light regime had a considerable effect on organogenesis. Within about 2 months after the seedlings had been transferred to photoperiods of cool white fluorescent light the protocorms were 5 mm long, and they had produced leaves that measured 50 mm and a number of roots. However, when seedlings were kept in darkness, the protocorms grew to a length of 30 mm producing absorbing hairs and scale leaves only – in other words the seedlings remained in the heterotrophic phase until subjected to light. On the other hand, Myers & Ascher (1982) were able to initiate the development of foliage leaves in the dark (on Norstog medium), the leaves unfolding when the seedlings had been transferred to light.

After the foliage leaves had developed and a good root system had been established, the small plants could either be transplanted to *Sphagnum* pots that were kept in the shade (Myers & Ascher, 1982), or kept *in vitro* until the leaves and roots died down, by which time a corm had been formed at the base of the leafy shoot (Carlson, 1935). The corm remains indefinitely dormant if the cultures are left at room temperature, but chilling for 4–6 weeks stimulates the production of a new shoot, initiating a new growth cycle. After vernalization (4 °C for 2 months) the corms can be planted in soil outdoors in spring (Anderson, 1990*a*).

The flowering stage is attained 2–3 years after the seeds have germinated (Riley, 1983) and this period can probably be further reduced by experimentally accelerating development to two growth cycles a year (Stoutamire, 1964).

13.30 **Bletilla**

Bletilla is a small genus distributed in eastern Asia. The species of *Bletilla* grow on grassy slopes and in the margins of woods and thickets, up to 3200 m, on loamy well-drained humus-rich soils (Ohwi, 1965; Cribb & Bailes, 1989). The plants produce leafy and flowering shoots in summer from a creeping, cormous rhizome; the perennial roots grow from May to September. In autumn numerous living pelotons are found in the roots but they decrease in number during the winter months until there is an increase again in April. Mycotrophic activity is greatest in spring and autumn, which are also the wettest periods (Masuhara *et al.*, 1988).

Endophytes

Bernard (1909) succeeded in establishing seedling mycorrhiza in *B. striata* with a strain of fungus that he called *Rhizoctonia repens*, on a substrate containing a decoction of salep. Masuhara and co-workers (1988) obtained 18 isolates from adult plants, all tentatively referred to the *Rhizoctonia repens* group, and two strains with more slender hyphae (1–3 μm) and small monilioid cells (4.0–8.0 $\times$ 2.5–7.5 μm) that did not conform to the description of *R. repens* Bernard as specified by Saksena & Vaartaja (1961). These isolates differed considerably in symbiotic capacity. With some of them, including those with slender hyphae, germination levels above 90% were obtained as compared with 8% in the asymbiotic control. Other isolates, although able to form pelotons in the embryos, did not promote germination or growth.

Germination in culture

The seeds of *B. striata* can be surface sterilized in saturated $Ca(OCl)_2$ for 15–20 minutes or in 0.5–0.8% NaOCl for 5 minutes (Arditti, 1982).

Several asymbiotic substrates are suitable, for instance, modified Knudson B (Arditti, 1982), Knudson C and Karasawa medium (Sawa *et al.*, 1979, according to Nishimura, 1982). A medium based on Hyponex with peptone and sucrose produced excellent results, *c.* 98% of the seeds germinating within 48 days. This is far superior to the asymbiotic germination results obtained on oat medium without additional sugars or nitrogen compounds (Masuhara & Katsuya, 1989). Glucose and trehalose are more suitable sources of carbohydrate than sucrose and mannitol (Arditti, 1982). The addition of ammonium to the substrate may somewhat inhibit germination (Ichihashi & Yamashita, 1977), but apart from that germination is not particularly sensitive to the mineral composition of the substrate (Ichihashi, 1978).

Masuhara & Katsuya (1989) succeeded in germinating seeds of *B. striata* symbiotically on oat medium, the germination varying from 41% to 93% depending on the strain. With one isolate no seeds germinated at all. The asymbiotic controls on oats resulted in 7% of the seeds germinating.

The seeds germinate in darkness as well as in low light intensities (400 lux, *c.* $1\,\mathrm{W\,m^{-2}}$: Arditti, 1982).

Culture of seedlings

Bernard (1909) found that in asymbiotic culture the protocorm meristem developed into a leafy shoot before any roots appeared. In an inoculated culture with a vigorous symbiont the growth rate was higher, the stem remained short and root development was better (Chapter 8).

The growth of seedlings is promoted by the addition of ammonium and nitrates up to the point where nitrogen compounds comprise 90% of all ions in the substrate (Ichihashi & Yamashita, 1977). Low mineral concentrations (20 mM) promoted the growth of roots rather than shoot development, but when concentrations reached above 60 mM shoot development was favoured instead (Ichihashi, 1979).

13.31 **Liparis, Malaxis, Hammarbya**

The genera *Liparis, Malaxis* and *Hammarbya* are morphologically similar and have been cultivated so little that they are treated together here. Both *Liparis* and *Malaxis* are circumboreal, growing in *Sphagnum* bogs, wet meadows and damp humus in coniferous or hardwood forest, in North America usually on an acid substrate (Wherry, 1918). *L. loeselii* prefers alkaline bogs with a pH up to about 9 (Summerhayes, 1951; Sundermann, 1962b). In contrast, *Hammarbya* occurs on a distinctly acid substrate. All species have leaves in the summer and are rhizomatous with a leafy and flowering shoot arising from a bulbous base (Correll, 1950; Williams & Williams, 1983; Buttler, 1986).

The plants reproduce vegetatively by bulbils that develop on the leaf margins (*H. paludosa*); alternatively buds are formed on the swollen stem inside the bulbs. These propagules are dispersed in autumn (Huber, 1921).

The roots of *L. loeselii* are fairly well developed, are only weakly mycotrophic, and form numerous root hairs. The persistent, partially decayed leaf bases and the old mycorhizome, which constitutes a hairy, spongy covering around the lower part of the plant, assist in the uptake of water (Fuchs & Ziegenspeck, 1927a).

Life history

The developmental history of *L. loeselii* was one of the first to be described (references in Fuchs & Ziegenspeck, 1927a), but the youngest stages were not observed until Fuchs & Ziegenspeck (1927a) made their extensive field study. They assumed that germination takes place in spring, which has recently been confirmed by Mrkvicka's (1990a) field observations. The seedlings they found suggested that only a scale leaf had been produced on the protocorm by the end of the first summer. The first juvenile bulb developed by the following (second) summer and consisted of a thick scale leaf sheathing a swollen internode, but no foliage leaf was formed and the rhizome continued to grow monopodially, producing a second scale leaf in autumn. The rhizoids on these parts occurred in groups on papillae, not evenly distributed as on the young protocorm.

According to Fuchs & Ziegenspeck (1927a), a fully developed bulb had formed by the third summer, consisting of a sheathing scale leaf, a swollen internode terminating in a rudimentary apical bud which is normally dormant but sometimes involved in vegetative reproduction, and at the base of the internode an axillary bud. From this stage onwards, the rhizome grows sympodially. The base of each swollen internode is delimited from the older part of the rhizome by a zone of lignified cortical tissue which the endophyte cannot penetrate (Fig. 7.2c, p. 126).

When the axillary bud develops into a new rhizome segment in the third autumn, an exogenous root is produced from its base. It pierces the sheathing scale leaf, grows down into the ageing rhizome segment, and forms long hairs that penetrate the dying tissue of the old rhizome cortex. Living hyphae remaining in the old shoot can thus pass into the young rhizome segment by way of the hairs and cortex of this internal root, the infection thereby being transferred from the old rhizome segment to the young one without passing through the swollen internode. The young rhizome is terminated by a bulb primordium that develops during winter and spring. The bulb that forms in the fourth year is sheathed by the base of a foliage leaf instead of a scale leaf, and external roots are produced in addition to the internal one. To some extent they are mycotrophic though mainly water conducting, according to Fuchs & Ziegenspeck (1927a). In fully grown plants two foliage leaves are produced each year on the rhizome segment.

Mrkvicka (1990a) observed a similar though much faster development in *L. loeselii*. When seeds had been sown in August under semi-natural conditions close to the roots of adult plants, young protocorms were observed in May the following year. They were pale, uniformly hirsute and had attained a length of about 0.8–1.2 mm. By August the same year the protocorms had produced the first foliage leaf and a small bulb. This suggests that Fuchs & Ziegenspeck were mistaken when they described the continuation of the mycorhizome after the first swelling as monopodial, which does indeed appear improbable since the node below the swelling is constricted and lignified. It is far more likely that growth immediately becomes sympodial. No internal root was noted by Mrkvicka (1990a), but a few ordinary roots developed the second spring, after which roots were formed each spring and functioned throughout the summer season.

The bulbs tend to enlarge from year to year. On plants growing in *Sphagnum* the succession of shoots from several years can clearly be seen (Fuchs & Ziegenspeck, 1927a). Both the tip of the swollen stem within the bulb and axillary buds may produce detachable propagules. When the plant reaches flowering size – according to Mrkvicka (1990a) after 4 years and to Fuchs & Ziegenspeck (1927a) after 15 years – an inflorescence terminates the swollen internode.

Fuchs & Ziegenspeck (1927a) point out that the leaf sheaths round the young bulb consist of living tissue with numerous rhizoids that may absorb considerable amounts of water. At first the mesophyll is filled with starch; later a kind of velamen develops, the tissue dies and the cell walls acquire secondary thickenings, forming a water-retentive, partially decayed layer around the old bulb where fungi live without being digested. If hyphae from the same mycelium enter living parts of the plant and are lysed there, the plant may actually recycle its own nutrients.

The stiff inflorescence persists in an upright position for a long time after flowering. According to Ziegenspeck (1936) the capsules do not open until autumn when the degree of moisture in the atmosphere rises, or as reported by Huber (1921) when the inflorescence is weighed down to the ground by the snow cover. The transfer of the endophyte from the mother plant to the seeds by way of the placenta, as suggested by Huber (1921), is unlikely since the swollen internode at the base of the inflorescence is not usually infected.

The arrangement with an internal root that carries the fungus across the barrier between one rhizome segment and the next has also been described in *Hammarbya paludosa* (Fig. 7.2c, p. 126; Irmisch, 1853) and *M. monophyllos* (Fuchs & Ziegenspeck, 1927a). *H. paludosa* does not produce any roots other than the internal one. Since this species grows in extremely wet habitats its requirements for water-absorbing organs are probably small. The hairy leaf bases may also absorb considerable amounts of water. According to Fuchs & Ziegenspeck (1927a) about 5 years elapse before typical bulbs are formed both in this species and in *M. monophyllos*, but these estimates may be subject to the same kind of objection as the estimate for *L. loeselii*.

Endophytes

Since the infection is transferred from one segment of the rhizome to the next, it is probable that a plant is infected with the original fungus throughout its entire life. The best time for isolating endophytes from adult plants should be winter and spring when the infection in the youngest segment of the rhizome is still new. Huber (1921) observed living pelotons that showed no signs of lysis for several weeks in spring. In summer the lysis of pelotons begins, and by September and October only a few scattered pelotons are living (Huber, 1921). The swollen stem within the bulb contains no fungus although the rest of the rhizome is filled with pelotons. The upper part of the rhizome, adjacent to the new bulb, is less heavily infected and the pelotons there remain alive. The transfer of the fungus is the primary function of the internal root that contains few hyphae and no real mycotrophic tissue, but the mechanical and conducting tissue is well developed so that the root probably has an absorptive function as well (Fuchs & Ziegenspeck, 1927a). The external roots that form in late spring are scarcely mycotrophic at all except at the base (Huber, 1921; Fuchs & Ziegenspeck, 1927a; Mrkvicka, 1990a). On the other hand, the bases of the leaves are usually infected, the hyphae passing in and out through epidermal hairs and forming pelotons in the mesophyll.

The propagules readily become infected through rhizoid papillae on the sheathing leaves, and a mycotrophic tissue is established in the lower part of the stem in these bulbils. Those originating from tissues within the bulb are probably infected with the same fungus as the mother plant, but those formed from leaf margins (*H. paludosa*) are not infected at the time of dispersal (Fuchs & Ziegenspeck, 1927a).

In these species mycotrophy thus takes place mainly in the rhizomatous tissue, in contrast to most other orchids in which infection is sooner or later transferred from the rhizome to the roots. The importance of mycotrophy is illustrated by an experiment that Huber (1921) performed on *Liparis loeselii* in which the plants were deprived of the fungus by cultivating the swollen internode separately. The plants died when the stored nutrients were exhausted.

It is a simple matter to isolate the endophyte from *L. loeselii*, according to Huber (1921). The hyphae are typically *Rhizoctonia*-like with chains of round monilioid cells and short divaricate side branches. In pure culture it prefers a neutral to slightly alkaline substrate with ammonium salts or organic nitrogen instead of nitrate (Huber, 1921). We have also isolated a *Rhizoctonia*-like strain from a presumed 2-year-old seedling of *Liparis loeselii*, but have not tested its symbiotic capacity. Isolates from *L. lilifolia* were typical *Rhizoctonia* with numerous monilioid cells and supported vigorous seedling development *in vitro* (Fig. 6.1*b*, p. 114; Rasmussen, 1993*b*).

Seed storage and survival

Mrkvicka (1990*a*) germinated seeds of *Liparis loeselii* that had overwintered in the infructescence and were sown about 6 months after maturation.

Germination in culture

Optimum sterilization was obtained with saturated $Ca(OCl)_2$ for 2 hours for seeds of *Liparis lilifolia* (Fig. 4.2, p. 42). Seeds of *Liparis nervosa* were surface sterilized in 0.5% available chlorine in NaOCl for 2 minutes by Tsutsui & Tomita (1988) and germinated on Knudson C and Karawawa medium. *L. kumokiro*, *L. odorata* and *Malaxis monophyllos* also germinated on these substrates (Sawa *et al.*, 1979, according to Nishimura, 1982). Henrich *et al.*, (1981) reported 25% germination of *L. loeselii* on Norstog medium. Arditti (1982) recommended Curtis solution 5 as a substrate for *M. unifolia*.

Huber's (1921) observation that the capsules of *L. loeselii* remain closed during winter suggests that cold stratification is necessary, but apparently a large portion of the seeds can germinate without chilling.

Huber (1921) attempted to germinate fresh seeds of *L. loeselii* with the endophyte under varying physical conditions (light, darkness, constant and fluctuating temperatures), but without success. This is surprising but no details as to the composition of the substrate used were given. Tsutsui & Tomita (1988) obtained symbiotic seedlings of *L. nervosa* with a *Rhizoctonia repens*-like fungus strain on oat medium. This species germinated in 16 hour photoperiods. I found that seeds of *L. lilifolia* reached a germination of about 45% on W2 medium when inoculated with either of two *Rhizoctonia*-like isolates from the species; seedling development was rapid, especially when they were transferred to the richer oat medium (Rasmussen, 1993*b* and unpublished data).

13.32 Calypso

Calypso is a monotypic, circumboreal genus. In North America there are two ecologically distinct forms of *C. bulbosa*: an eastern form that occurs in areas with damp, approximately neutral humus-rich soils and cool summers (< 15 °C), and a western form that grows in moderately dry subacid coniferous litter in areas with warmer summers (Wherry in Correll, 1950). After flowering (May to early July), the sympodial rhizome develops a new cormous segment with a solitary foliage leaf soon after the mature leaf on the preceding segment has withered. Thus, in some areas the plant may be virtually evergreen, with a brief leafless period in late summer (Lindén, 1980; Williams & Williams, 1983; Currah *et al.*, 1988).

Life history

There are no reports on germination in the field. However, Correll (1950) and MacDougal (1899*b*) have illustrations of the flowering plant that may give a clue to the structure of the seedling. In both drawings there is a short, branched ('coralloid') rhizome system below the corm, which probably represents a mycorhizome system. This interpretation is supported by the observation of similar structures in cultivated seedlings, and it also seems likely in comparison with the juvenile stages of the related genus *Tipularia*.

Mousley (1925) observed coralloid structures but interpreted them as being side shoots from the adult rhizome, and as representing a shift in nutrition towards more intensive mycotrophy. Failing to find more coralloid side shoots on the *Calypso* rhizome some authors have suggested that they form very rarely (Ames according to Mousley, 1925; Currah *et al.*, 1988). However, the photograph Mousley presented is not very clear, so that it is difficult to judge whether the coralloid branch he observed could not have been the remains of the mycorhizome. Smreciu & Currah (1989*b*) recorded having seen a coralloid structure once and interpreted it as a juvenile stage.

If development proceeds as in seedlings of *Tipularia*, a leafy shoot develops on the mycorhizome and the base of this shoot swells to produce the first corm. The rhizome then continues to grow sympodially from a renewal bud on the corm.

Endophytes

At flowering time, the corm has two or three short roots with numerous rhizoids in the proximal zone that is covered by a sheath consisting of a coarse brown mycelium (Currah *et al.*, 1988). Endotrophic hyphae are found in rhizoids and in the outer cortex of the roots, as well as in the tissues of the corm.

MacDougal (1899*b*) described a thick-walled type of hyphae, with monilioid cells, isolated from the outer cortex of the roots. A strain with rather broad multinucleate cells and sclerotia consisting of monilioid cells was isolated from two different collections of *Calypso* and described as *Thanetephorus pennatus* (Currah, 1987). Although this fungus formed pelotons in the tissue its mycorrhizal status is uncertain and it did not induce germination in *Calypso bulbosa* nor did it stimulate germination in any of a number of other terrestrial orchids tested (Smreciu & Currah, 1989*a*). Currah and colleagues (1988) later isolated both a *Rhizoctonia*-like mycelium with slender hyphae (1.2–2.5 μm) and a clamp-forming mycelium with coarse hyphae (6–10 μm) from pelotons in *Calypso bulbosa*.

Germination in culture

Seeds were surface sterilized in saturated $Ca(OCl)_2$ for 10 minutes by Arditti *et al.*, (1985) and germinated at 22 °C, either in darkness or in 16 hour photoperiods. In American material extremely low germination levels have been noted (estimated at 1–5%) with both mature and immature seeds. Germination was slow (4–6 months) and seedling survival low (Arditti, 1982; Arditti *et al.*, 1985; Smreciu & Currah, 1989*b*). Lindén (1980) obtained similarly poor results with American seeds, but in contrast 25–50% of the seeds of Scandinavian origin germinated.

A number of asymbiotic substrates have been tested, i.e. variations of Curtis (Arditti *et al.*, 1985), Lucke medium (Stoutamire, 1983), and potato-dextrose agar that Harvais (1974) found produced only a slight sign of germination. The culture media used by Lindén (1980) with reasonable success were Eg1 and Fast supplemented with microelements, vitamins, kinetin and auxin. The usefulness of Fast medium is corroborated by Smreciu & Currah (1989*a*).

Culture of seedlings

The seedlings form a leafless and branched rhizome system, 'a coralloid mass' like those that are sometimes preserved at the base of flowering individuals in the field (Stoutamire, 1983). On substrates containing both cytokinin and auxin Lindén (1980) also observed proliferation from the nodes which either gave rise to new plantlets or assumed coralloid shapes.

In vitro the succession of rhizome segments and renewal of leaves occurred at intervals of roughly 12 months without requiring any external changes in growth conditions (Lindén, 1980).

13.33 **Aplectrum with notes on Cremastra**

Aplectrum is usually treated as a monotypic genus distributed in eastern North America, in moist woodland and bogs, but *Cremastra*, which is a small, closely related genus in temperate eastern Asia (Lund, 1988), is sometimes included. The North American species *Aplectrum hyemale* occurs on acid to approximately neutral humus-rich woodland soil (Wherry, 1918; Correll, 1950; Williams & Williams, 1983). Solitary aerial shoots with a single foliage leaf appear in autumn and persist throughout the winter. In spring the rhizome produces an inflorescence, and the plant flowers in May with a withered leaf remaining (MacDougal, 1898). Each rhizome segment consists of a proximal slender part and a distal globose corm (cf. Correll, 1950; Williams & Williams, 1983).

Life history

Citing an unpublished manuscript written by T. Holm, Correll (1950) reported that the sympodial, cormous rhizome of *A. hyemale* is 'preceded by a coral-like, creeping rhizome, destitute of roots, but covered with hairy papillae', this rhizome system being 'of the same structure as that of *Corallorhiza*, *Calypso* and *Hexalectris*'. The word 'preceded' seems to indicate that the rhizome system in question is regarded as being a development of the mycorhizome, and the description at the same time disregards any controversy concerning similar structures in *Calypso* (see above).

Other observations of coralloid structures in *A. hyemale* have been described as lateral shoots on the adult rhizome (MacDougal, 1898; MacDougal & Dufrenoy, 1944), and interpreted as evidence that the plant can revert to a high degree of mycotrophy by producing a mycotrophic side shoot, a 'mutative coralloid offset'. Judging from the illustration in MacDougal & Dufrenoy (1944, fig. 3) it could be the initial rhizome system of the seedling that was preserved in the adult plant, but on one occasion MacDougal (1899*a*) was able to provoke the formation of a coralloid side shoot from a mature rhizome by disconnecting it from the young rhizome segment (see Fig. 8.3, p. 171).

Endophytes

MacDougal (1898) observed that the basal part of the corms and the short, straight unbranched roots that extend from it were all infected, the hyphae growing out through rhizoids and ramifying in the soil. The upper part of the corms was not infected and produced no rhizoids.

Curtis (1939) isolated a fungus from *A. hyemale* that he identified as '*R. neottiae*'. It was a coarse dark mycelium with dark brown, open textured sclerotia, but other plants examined later contained a strain with finer, hyaline hyphae (MacDougal & Dufrenoy, 1944). I have isolated a typical *Rhizoctonia* from *A. hyemale* which was effective in germinating seeds of *Liparis lilifolia* and supported seedling development in *Goodyera pubescens* (Rasmussen, unpublished data), but it was not tested with seeds of *A. hyemale*.

Some ectotrophic fungi that occur on the roots of oaks and maples have been observed to adhere to roots of *A. hyemale* when it grows adjacent to the tree roots (MacDougal, 1898). This observation could be a sign of triple symbiosis or epiparasitism, but no further investigations have been undertaken.

Germination in culture

The seeds of *A. hyemale* can be surface sterilized by rinsing them in 70% ethanol for 1 minute, followed by 10 minutes in 1% NaOCl with Tween (Coke, 1990).

Henrich and co-workers (1981) obtained about 25% germination on a Norstog medium while keeping the seeds at room temperature, whereas Coke (1990) succeeded in germinating 60% of the seeds on Knudson C medium after the seeds had been chilled at 5 °C for 3–5 months after sowing and then incubated at 25 °C for a further 3–5 months. The difference between the results of the two experiments could be a result of the chilling, but the differences between the substrates and possible differences in seed quality must also be taken into account.

Seeds of *Cremastra unguiculatum* germinate on Knudson C and Karasawa medium (Sawa *et al.*, 1979, according to Nishimura, 1982). The protocorms become enlarged and produce branched leafless rhizome systems in asymbiotic culture (Stoutamire, 1974).

Culture of seedlings

Asymbiotic seedlings can be transferred to soil when the corms measure 0.5–1 cm in diameter (Coke, 1990) but the survival rate is not reported.

13.34 **Corallorhiza**

Corallorhiza is a minor genus distributed in North America, Europe and Asia, from the arctic zone and southwards, occurring at higher elevations in the southern part of its range. As natives of coniferous or deciduous forest or tundra, these plants grow in deep layers of humus or moss, often in damp depressions and shady places (Summerhayes, 1951; Buttler, 1986; Williams & Williams, 1983). The soil may range from strongly acid to alkaline (Wherry, 1927; Correll, 1950).

The creeping underground rhizome is much branched and mycotrophic. The rhizomes of *C. trifida* have been found with a soil cover of a few centimetres (Wolff, 1927), and I have found *C. odontorhiza* less than 1 cm below the surface of the soil but covered with additional layers of leaf litter.

No foliage leaves are produced. Although the inflorescence bears some scale leaves, the chlorophyll is mainly contained within the tissues of the stem and ovaries. *C. trifida* contains sufficient amounts of photosynthetic pigments for diffuse daylight to bring the aerial shoot above compensation point, i.e. the point where the production of oxygen by photosynthesis exceeds the consumption of oxygen in the respiration process (Montfort & Küsters, 1940). In *C. maculata* the experimental exclusion of light caused no excess elongation of the inflorescence, but there was a loss of green and red pigments. Specimens that were transplanted to boxes with soil and placed in darkness were able to flower and produce seeds 9 months later in total darkness (Reed & MacDougal, 1945), which means that either the plants had adequate supplies of stored nutrients or they obtained them from the fungi.

Life history

The seeds of *C. trifida* presumably germinate in autumn when a round, uniformly hirsute protocorm is produced (Fuchs & Ziegenspeck, 1927a; Weber, 1981). Fuchs & Ziegenspeck (1924a) found protocorms of *C. trifida* in August in an alpine locality, usually in pockets of deep coniferous litter. By spring a stele had developed in the protocorms and the youngest rhizoids on the upper part of the protocorm were clustered on hair cushions. They reported that the mycorhizome elongated monopodially but a side shoot differentiated from the axil of a scale leaf in the third autumn, and in the fourth autumn a second developed higher up on the mycorhizome. Still according to Fuchs & Ziegenspeck (1927a) two more lateral shoots appeared in the fifth autumn. The side shoots were short and with determinate development, branching perhaps one or more times before they ceased to grow, and they were densely infected except at the tip. An inflorescence was differentiated at the apex of the long shoot 1 year prior to flowering, which according to Fuchs & Ziegenspeck (1927a) occurs in the seventh summer or later. However, rhizomes that Reinke (1873, fig. 1) found were *c.* 15 mm long, with two or three branches, and they had already produced a primordium for the inflorescence. He regarded the plants as being 2 years old, but they resembled plants that Ziegenspeck would refer to the fifth year.

C. odontorhiza germinated in June in a locality in Maryland, USA (Rasmussen & Whigham, 1993) and in November the same year the largest seedling was already almost 2 cm long and branched. Most of the youngest seedlings were turnip-shaped and produced no rhizoids at the base. Higher up there were hair cushions with short papilla-like hairs (illustrated in Rasmussen, 1994a). There were great differences between seedlings of the same cohort, some being still quite small 5 months after germination, but these observations nevertheless cast doubt on the slow development described above for *C. trifida*.

The long shoot is generally infected, but the inflorescence and its base are not. After *C. trifida* has flowered an axillary bud may replace the long shoot, but some of the short shoots may also begin to elongate, particularly if they happen to become detached from the rest of the rhizome system (Fuchs & Ziegenspeck, 1924a). Correll (1950) reported that *C. odontorhiza* is monocarpic, which also applies to the populations that I have investigated.

No roots are formed, and water must be absorbed through rhizoids and through the hyphal connections with the soil. The bases of old hair cushions suberize, and the old rhizoids are shed, which presumably explains some previous reports that the rhizomes are glabrous (Fuchs & Ziegenspeck, 1927a).

Endophytes

Theoretically, living hyphae should be present somewhere in the rhizome system of these plants at any time. Thomas (1893) indicated that living hyphae occur 3–4 mm below the shoot tip of the rhizome. Living pelotons are located in the outer cortex, in particular below the hair cushions, whereas in the inner cortex the hyphae become lysed (Thomas, 1893; Fuchs & Ziegenspeck, 1924a; Nieuwdorp, 1972; Freudenstein, 1992). In the cortex of *C. odontorhiza*, however, the living hyphae are found in the central part of the cortex, the outer one to three layers lacking infection, and the innermost part comprising storage tissue adjacent to the stele (Freudenstein, 1992). This species contains numerous living pelotons during the flowering period, which occurs in autumn and continues until late in the year. According to Harvais (1974) *C. maculata* also contains quite numerous living pelotons during the growing season (by which he probably meant the above-ground, flowering season), but in *C. trifida* it is difficult to isolate the endophyte at flowering time (Fuchs & Ziegenspeck, 1924*a*).

Wolff's (1933) isolates from *C. trifida* resembled a *Rhizoctonia*, but Burgeff's (1936) isolations from the same species consistently yielded clamp-bearing mycelia, and Jennings & Hanna (1898) described the endophyte of *C. trifida* as a septate, repeatedly branching mycelium with clamp connections. The ultrastructural study by Nieuwdorp (1972) shows dolipores with perforate parenthesome. Campbell (1970*a*) found two different fungi in *C. trifida*, both with clamps, one of them tentatively identified as *Mycena thujina*.

In *C. striata* Campbell (1970*a*) found two clampless fungi that also appeared to produce sheaths and Hartig net on adjacent roots of *Pinus strobus* and various hardwood trees. Campbell (1970*a*) suspected a kind of association between *C. maculata* and *Armillaria mellea*, since rhizomorphs were found in the vicinity of the rhizome of 17 of 20 randomly chosen individuals, but in this species MacDougal & Dufrenoy (1944) observed an endophyte that they regarded as being a *Rhizoctonia*. The strains isolated by Harvais (1974) from *C. maculata* comprised a *Racodium* sp. and two hyaline, clampless and non-sporulating mycelia. In germination tests with the seeds none of Harvais's fungi formed a symbiotic association.

I have isolated the endophyte of *C. odontorhiza* on two occasions during the winter months, and have succeeded in growing the cultures on substrates with glucose, i.e. potato–dextrose agar or dextrose–peptone medium, on which they developed only slowly. The isolates originated from different populations that were separated by about 1 km, and they had similar growth patterns, including clamp connections. One of these isolates was used for producing symbiotic seedlings *in vitro*; the plants developed poorly and eventually died, but this could also have been owing to unsuitable culture conditions. A third isolate with the same characteristics was obtained from a seedling that was produced after field sowing (Rasmussen & Whigham, 1993).

These studies indicate that the species of *Corallorhiza* when mature can form associations with a variety of fungal partners, and epiparasitism on neighbouring trees can be considered a possibility. Except for one example, however, little is known about the capacity of these fungi as a germination partner, and some of the isolated strains were perhaps the result of fortuitous contamination.

Germination in *C. trifida* has been observed on one occasion, when the seeds were incubated with an unidentified clamp-bearing fungus and sown on a substrate containing Pfeffer's salts and potato extract (Downie, 1943*a*). A later attempt to establish symbiosis with a strain of *Rhizoctonia solani* proved unsuccessful (Downie, 1959*a*).

Germination in culture

Seeds of *Corallorhiza* have failed repeatedly to germinate asymbiotically on a variety of substrates (Downie, 1941; Henrich *et al.*, 1981; Riether, 1990). According to Arditti (1982), Harvais (1974) managed to germinate the seeds of *C. maculata*, but in actual fact this amounted to little more than an indication of seed imbibition, i.e. about 5% of the embryos showed signs of converting their reserve nutrients into discernible amounts of starch.

Downie (1943*a*) succeeded in germinating seeds of *C. trifida*. The seeds had failed to germinate in water and in any of the substrates containing combinations of Pfeffer's mineral salts, glucose and potato extract (Downie, 1941). However, after inoculation with a fresh isolate of an endophyte with clamp connections they germinated on a substrate containing Pfeffer's mineral salts and potato extract. Downie (1943*a*) described the level of germination and the subsequent growth of the seedlings as good, but after 9 months the seedlings had only attained a length of 0.2–1.6 mm.

I have been partially successful in germinating *C. odontorhiza in vitro* after cold treatment. The seeds had been incubated on oat medium at 20 °C for 6 months, during the last 40 days of which they had been inoculated with an isolate, described above. They were then subjected to chilling for 3, 6 or 9 weeks at 5 °C. The highest germination percentage, recorded 3.5 months later, was *c.* 20% and occurred after the longest cold treatment. There seemed to be a positive correlation between length of cold treatment and germination but the results were so variable that none of the treatments differed significantly from the control (Rasmussen, unpublished data).

According to Wherry's notes on culture in Correll (1950), spontaneous seedlings of *C. odontorhiza* can often be observed near cultured plants when soil conditions are suitable. Seedlings flower within 5–6 years.

13.35 Tipularia

Tipularia is a small genus distributed in North America and Japan. *T. discolor* grows in deciduous forest on subacid, humus-rich soils (Wherry, 1918; Correll, 1950). The solitary foliage leaf appears in September and lasts until May, so that the leafy season alternates with the flowering season from June to September. At the base of the foliage leaf a corm begins to develop in September, attains a maximum weight immediately before flowering and loses *c.* 50% of its biomass again during flowering and fructification (Whigham, 1984). Roots appear in autumn at the base of the youngest corm (Snow & Whigham, 1989, fig. 1).

Life history

It is not known when the seeds of *Tipularia discolor* germinate, field sowings having so far been without result (Rasmussen & Whigham, 1993). However, I found spontaneous seedlings in considerable numbers in the forest around the laboratory in Maryland, USA, and the smallest, entirely underground ones were found in autumn. Slightly larger protocorms, probably belonging to the same cohort, were producing a small leaf in late autumn. The protocorms assumed a variety of shapes, often lobed, and usually broader than long (Fig. 7.1*a*, *b*, p. 124). By March the following year a little corm had developed at the base of the leafy shoot; the protocorm was still functional but withered during the spring months. The first corm was considerably smaller than those produced the following seasons.

Germination in culture

Coke (1990) obtained 95% germination in *T. discolor* on Knudson C medium, in Ballard's modification with potato, pH 5.5. The seeds were kept at 5 °C for 3–5 months and then at 25 °C for 3–5 months before the germination was recorded. The seedlings *in vitro* developed a fairly large mycorhizome and then either produced leaves and roots or continued to branch to form a coralloid system (Stoutamire, 1983). The mycorhizome bears solitary rhizoids or clusters of rhizoids.

Culture of seedlings

The seedlings can be grown on Knudson C until they have produced corms about 5–10 mm in diameter, and can then be transferred to leaf mould for further, non-sterile cultivation (Coke, 1990).

13.36 **Epipogium**

The holomycotrophic genus *Epipogium* contains 2 or 3 species, among these *E. aphyllum* which is widely distributed in the northern parts of Eurasia and Japan but not found in the Mediterranean and continental areas (Buttler, 1986). It inhabits deciduous or, more rarely, coniferous forest, in thick humus with constant conditions of moisture and shade (Summerhayes, 1951). The plant consists of a branched rhizomatous system with scale leaves. The pale inflorescences appear in summer at irregular intervals, usually following a wet spring, and seed production is low (Summerhayes, 1951; Geitler, 1956). Vegetative reproduction takes place by means of bulbils formed on slender underground stolons (Irmisch, 1853).

The inflorescences are without apparent chlorophyll but there are no physiological data to show whether the aerial shoots photosynthezise at all. However, the species is obviously highly dependent on mycotrophy, since its appearance above ground is interrupted by intervals of several years, possibly up to 20 years (Summerhayes, 1951). However, there are no long-term analyses of marked individuals and it should be noted that the aerial shoots are easily overlooked.

Life history

Irmisch (1853) observed protocorms of E. aphyllum with numerous rhizoids in deep beech humus but did not mention the time of year. Fuchs & Ziegenspeck (1927a) found the smallest stages in autumn. The following summer, one or more short shoots had formed in the axils of the scale leaves, and often a slender stolon was produced as well. At the nodes of the stolon bulbils develop, a few of which become detached as propagules. The small plants developing from bulbils, in autumn, closely resemble seedlings. They form rhizoids at their base and are infected from the soil. The stolons are apparently not infected.

By the production of short side shoots, the main shoot is bent to an angle of about 90°, so that false dichotomies appear. Development resembles that in Corallorhiza, except for the existence of stolons (Fuchs & Ziegenspeck, 1924a). When it is about 10 years old the rhizome system may be ready to flower. The primordium of the flowering shoot is observed in late autumn and winter (Fuchs & Ziegenspeck, 1927a) and appears above ground in the subsequent summer if the spring weather has been sufficiently wet (Summerhayes, 1951; Mossberg & Nilsson, 1977). Most reserve nutrients in the rhizome are apparently translocated to the base of the young inflorescence, which is bulbous and filled with starch before flowering. Usually only the stolons survive flowering. Fuchs & Ziegenspeck (1927a) and Niewiezerzalowna (1933) described the paucity of conducting xylem in the rhizomes and the effect that this must have on water transport to the inflorescence. Stringent water economy is suggested by the presence of water storage tissues at the base of the inflorescence and the lack of stomata on the aerial parts, and also explains the necessity for wet weather during inflorescence development.

Endophytes

The endophyte observed by Marcuse (1902, according to Burgeff, 1909) was clampbearing, resembling the strain described by Mac-Dougal (1899*a*) a few years earlier from *Hexalectris spicata*.

A

Appendix. Nutrient substrates mentioned in the text

Blumenfeld medium (Blumenfeld, 1935)
for endophytes of heterotrophic orchids. Used for endophytes of
Corallorhiza, Neottia, Limodorum

KH_2PO_4	10 g	
$MgSO_4$	2 g	
NH_4NO_3		200 mg
Glucose	20 g	

No agar (fluent medium)

BM medium (Van Waes, 1984)

KH_2PO_4	300 mg
KNO_3	400 mg
$MgSO_4 \cdot 7H_2O$	100 mg
$(NH_4)_2SO_4$	60 mg
NH_4NO_3	370 mg
$CuSO_4 \cdot 5H_2O$	0.025 mg
$FeSO_4 \cdot 7H_2O$	27.85 mg
Na_2EDTA	37.25 mg
$Na_2MoO_4 \cdot 2H_2O$	0.25 mg
H_3BO_3	10 mg
$MnSO_4 \cdot 4H_2O$	25 mg
$CoCl_2 \cdot 6H_2O$	0.025 mg
$ZnSO_4 \cdot 4H_2O$	10 mg
Biotin	0.05 mg
Folic acid	0.5 mg
L-glycine	2.0 mg
Myo-inositol	100.0 mg
Nicotinic acid	5.0 mg
Pyridoxine	0.5 mg
Thiamine	0.5 mg
Sucrose	20 g
Agar	6 g

BM1
Contains no inorganic nitrogen but 500 mg casein hydrolysate and 100 mg L-glutamine is added.

BM2
BM1 + 0.2 mg benzyladenine.

BM4
BM2 + 1 g myo-inositol.

Chang (Harbeck, 1963)

Peptone	2 g	
Atlas Fish Emulsion		6 ml
Sucrose	20 g	
Agar	12 g	

Clements' isolation medium (Clements, 1982*a*)
for Australian orchid fungi

$NaNO_3$		300 mg
KH_2PO_4		200 mg
$MgSO_4 \cdot 7H_2O$		100 mg
KCl		100 mg
Streptomycin sulphate		50 mg
Sucrose	5 g	
Yeast extract		100 mg
Agar	10 g	
pH 4.5–5.0		

CM₁
for European orchid fungi (Rasmussen *et al.*, 1990*a*), modified
with novobiocin after Clements *et al.* (1986)

$Ca(NO_3)_2 \cdot 4H_2O$		200 mg
KH_2PO_4		200 mg
$MgSO_4 \cdot 7H_2O$		100 mg
KCl		100 mg
Novobiocin*		50 mg
Sucrose	4 g	
Yeast extract		100 mg
Agar	12 g	

pH 7.0; also used for American orchid fungi, with pH adjusted to
5.8

* Remains antibiotic after autoclaving.

Curtis medium (Oliva & Arditti, 1984)

$CaCO_3$		148 mg
KH_2PO_4		240 mg
$MgSO_4 \cdot 7H_2O$		520 mg
Urea		250 mg
$AlCl_3$		0.03 mg
$CoCl_2 \cdot 6H_2O$		0.025 mg
$CuSO_4 \cdot 5H_2O$		25–0.03 mg
$FeCl_3 \cdot 6H_2O$		1 mg
$FeSO_4 \cdot 7H_2O$		11 mg
H_3BO_3		1.0–6.2 mg
KI		0.01–0.83 mg
$MnSO_4 \cdot 4H_2O$		1.0–22.3 mg
$Na_2MoO_4 \cdot 2H_2O$		0.25 mg
$NiCl_2 \cdot 6H_2O$		0.03 mg
$ZnCl_2$		3.93 mg
$ZnSO_4 \cdot 4H_2O$		1.0 mg
'Wuchsstoff 66_f'*		0.1 ml
Coconut milk		100 ml
Banana homogenate	75 g	
Glucose	10 g	
Graphite	2 g	
Agar	14 g	
pH 5.0		

* Available from Gerlach GmbH, Germany.

Curtis solution 5 (Curtis, 1936)

$Ca(NO_3)_2 \cdot 4H_2O$		350 mg
$FePO_4$		3 mg
KH_2PO_4		120 mg
$MgSO_4 \cdot 7H_2O$		260 mg
NH_4NO_3		220 mg
Glucose	10 g	
Agar	14 g	

Czapek-Dox agar (Warcup, 1971)

$NaNO_3$	2 g
K_2HPO_4	1 g
$MgSO_4 \cdot 7H_2O$	0.5 g
KCl	0.5 g
$FeSO_4$	0.01 g
$ZnSO_4$ in a 1% aqueous solution	1 ml
$CuSO_4$ in 0.05% aqueous solution	1 ml
Yeast extract (Difco)	0.5 g
Cellulose powder (Whatman Column Chromedia CF 11)	20 g
Agar	10–12 g

Dextrose–peptone medium

Glucose	5 g
Peptone	1 g
Agar	12 g

Eg1 (Burgeff, 1936)

$Ca(NO_3)_2 \cdot 4H_2O$	1000 mg
$Fe_3(PO_4)_2$	25 mg
KH_2PO_4	400 mg
$MgSO_4 \cdot 7H_2O$	250 mg
$(NH_4)_2SO_4$	500 mg
Glucose	10 g
Fructose	10 g
Agar	15 g

Eg1 (Borris, 1970)

$Ca(NO_3)_2 \cdot 4H_2O$	500 mg

$FeSO_4 \cdot 7H_2O$		20 mg
K_2HPO_4		250 mg
KH_2PO_4		250 mg
$MgSO_4 \cdot 7H_2O$		250 mg
$(NH_4)_2SO_4$		250 mg
Nicotinic acid		2 mg
Pyridoxine		2 mg
Thiamine		2 mg
Optional: Coconut milk	20 g	
Sucrose	20 g	
Agar	18 g	

Eg1 (Fast, 1978)

$Ca(NO_3)_2 \cdot 4H_2O$		1000 mg
$FeSO_4 \cdot 7H_2O$		20 mg
K_2HPO_4		250 mg
KH_2PO_4		250 mg
$MgSO_4 \cdot 7H_2O$		250 mg
$(NH_4)_2SO_4$		250 mg
Glucose	10 g	
Fructose	10 g	
Agar	10 g	

Eg1 (Lindén, 1980)

$Ca(NO_3)_2 \cdot 4H_2O$		1000 mg
$FeSO_4 \cdot 7H_2O$		20 mg
K_2HPO_4		250 mg
KH_2PO_4		250 mg
$MgSO_4 \cdot 7H_2O$		250 mg
$(NH_4)_2SO_4$		250 mg
Sucrose	10 g	
Coconut milk (Gibco)		20 ml
Agar	10 g	

Eg1 (Riether, 1990)

$Ca(NO_3)_2 \cdot 4H_2O$		500 mg
Fe-citrate		50 mg
K_2HPO_4		250 mg
KH_2PO_4		250 mg
$MgSO_4 \cdot 7H_2O$		200 mg
$(NH_4)_2SO_4$		250 mg
A–Z trace-element solution		1 ml
Vitamin B complex		1 ml
Activated charcoal	1 g	
Sucrose	10 g	
Agar-agar	5 g	
Birch phloem sap		250 ml
Distilled water		750 ml

Fast medium (Fast, 1974)
recommended for *Cypripedium calceolus*

$Ca(NO_3)_2 \cdot 4H_2O$		83 mg
KCl		167 mg
KH_2PO_4		83 mg
$MgSO_4 \cdot 7H_2O$		83 mg
NH_4NO_3		167 mg
$FeNa_2EDTA$		17 mg
Heller's trace elements		0.8 ml
Biotin		0.01 mg
Nicotinic acid		0.1 mg
Peptone	1.67 g	
Fresh brewer's yeast	8.3 g	
Sucrose	11.67 g	
Fructose	5 g	
Agar	8.3 g	

pH adjusted to 5.5; concentration of mineral salts 0.06%

Fast medium (Lindén, 1980)

$Ca(NO_3)_2 \cdot 4H_2O$		83 mg

KCl		167 mg
KH_2PO_4		83 mg
$MgSO_4 \cdot 7H_2O$		83 mg
NH_4NO_3		167 mg
$NaFe_2EDTA$		8 mg
Waris' trace elements*		2 ml
Nicotinic acid amide		2 mg
Pyridoxine		2 mg
Thiamine		2 mg
Yeast extract	2 g	
Kinetin		1–2 mg
Indoleacetic acid		0.1–0.2 mg
Sucrose	12 g	
Fructose	5 g	
Agar	6 g	
pH adjusted to 5.5		

* Contains 1.5 g boric acid, 0.15 g ammonium molybdate, 0.1 g sequestrene Na_2Cu, 0.1 g sequestrene Na_2Co, 0.3 g sequestrene Na_2Zn per litre solution.

Fast medium (Anderson, 1990*a*)

$Ca(NO_3)_2 \cdot 4H_2O$		65 mg
KCl		33 mg
KH_2PO_4		33 mg
$MgSO_4 \cdot 7H_2O$		33 mg
NH_4NO_3		65 mg
Ferric ammonium citrate		25 mg
Biotin		0.25 mg
Calcium pantothenate		5 mg
Nicotinic acid		10 mg
Pyridoxine		2 mg
Thiamine		5 mg
Peptone	1 g	
Glucose	7 g	
Agar	6 g	
pH adjusted to 5.5		

FIM (Clements *et al.*, 1986)
fungus isolating medium

$Ca(NO_3)_2 \cdot 4H_2O$		500 mg
KH_2PO_4		200 mg
$MgSO_4 \cdot 7H_2O$		100 mg
KCl		100 mg
Sucrose	5 g	
Yeast extract		100 mg
Agar	10 g	

FN (Fast, 1978)
recommended for a range of European species

$Ca(NO_3)_2 \cdot 4H_2O$		65 mg
KCl		33 mg
K_2HPO_4		33 mg
$MgSO_4 \cdot 7H_2O$		33 mg
NH_4NO_3		65 mg
$FeNA_2EDTA$		13 mg
Sodium deoxyribonucleinate		130 mg
Peptone	1.3 g	
Sucrose	7 g	
Agar	8 g	

pH adjusted to 5.5; mineral salts concentration 0.06%

GD medium (Thomale, 1957)

$FeSO_4 \cdot 7H_2O$		20 mg
KH_2PO_4		300 mg
KNO_3		400 mg
$Mg(NO_3)_2 \cdot 6H_2O$		110 mg
$(NH_4)_2SO_4$		60 mg
NH_4NO_3		370 mg
Nicotinic acid amide		1 mg
'Wuchsstoff 66$_f$'*		10 mg
Glucose	10 g	
Fructose	10 g	

Agar	13 g

* Available from Gerlach GmbH, Germany.

H1 oat medium (Clements *et al.*, 1986, modified by Rasmussen *et al.*, 1990*a*)
for symbiotic cultures

$Ca(NO_3)_2 \cdot 4H_2O$	200 mg
KCl	100 mg
KH_2PO_4	200 mg
$MgSO_4 \cdot 7H_2O$	100 mg
Yeast extract (Difco)	100 mg

Sucrose	2 g
Finely ground oats	3 g
Agar	12 g

H2 oat medium
as H1 but with 2 g glucose instead of sucrose.

H4 oat medium
as H2 but with 72 mg $NH_4NO_3 \cdot 4H_2O$
and no $Ca(NO_3)_2$.

Harvais medium (Harvais, 1974)

$Ca(NO_3)_2 \cdot 4H_2O$	800 mg
KCl	100 mg
KH_2PO_4	200 mg
KNO_3	200 mg
$MgSO_4 \cdot 7H_2O$	200 mg
$CuSO_4 \cdot 5H_2O$	0.5 mg
H_3BO_3	0.5 mg
$MnSO_4 \cdot 4H_2O$	0.5 mg
$ZnSO_4 \cdot 7H_2O$	0.5 mg
Ferric ammonium citrate	1.0 mg

Glucose	10 g
Agar	10 g

HPZ (Reichter, 1990)

A–Z trace-element solution	1 ml
Yeast extract (Bramsch)	2 g
Peptone	4 g
Agar	5 g

Hyponex medium (Nishimura, 1982)

Hyponex*	3 g
Sucrose	20, 30 or 35 g
Peptone or tryptone	2 g
Activated charcoal	2 g
Agar	8, 10, 12 or 15 g

* Hydroponic Chemicals Co., Inc., Copley, OH, USA.

Karasawa medium (Karasawa, 1966)

Hyponex (macronutrients)	3 g
Peptone	2 g
Banana	15 g
Bee honey	30 g
Sucrose	20 g
Agar	15 g
pH 5.2	

Knudson B solution (Knudson, 1925/1946)

$Ca(NO_3)_2 \cdot 4H_2O$	1000 mg
KH_2PO_4	250 mg
$MgSO_4 \cdot 7H_2O$	250 mg
$(NH_4)_2SO_4$	500 mg
$FePO_4 \cdot 4H_2O$	50 mg (1925) or 25 mg (1946)
Sucrose	20 g
Agar	12 g (1925) or 15 g (1946)

Knudson C solution (Knudson, 1946)

$Ca(NO_3)_2 \cdot 4H_2O$		1000 mg
KH_2PO_4		250 mg
$MgSO_4 \cdot 7H_2O$		250 mg
$(NH_4)_2SO_4$		500 mg
$FeSO_4 \cdot 7H_2O$		25 mg
$MnSO_4 \cdot 4H_2O$		7.5 mg
Sucrose	20 g	
Agar	12–15 g	

Knudson C (Stoutamire, 1964)

$Ca(NO_3)_2 \cdot 4H_2O$		1000 mg
KH_2PO_4		250 mg
$MgSO_4 \cdot 7H_2O$		250 mg
$(NH_4)_2SO_4$		500 mg
Fe-EDTA		25 mg
$MnSO_4 \cdot 4H_2O$		7.5 mg
Sucrose	20 g	
Peptone	2 g	
Agar	8 g	

Knudson C (Arditti, 1982)

$Ca(NO_3)_2 \cdot 4H_2O$		1000 mg
KH_2PO_4		228 mg
K_2HPO_4		7.8 mg
$MgSO_4 \cdot 7H_2O$		250 mg
$(NH_4)_2SO_4$		500 mg
$FeSO_4 \cdot 7H_2O$		25 mg
$MnSO_4 \cdot 4H_2O$		7.5 mg
$CuSO_4$		0.040 mg
H_3BO_3		0.056 mg
MoO_3		0.016 mg
$ZnSO_4$		0.331 mg
Ripe banana*	100–150 g	

Vegetable charcoal
(Nuchar C)   2 g

Sucrose   20 g

Agar   12–15 g

* Coke (1990) used potato instead (unspecified amounts)

Knudson C (Anderson, 1990*a*)

$Ca(NO_3)_2 \cdot 4H_2O$	1000 mg
KH_2PO_4	228 mg
K_2HPO_4	7.8 mg
$MgSO_4 \cdot 7H_2O$	125 mg
$(NH_4)_2SO_4$	250 mg
Ferric ammonium citrate	25 mg
Biotin	0.25 mg
Calcium pantothenate	5 mg
Nicotinic acid	10 mg
Pyridoxine	2 mg
Thiamine	5 mg
Potato extract	50 ml

Glucose   16 g

Agar   6 g

Lucke O₄ (Lucke, 1971*a*)

KH_2PO_4	300 mg
KNO_3	400 mg
$Mg(NO_3)_2 \cdot 6H_2O$	110 mg
NH_4NO_3	370 mg
$(NH_4)_2SO_4$	60 mg
$FeSO_4 \cdot 7H_2O$	20 mg
Biotin	2 mg
Nicotinic acid amide	50 mg
Pyridoxine	10 mg
Thiamine	10 mg
Coconut milk	50 mg

Glucose   10 g

Fructose	10 g
Activated charcoal	1 g
Agar	10 g

Lucke modified (Anderson, 1990*a*)

KH_2PO_4	300 mg
KNO_3	400 mg
$Mg(NO_3)_2 \cdot 6H_2O$	110 mg
NH_4NO_3	370 mg
$(NH_4)_2SO_4$	60 mg
Ferric ammonium citrate	25 mg
Biotin	0.25 mg
Calcium pantothenate	5 mg
Nicotinic acid	10 mg
Pyridoxine	2 mg
Thiamine	5 mg
Coconut milk	50 ml

Glucose	10 g
Fructose	10 g
Activated charcoal	1 g
Agar	6 g

Mead & Bulard medium (Mead & Bulard, 1975)

$CaH_4(PO_4)_2 \cdot H_2O$	100 mg
$CaSO_4 \cdot 2H_2O$	80 mg
KH_2PO_4	270 mg
$MgSO_4 \cdot 7H_2O$	240 mg
NH_4NO_3	80 mg
$AlCl_3$	0.03 mg
$CuSO_4 \cdot 5H_2O$	0.03 mg
$FeNa_2EDTA$	35 mg
H_3BO_3	1 mg
KI	0.01 mg
$MnSO_4 \cdot H_2O$	0.1 mg
$NiCl_2 \cdot 6H_2O$	0.03 mg
$ZnSO_4 \cdot 7H_2O$	1 mg

Casein hydrolysate*	5 g
Sucrose	5–80 g
Agar	8 g

* This undefined additive can be replaced by a combination of the following amino acids, totalling $2.425 \, g l^{-1}$ medium: arginine 100 mg, aspartic acid 250 mg, glutamic acid 750 mg, glycine 50 mg, histidine 75 mg, isoleucine 200 mg, leucine 250 mg, lysine 250 mg, methionine 50 mg, phenylalanine 100 mg, threonine 100 mg, tyrosine 50 mg, valine 200 mg. Casein hydrolysate is usually almost free from vitamins and growth factors, due to the severe conditions under which hydrolysis takes place.

N_3f (Burgeff, 1936, according to Withner, 1953)

Should be dissolved in sequence as indicated in brackets, thereby avoiding precipitation.

$Ca(NO_3)_2 \cdot 4H_2O$ (6)		1000 mg
$FeSO_4$ (5)		20 mg
K_2HPO_4 (2)		250 mg
KCl (4)		250 mg
$MgSO_4 \cdot 7H_2O$ (3)		250 mg
$(NH_4)_2SO_4$ (7)		250 mg
Citric acid (1)		90 mg
Sucrose (8)	20 g	
(or 10 g glucose + 10 g fructose)		
Agar	12–15 g	

NB-1 (Vöth, 1976)

Medical yeast	5 g
Peptone	3 g
Activated charcoal	2 g
Sucrose	10 g
Raw sugar (i.e. uncleaned)	10 g
Agar	10 g

NB 8X (Vöth, 1976)

$Ca(NO_3)_2$	500 mg
KH_2PO_4	125 mg
K_2HPO_4	125 mg
$MgSO_4 7H_2O$	125 mg
$(NH_4)_2SO_4$	125 mg
$FeSO_4 7H_2O$	10 mg

1 Multivit B tablet containing 1.5 mg thiamine, 1 mg riboflavin, 10 mg nicotinic acid amide, 1 mg calcium pantothenate, 0.5 mg pyridoxine

Sucrose	20 g
Medicinal yeast	5 g
Activated charcoal	2 g
Agar	10 g

Norstog medium (Norstog, 1973)

$CaCl_2 \cdot 2H_2O$	740 mg
KCl	750 mg
K_2HPO_4	910 mg
$MgSO_4 \cdot 7H_2O$	740 mg
Ferric citrate	10 mg
$CoCl \cdot 6H_2O$	0.025 mg
$CuSO_4 5H_2O$	0.025 mg
H_3BO_3	0.5 mg
$NaMoO_4 \cdot H_2O$	0.025 mg
$MnSO_4 \cdot H_2O$	3 mg
$ZnSO_4 \cdot 7H_2O$	0.5 mg
Calcium pantothenate	0.25 mg
Inositol (meso-)	50 mg
Pyridoxine	0.25 mg
Thiamine	0.25 mg
Alanine	50 mg
Arginine	10 mg
Cysteine	20 mg
Glutamine	400 mg
Leucine	10 mg
Phenylalanine	10 mg

Tyrosine   10 mg

Malic acid   1.0 g
Sucrose   34.2 g
Agar (purified)   6 g

PDA
potato–dextrose agar

Potato   200 g
Glucose   10 g

Agar   10 g

Penningsfeld medium (Penningsfeld, 1990*a*, *b*)
for symbiotic culture of European species

$Ca(NO_3)_2 \cdot 4H_2O$ 80 mg
KCl   160 mg
KH_2PO_4 80 mg
$MgSO_4 \cdot 7H_2O$ 80 mg

NaFeEDTA   25 mg

Starch   10 g
Agar-agar   10 g

pH adjusted to 6.0

Pfeffer medium (Hadley, 1982*b*)

$Ca(NO_3)_2$ 800 mg
KCl   100 mg
K_2HPO_4 200 mg
KNO_3 200 mg
$MgSO_4 \cdot 7H_2O$ 200 mg
$FeSO_4 \cdot 7H_2O$ 0.97 mg
$CuSO_4 \cdot 5H_2O$ 0.16 mg
$MnSO_4 \cdot 4H_2O$ 0.08 mg
$(NH_4)_6Mo_7O_{24} \cdot 4H_2O$ 0.26 mg
$ZnSO_4 \cdot 7H_2O$ 0.74 mg

Glucose   20 g

Agar   10–15 g

Optional: 1 g potato extract, 1 g yeast extract or 100 ml coconut milk

Pfeffer salt solution (Knudson, 1925)

$Ca(NO_3)_2$	800 mg
KCl	100 mg
K_2HPO_4	200 mg
KNO_3	200 mg
$MgSO_4 \cdot 7H_2O$	200 mg
$FeCl_3$	8 mg
Soluble starch	5 g

RAGA B (Galunder, 1986)
Ramin medium with brewer's yeast instead of yeast extract.

Ramin medium (Galunder, 1986)
asymbiotic medium for a range of difficult species: *Cypripedium calceolus, Anacamptis pyramidalis, Ophrys* spp., *Orchis militaris, O. morio*

Peptone	4 g
Yeast extract	2 g
Polybion	10 drops
Sucrose	10 g
Agar-agar	6 g

SM-organic (Malmgren, 1994)

$Ca_3(PO_4)_2$	75 mg
KH_2PO_4	75 mg
$MgSO_4 \cdot 7H_2O$	75 mg
$FeSO_4$	10 mg
Solu-Vit* (vitamin B complex)	1 ampoule

Vaminolac* (amino acid mixture)		0.5 ml†
Kinetin	5 mg	
Unsweetened pineapple juice		25 ml
Activated charcoal	1 g	
Sucrose	10 g	
Agar	8 g	

pH adjusted to 5.5 after addition of pineapple
* Cf. SM-spec.
† Rising to 0.75 or 1.0 ml when the seedlings are larger.

SM-spec (Malmgren, 1989a)

$Ca_3(PO_4)_2$		70 mg
KNO_3		150 mg
KH_2PO_4		60 mg
$MgSO_4 \cdot 7H_2O$		60 mg
$(NH_4)_2SO_4$		125 mg
Ferric tartrate		7 mg
$MnSO_4 \cdot 4H_2O$		2 mg
Solu-Vit* (vitamin B complex)		1 ampoule
Vaminolac† (amino acid mixture)		5 ml
Kinetin		5 mg
Unsweetened pineapple juice		25 ml
Activated charcoal	1 g	
Sucrose	10 g	
Agar	6–8 g	

pH adjusted to 5.5 after addition of pineapple

* Contains 3 mg thiamine, 3.6 mg riboflavine, 15 mg sodium pantothenate, 4 mg pyridoxine, 40 nicotinic acid amide, 5 mg cyanocobalamin, 60 mg biotin, 100 mg ascorbic acid and 0.4 folic acid.
† This mixture contains 18 different amino acids (g l^{-1}): 6.3 alanine, 2.1 glycine, 4.1 arginine, 4.1 aspartic acid, 1 cysteine and cystine, 2.7 phenylalanine, 7.1 glutamic acid, 2.1 histidine,

3.1 isoleucine, 7 leucine, 5.6 lysine, 1.3 methionine, 5.6 proline,
3.8 serine, 3.6 threonine, 1.4 tryptophan, 0.5 tyrosine, 3.6
valine.

Stephan medium (Stephan, 1988)
used for *Spiranthes spiralis*

KH_2PO_4		250 mg
$MgSO_4 \cdot 7H_2O$		125 mg
NH_4NO_3		125 mg
Peptone	2 g	
Yeast extract	1 g	
Activated charcoal	1.5 g	
Sucrose	10 g	
Agar	10 g	

Tashima medium (Tashima *et al.*, 1978)
for symbiotic culture of *Gastrodia*

$Ca(NO_3)_2 \cdot 4H_2O$		117.0 mg
KH_2PO_4		38.3 mg
$MgSO_4 \cdot 7H_2O$		120.0 mg
NH_4NO_3		57.5 mg
$CuSO_4 \cdot 5H_2O$		0.05 mg
H_3BO_3		0.6 mg
$H_2MoO_4 \cdot H_2O$		0.02 mg
$MnCl_2 \cdot 4H_2O$		0.4 mg
$ZnSO_4 \cdot 7H_2O$		0.05 mg
Ferric citrate		10 mg
Water melon juice		70 ml
Sucrose	20 g	
Meal of withered bamboo leaves*		

pH adjusted to 5.4–5.5 after autoclaving

* The leaves were rinsed in tap water, boiled at 100 °C for 60
minutes in deionized water, dried in 80 °C with hot air, and finely
ground. It functions as a carbon source for the fungus and gives
texture to the substrate (no agar!)

TGZ-N (Malmgren, 1989*a*)

Inorganic salts	0.5 g
Sucrose	10 g
Peptone	3 g
Activated charcoal	1 g
Agar	10 g

Thompson medium (Thompson, 1974)

$Ca(CH_3COO)_2$	79 mg
KCH_3COO	392 mg
$MgSO_4 \cdot 7H_2O$	369 mg
$NH_4H_2PO_4$	345 mg
Urea	540 mg
$FeSO_4 \cdot 7H_2O$	25 mg
Na_2-EDTA	37 mg
$CuSO_4 \cdot 5H_2O$	0.24 mg
H_3BO_3	1.86 mg
$MnCl_2 \cdot 4H_2O$	2.23 mg
$(NH_4)_6Mo_7O_{24} \cdot 4H_2O$	0.035 mg
$ZnSO_4 \cdot 7H_2O$	0.29 mg
Sucrose	30 g
Agar	10 g

Vacin & Went medium (Clements, 1982*c*)

$Ca_3(PO_4)_2$	200 mg
KH_2PO_4	250 mg
KNO_3	525 mg
$MgSO_4 \cdot 7H_2O$	250 mg
$(NH_4)_2SO_4$	500 mg
Ferric tartrate	28 mg
$MnSO_4 \cdot 4H_2O$	7.5 mg
Sucrose	20 g
Agar	8 g

Veyret medium (Veyret, 1969)

$Ca(NO_3)_2 \cdot 4H_2O$	1000 mg
KH_2PO_4	250 mg
$MgSO_4 \cdot 7H_2O$	250 mg
$(NH_4)_2SO_4$	500 mg
$FeCl_3 \cdot 6H_2O$	1 mg
$AlCl_3$	0.03 mg
$CuSO_4$	0.03 mg
H_3BO_3	1 mg
KI	0.01 mg
$MnSO_4 \cdot 4H_2O$	0.01 mg
MoO_3	0.02 mg
$NiCl_2 \cdot 6H_2O$	0.03 mg
$ZnSO_4 \cdot 7H_2O$	1 mg
Sucrose	10 g
Glucose	15 g
Agar	6 g

pH adjusted to 5.2

W2 medium

As H2 except that oat is replaced by an equal amount of finely granulated (< 2 mm) decayed wood.

ZAK medium (Borris, 1969)

Coconut milk	20 ml
Sucrose	10 g
Agar	15 g

B

Appendix. Names and synonyms

Latin names used in this work, including synonyms employed in cited literature (the latter not in italics).

Aceras anthropophorum (L.) Aiton f.
Acianthus reniformis (R. Br.) Schltr. = *Cyrtostylis reniformis*
Amerorchis rotundifolia (Banks) Hultén
Amitostigma gracile (Bl.) Schltr.
Amitostigma keiskei (Maxim.) Schltr.
Anacamptis pyramidalis (L.) L.C.M. Richard
Aplectrum hyemale (Mühlenb. & Willd.) Nuttall
Aplectrum unguiculatum (Finet) Maekawa = *Cremastra unguiculatum* = Oreorchis unguiculatus Finet
× *Aranda* hort. (Arachnis × Vanda; × Vandachnantha hort., × Vandarachnis hort.)
Arethusa bulbosa L.
Arethusa japonica A. Gray
Arundina chinensis Bl. = *A. graminifolia* (D. Don) Hochr.
Arundina graminifolia (D. Don) Hochr.

Barlia longibracteata (Biv.) Parl. = *Barlia robertiana*
Barlia robertiana (Loisel.) W. Greuter
Bletilla hyacinthina (J.E.Smith) Rchb. f. = *B. striata*
Bletilla striata (Thunb.) Rchb. f.

Calanthe discolor Lindl.
Calopogon pulchellus (Salisb.) R. Br. = *C. tuberosus*
Calopogon tuberosus(L.) Britton, Sterns & Poggenberg
Calypso bulbosa (L.) Oakes

Cattleya aurantiaca (Batem. ex Lindl.) P.N. Don
Cephalanthera alba (Crantz) Simonkai = *C. damasonium*
Cephalanthera austinae (A. Gray) Heller
Cephalanthera damasonium (Miller) Druce
Cephalanthera ensifolia (Murray) L.C.M. Richard = *C. longifolia*
Cephalanthera grandiflora S.F. Gray = *C. damasonium*
Cephalanthera longifolia (L.) Fritsch
Cephalanthera pallens (Jundz.) L.C.M. Richard = *C. damasonium*
Cephalanthera rubra (L.) L.C.M. Richard
Cleistes divaricata (L.) Ames
Cleistes divaricata (L.) Ames var. *bifaria* Fern.
Coeloglossum viride (L.) Hartman
Corallorhiza arizonica S. Wats. = *Hexalectris spicata*
Corallorhiza innata R. Br. = *C. trifida*
Corallorhiza maculata (Rafin.) Rafin.
Corallorhiza odontorhiza (Willd.) Nuttall
Corallorhiza striata Lindl.
Corallorhiza trifida Chat.
Cremastra unguiculatum (Finet) Finet
Cymbidium ensifolium (L.) Sw.
Cymbidium floribundum Lindl.
Cymbidium goeringii (Rchb. f.) Rchb. f.
Cymbidium pumilum Rolfe = *C. floribundum*
Cymbidium virescens Lindl. = *C. goeringii*
Cypripedium acaule Aiton
Cypripedium arietinum R. Br.
Cypripedium calceolus L.
Cypripedium californicum A. Gray
Cypripedium candidum Mühlenb. ex Willd.
Cypripedium debile Rchb. f.
Cypripedium japonicum Thunb.
Cypripedium parviflora Salisb. = *C. calceolus* var. *parviflorum* (Salisb.) Fern.
Cypripedium pubescens = *C. calceolus* var. *pubescens* (Willd.) Correll
Cypripedium reginae Walter
Cyrtosia septentrionalis Garay = *Galeola septentrionalis*
Cyrtostylis reniformis R. Br.

Dactylorhiza elata (Poir.) Soó
Dactylorhiza fistulosa (Moench) H. Baumann & S. Künkele = *D. majalis* (nomenclaturally unsettled)

Dactylorhiza foliosa (Vermeulen) Soó
Dactylorhiza fuchsii (Druce) Soó
Dactylorhiza incarnata (L.) Soó
Dactylorhiza maculata (L.) Soó
Dactylorhiza majalis (Rchb. f.) Hunt & Summerh.
Dactylorhiza praetermissa (Druce) Soó
Dactylorhiza purpurella (T. & T.A.Steph.) Soó
Dactylorhiza romana Sebast. & Mauri
Dactylorhiza saccifera (Brong.) Soó
Dactylorhiza sambucina (L.) Soó
Dactylorhiza sesquipedalis Willd. = *D. elata*
Dactylorhiza traunsteineri (Rchb.) Soó
Dendrobium Sw.
Disa uniflora Berg.
Diuris Sm.

Eleorchis japonica (A.G ray) Maekawa = *Arethusa japonica*
Epidendrum ibaguense H.B.K.
Epipactis atrorubens(Bernh.) Besser
Epipactis gigantea Dougl. ex Hooker
Epipactis helleborine (L.) Crantz
Epipactis latifolia (L.) Allioni = *E. helleborine*
Epipactis microphylla (Ehrh.) Sw.
Epipactis palustris (L.) Crantz
Epipactis purpurata J.E. Sm.
Epipactis royleana Lindl.
Epipactis rubiginosa Crantz = *E. atrorubens*
Epipactis violacea (Dur. Duq.) Bor. = *E. purpurata*
Epipogium aphyllum Sw.
Epipogium gmelinii L.C. Richard = *E. aphyllum*
Eulophia alta (L.) Fawcett & Rendle
Eulophidium maculatum (Rchb. f.) Pfitz.

Galearis spectabilis (L.) Rafin.
Galeola altissima (Blume) Rchb. f.
Galeola septentrionalis Rchb. f.
Gastrodia elata Bl.
Gastrodia sesamoides R. Br.
Goodyera macrantha Maxim.
Goodyera maximowicziana Makino.
Goodyera oblongifolia Rafin.
Goodyera pubescens (Willd.) R. Br.

Goodyera repens (L.) R. Br.
Goodyera schlechtendaliana Rchb. f.
Goodyera tessellata Loddiges
Goodyera velutina Maxim.
Govenia liliacea (Llave & Lex) Lindl.
Gymnadenia albida (L.) L.C.M. Richard
Gymnadenia conopsea (L.) R. Br.
Gymnadenia odoratissima (L.) L.C.M. Richard

Habenaria dentata Schltr.
Hammarbya paludosa (L.) O.Kuntze
Herminium monorchis (L.) R. Br.
Hexalextris spicata (Walt) Barn.
Himantoglossum hircinum (L.) Spreng.

Isotria medeoloides (Pursh) Raf.
Isotria verticillata (Muhl. ex Willd.) Raf.

Laelia Lindl.
× *Laeliocattleya* Rolfe (Cattleya × Laelia; Catlaelia hort.)
Lecanorchis javanica Bl.
Leucorchis albida (L.) E.H.F. Meyer = *Gymnadenia albida*
Limodorum abortivum (L.) Swartz
Liparis kumokiro F. Maekawa
Liparis lilifolia (L.) L.C.M. Rich. ex Lindl.
Liparis loeselii (L.) L.C.M. Rich.
Liparis nervosa (Thunb.) Lindl.
Liparis odorata (Willd.) Lindl.
Listera borealis Morong
Listera cordata (L.) R. Br.
Listera ovata (L.) R. Br.
Loroglossum hircinum L.C.M. Rich. = *Himantoglossum hircinum*

Malaxis monophyllos (L.) Sw.
Malaxis paludosa L. = *Hammarbya paludosa*
Malaxis unifolia Michaux

Neotinea intacta (Link) Rchb. f. = *Neotinea maculata*
Neotinea maculata (Desf.) Stearn
Neottia nidus-avis (L.) L.C.M. Richard
Nigritella nigra (L.) Rchb. f.
Nigritella miniata (Crantz) Janchen = (?) *N. rubra*

Nigritella rubra (Wettst.) K. Richter

Ophrys apifera Huds.
Ophrys arachnites (L.) Reich. = *O. holoserica*
Ophrys bombyliflora Link
Ophrys ciliata Biv.
Ophrys fuciflora (F.W. Schmidt) Moench. = *O. holoserica*
Ophrys fusca Link
Ophrys holoserica (Burm. f.) W. Greut.
Ophrys insectifera L.
Ophrys lutea Cav.
Ophrys myodes Jacq. = *O. insectifera* L.
Ophrys speculum Link p.p. = *O. ciliata* Biv.
Ophrys sphegodes Mill.
Orchis coriophora L.
Orchis ericetorum = *Dactylorhiza maculata* ssp. *ericetorum* (E.F.
    Lipton) Hunt & Summerhayes = *Dactylorhiza maculata p.p.*
Orchis globosa L. = *Traunsteinera globosa*
Orchis latifolia auct. = *Dactylorhiza majalis*
Orchis laxiflora Lam.
Orchis maculata L. = *Dactylorhiza maculata*
Orchis mascula L.
Orchis militaris L.
Orchis morio L.
Orchis pallens L.
Orchis palustris Jacq.
Orchis papilionacea L.
Orchis purpurea Hudson
Orchis rotundifolia Banks ex Pursh = *Amerorchis rotundifolia*
Orchis sancta L.
Orchis sesquipedalis Willd. = *Dactylorhiza elata*
Orchis simia Lam.
Orchis spectabilis L. = *Galearis spectabilis*
Orchis spitzelii Sauter
Orchis tridentata Scop.
Orchis ustulata L.
Oreorchis patens (Lindl.) Lindl.

Paphiopedilum Pfitz.
Phalaenopsis Bl.
Piperia Rydb.
Platanthera bifolia (L.) L.C.M. Richard

Platanthera blephariglottis (Willd.) Lindl.
Platanthera chlorantha (Custer) Reichenb.
Platanthera ciliaris (L.) Lindl.
Platanthera dilatata (Pursh) Lindl.
Platanthera hachijoensis Honda
Platanthera hyperborea (L.) Lindl.
Platanthera integrilabia (Correll) Luer
Platanthera lacera (Michaux) G. Don
Platanthera leucophaea (Nuttall) Lindl.
Platanthera minor (Miq.) Rchb. f.
Platanthera montana Rchb.f. = *P. chlorantha*
Platanthera obtusata (Banks & Pursh) Lindl.
Platanthera orbiculata (Pursh) Lindl.
Platanthera platycorys Schltr.
Platanthera psycodes (L.)Lindl.
Platanthera saccata = (?) Habenaria saccata Greene = *Platanthera stricta*
Platanthera sparsiflora (S. Watson) Schltr.
Platanthera stricta Lindl.
Platanthera takedai Makino
Platanthera tipuloides Lindl.
Pogonia affinis Austin ex A. Gray = *Isotria medeoloides*
Pogonia divaricata (L.) R. Br. = *Cleistes divaricata*
Pogonia japonica Rchb. f.
Pogonia ophioglossoides (L.) Juss.
Pogonia pendula (Muhl. ex Willd.) = *Triphora trianthophora*
Pogonia verticillata (Muhl. ex Willd.) Nut. = *Isotria verticillata*
Poneorchis graminifolia Rchb. f.
Ponthieva racemosa (Walter) Mohr

Rhizanthella gardneri R.S. Rogers

Serapias lingua L.
Spiranthes aestivalis (Poir.) L.C.M. Richard
Spiranthes autumnalis L.C.M. Richard = *S. spiralis*
Spiranthes cernua (L.) L.C.M. Richard
Spiranthes gracilis (Bigel.) Beck
Spiranthes magnicamporum Sheviak
Spiranthes ochroleuca (Rydb.) Rydb.
Spiranthes ovalis Lindl. var. *erostellata*
Spiranthes romanzoffiana Cham.
Spiranthes sinensis (Pers.) Ames

Spiranthes spiralis (L.) Chevall.

Thunia Rchb .f.
Tipularia discolor (Pursh) Nuttall
Traunsteinera globosa (L.) Rchb.
Triphora trianthophora (Sw.) Rydb.

Vanda Jones

Wullschlaegelia aphylla Rchb. f.

Yoania australis Hatch

Fungi

Armillaria mellea (Vahl ex Fries) Karten
Armillaria tabescens (Scop. ex Fr.) Dennis, Orton & Hora
Ascorhizoctonia Yang & Korf

Ceratobasidium angustisporum Warcup & Talbot
Ceratobasidium cereale Murray & Burpee
Ceratobasidium cornigerum (Bourdot) Rogers
Ceratobasidium globisporum Warcup & Talbot
Ceratobasidium obscurum Rogers
Ceratobasidium papillatum Warcup & Talbot
Ceratobasidium sphaerosporum Warcup & Talbot
Ceratobasidium stevensii (Burt) Talbot
Ceratorhiza Moore

Epulorhiza Moore

Favolaschia dybowskyana (Singer) Singer
Fusarium Link ex. Fr.

Helicobasidium purpureum Pat.

Leptodontidium orchidicola Currah & Sigler

Moniliopsis Ruhland
Moniliopsis solani (Kühn) Moore
Mycena thujina Smith

Nectriopsis solani (Reinke & Berth.) Booth

Rhacodium auct. = *Racodium* auct. [non Pers.], Hyphomycetes
Rhizoctonia DC
Rhizoctonia anaticula Currah = *Epulorhiza anaticula* (Currah)
    Currah
Rhizoctonia anomala Burgeff ex Currah = *Moniliopsis anomala*
    (Burgeff ex Currah) Currah
Rhizoctonia crocorum (Pers. ex Fr.) DC
Rhizoctonia goodyerae-repentis Costantin & Dufour. Doubtful
    taxon
Rhizoctonia lanuginosa Bernard. Doubtful taxon
Rhizoctonia monilioides J.T. Curtis. Nom. inval.
Rhizoctonia neottiae Wolff ex Burgeff. Nom. inval.
Rhizoctonia repens Bernard. Doubtful taxon
Rhizoctonia solani Kühn   (*Moniliopsis solani* (Kühn) Moore)
Rhizoctonia versicolor E. Müller & Nüesch. Nom. inval.

Sebacina vermifera Oberw.
Sistotrema Pers. ex Nocca & Balbis

Thanatephorus cucumeris (Frank) Donk
Thanatephorus gardneri Warcup
Thanatephorus orchidicola Warcup & Talbot
Thanatephorus pennatus Currah
Thanatephorus sterigmaticus (Bourdot) Talbot
Tulasnella allantospora Wakefield & Pearson
Tulasnella asymmetrica Warcup & Talbot
Tulasnella calospora (Boudier) Juel
Tulasnella cruciata Warcup & Talbot
Tulasnella irregularis Warcup & Talbot
Tulasnella violacea (Quélet) Bourdot & Galzin

Waitea Warcup & Talbot

References

Abeles, F. B. (1985). Sources of ethylene of horticultural significance. In *Ethylene and plant development*, ed. J. A. Roberts & G. A. Tucker, pp. 287–96. London: Butterworth.

Afzelius, K. (1916). Zur Embryosacentwicklung der Orchideen. *Svensk Botanisk Tidsskrift*, **10**, 183–227.

Alexander, C. & Alexander, I. J. (1984). Seasonal changes in populations of the orchid *Goodyera repens* Br. and its mycorrhizal development. *Transactions of the Botanical Society of Edinburgh*, **44**, 219–27.

Alexander, C., Alexander, I. J. & Hadley, G. (1984). Phosphate uptake in relation to mycorrhizal infection. *New Phytologist*, **97**, 401–11.

Alexander, C. & Hadley, G. (1983). Variation in symbiotic activity of *Rhizoctonia* isolates from *Goodyera repens* mycorrhizas. *Transactions of the British Mycological Society*, **80**, 99–106.

Alexander, C. & Hadley, G. (1984). The effect of mycorrhizal infection of *Goodyera repens* and its control by fungicide. *New Phytologist*, **97**, 391–400.

Alexander, C. & Hadley, G. (1985). Carbon movement between host and mycorrhizal endophyte during the development of the orchid *Goodyera repens* Br. *New Phytologist*, **101**, 657–65.

Alexander, M. (1977). *Soil microbiology*, 2nd edn. New York: Wiley.

Allenberg, H. (1976). Notizen zur Keimung, Meristemkultur und Regeneration von Erdorchideen. *Die Orchidee*, **27**, 28–31.

Allen, M. F. (1991). *The ecology of mycorrhizae*. Cambridge: Cambridge University Press.

Alvarez, M. R. (1968). Quantitative changes in nucleus DNA accompanying postgermination embryonic development in *Vanda* (Orchidaceae). *American Journal of Botany*, **55**, 1036–41.

Ames, O. (1921a). Notes on New England orchids. I. *Spiranthes*. *Rhodora*, **23**, 73–85.

Ames, O. (1921*b*). Seed dispersal in relation to colony formation in *Goodyera pubescens*. *Orchid Review*, **29**, 105–7.

Ames, O. (1922). Notes on New England orchids. II. The mycorrhiza of *Goodyera pubescens*. *Rhodora*, **24**, 37–46.

Andersen, T. F. (1990*a*). A study of hyphal morphology in the form genus *Rhizoctonia*. *Mycotaxon*, **37**, 25–46.

Andersen, T. F. (1990*b*). *Contributions to the taxonomy and nomenclature in the form genus* Rhizoctonia DC. PhD thesis. Denmark: University of Copenhagen.

Andersen, T. F. & Stalpers, J. A. (1994). A checklist of *Rhizoctonia* epithets. *Mycotaxon*, **51**, 437–57.

Anderson, A. B. (1990*a*). Asymbiotic germination of seeds of some North American orchids. In *North American native terrestrial orchid propagation and production*, ed. C. E. Sawyers, pp. 75–80. Chadds Ford, Pennsylvania: Brandywine Conservancy.

Anderson, A. B. (1990*b*). Improved germination and growth of rare native Ontario orchid species. In *Conserving Carolinian Canada*, ed. G. M. Allen, G. M. Eagles & P. F. J. Price, pp. 65–73. Waterloo, Canada: University of Waterloo Press.

Anderson, A. B. (1991). Symbiotic and asymbiotic germination and growth of *Spiranthes magnicamporum* (Orchidaceae). *Lindleyana*, **6**, 183–6.

Arber, A. (1925). *Monocotyledons: a morphological study*. Cambridge: Cambridge University Press.

Arditti, J. (1967). Factors affecting the germination of orchid seeds. *Botanical Review*, **33**, 1–97.

Arditti, J. (1979). Aspects of the physiology of orchids. *Advances in Botanical Research*, **7**, 421–655.

Arditti, J. (1982). Introduction, North American Terrestrial Orchids, etc. In *Orchid biology. II. Reviews and perspectives. Orchid seed germination and seedling culture – a manual*, ed. J. Arditti, pp. 245–73, 278–93. Ithaca, New York: Cornell University Press.

Arditti, J. (1990). Lewis Knudson (1884–1958): his science, his times, and his legacy. *Lindleyana*, **5**, 1–79.

Arditti, J., Michaud, J. D. & Healey, P. L. (1979). Morphometry of orchid seeds. I. *Paphiopedilum* and native California and related species of *Cypripedium*. *American Journal of Botany*, **66**, 1128–37.

Arditti, J., Michaud, J. D. & Healey, P. L. (1980). Morphometry of orchid seeds. II. Native California and related species of *Calypso*, *Cephalanthera*, *Corallorhiza*, and *Epipactis*. *American Journal of Botany*, **67**, 347–60.

Arditti, J., Michaud, J. D. & Oliva, A. P. (1981). Seed germination of North American orchids. I. Native Californian and related species of

Calypso, Epipactis, Goodyera, Piperia, and *Platanthera. Botanical Gazette,* **142,** 442–53.

Arditti, J., Michaud, J. D. & Oliva, A. P. (1982a). Practical germination of North American and related orchids. I. *Epipactis atrorubens* and *E. gigantea* and *E. helleborine. American Orchid Society Bulletin,* **51,** 162–71.

Arditti, J., Oliva, A. P. & Michaud, J. D. (1982b). Practical germination of North American and related orchids. II. *Goodyera oblongifolia* and *G. tessellata. American Orchid Society Bulletin,* **51,** 394–7.

Arditti, J., Oliva, A. & Michaud, J. D. (1985). Practical germination of North American and related orchids. III. *Calopogon tuberosus, Calypso bulbosa, Cypripedium* species and hybrids, *Piperia elegans* var. *elata, P. maritima, Platanthera hyperborea,* and *P. saccata. American Orchid Society Bulletin,* **54,** 859–66.

Auclair, A. N. (1972). Comparative ecology of the orchids *Aplectrum hyemale* and *Orchis spectabilis. Bulletin of the Torrey Botanical Club,* **99,** 1–10.

Bailes, C., Clements, M., Cribb, P., Muir, H. & Tasker, S. (1987). The cultivation of European orchids. *Orchid Review,* **95,** 19–24.

Baker, K. M., Mathes, M. C. & Wallace, B. J. (1987). Germination of *Ponthieva* and *Cattleya* seeds and development of *Phalaenopsis* protocorms. *Lindleyana,* **2,** 77–83.

Ballard, W. W. (1987). Sterile propagation of *Cypripedium reginae* from seeds. *American Orchid Society Bulletin,* **56,** 935–46.

Ballard, W. W. (1990). Further notes on *Cypripedium* germination. In *North American native terrestrial orchid propagation and production,* ed. C. E. Sawyers, pp. 87–9. Chadds Ford, Pennsylvania: Brandywine Conservancy.

Barabé, D., Saint-Arnaud, M. & Lauzer, D. (1993). Sur la nature des protocormes d'Orchidées (Orchidaceae). *Compte Rendu de l'Academie des Sciences, Paris, Sér. III,* **316,** 139–44.

Barmicheva, K. M. (1990). Ultrastructure of *Neottia nidus-avis* mycorrhizas. *Agriculture, Ecosystems and Environment,* **29,** 23–7.

Barroso, J. (1988). *Micorrizas em* Ophrys lutea *Cav. Aspectos citológicos, citoquímicos e bioquímicos da interacçâo hóspede/hospedeiro.* Doctoral thesis. Lisbon.

Barroso, J., Casimiro, A., Carrapiço, F. & Pais, M. S. S. (1988). Localization of uricase in mycorrhizas of *Ophrys lutea* Cav. *New Phytologist,* **108,** 335–40.

Barroso, J., Chaves Neves, H. & Pais, M. S. (1986a). Ultrastructural, cytochemical and biochemical aspects related to the formation of *O. lutea* endomycorrhizae. In *Physiological and genetical aspects of mycorrhizae*, ed. V. Gianinazzi-Pearson & S. Gianinazzi, pp. 265–8. Paris: INRA.

Barroso, J., Chaves Neves, H. & Pais, M. S. S. (1986*b*). Production of indole-3–ethanol and indole-3–acetic acid by the mycorrhizal fungus of *Ophrys lutea* (Orchidaceae). *New Phytologist*, **103**, 745–9.

Barroso, J., Chaves Neves, H. & Pais, M. S. S. (1987). Production of free sterols by infected tubers of *Ophrys lutea* Cav.: identification by gas chromatography-mass spectrometry. *New Phytologist*, **106**, 147–52.

Barroso, J., Fevereiro, P., Oliveira, M. M. & Pais, M. S. S. (1990). *In vitro* seed germination, differentiation and production of minitubers from *Ophrys lutea* Cav., *Ophrys fusca* Link and *Ophrys speculum* Link. *Scientia Horticulturae*, **42**, 329–37.

Barroso, J. & Pais, M. S. S. (1985). Cytochimie. Caractérisation cytochimique de l'interface hôte/endophyte des endomycorrhizes d'*Ophrys lutea*. Rôle de l'hôte dans la synthèse des polysaccharides. *Annales des Sciences Naturelles, Botanique*, **13**, 237–44.

Barroso, J. & Pais, M. S. S. (1987). Coated vesicles in the cytoplasm of the host cells in *Ophrys lutea* Cav. mycorrhizas (Orchidaceae). *New Phytologist*, **105**, 67–70.

Barosso, J. & Pais, M. S. S. (1990). Nuclear features in infected roots of *Ophrys lutea* Cav. (Orchidaceae). *New Phytologist*, **115**, 93–8.

Batygina, T. B. & Andronova, E. V. (1988). Is there a cotyledon in orchids?. *Doklady Academii Nauk SSSR*, **302**, 1017–19.

Beau, C. (1913). Sur les rapports entre la tuberisation et l'infestation des racines par des champignons endophytes au cours de développement du *Spiranthes autumnalis*. *Compte Rendu Hebdomadaire des Séances de l'Académie des Sciences Paris*, **157**, 512–15.

Beau, C. (1920*a*). Sur la germination de quelques orchidees indigenes. *Bulletin de la Société Histoire Naturelle Afrique du Nord*, **11**, 54–6.

Beau, C. (1920*b*). Sur le rôle trophique des endophytes d'orchidées. *Compte Rendu Hebdomadaire des Séances de l'Académie des Sciences Paris*, **171**, 675–7.

Beer, J. G. (1863). *Beiträge sur Morphologie und Biologie der Familie der Orchideen*. Wien: Carl Gerald's Son.

Benzing, D. H. (1981). Why is Orchidaceae so large, its seeds so small, and its seedlings mycotrophic. *Selbyana*, **5**, 241–2.

Bernard, N. (1899). Sur la germination du *Neottia nidus-avis*. *Compte Rendu Hebdomadaire des Séances de l'Académie des Sciences, Paris*, **128**, 1253–5.

Bernard, N. (1900). Sur quelques germination difficiles. *Revue Générale de Botanique*, **12**, 108–20.

Bernard, N. (1902). Études sur la tubérisation. *Revue Générale de*

Botanique, **14**, 5–25, 58–71, 101–19, 170–83, 219–34, 269–79, Tables 1–3.

Bernard, N. (1904). Recherches expérimentales sur les orchidées. La germination des orchidées. *Revue Générale de Botanique*, **16**, 405–51 458–75.

Bernard, N. (1909). L'evolution dans la symbioise des orchidées et leur champignons commensaux. *Annales des Sciences Naturelle Paris, 9. sér.*, **9**, 1–196.

Beyrle, H., Penningsfeld, F. & Hock, B. (1985). Orchideenmykorrhiza: symbiotische Anzucht einiger *Dactylorhiza*-arten. *Zeitschrift für Mykologie*, **51**, 185–98.

Beyrle, H., Penningsfeld, F. & Hock, B. (1987). Die gärtnerische Aufzucht von feuchtigkeitsliebenden *Dactylorhiza*-arten. *Die Orchidee*, **38**, 302–6.

Beyrle, H., Penningsfeld, F. & Hock, B. (1991). The role of nitrogen concentration in determining the outcome of the interaction between *Dactylorhiza incarnata* (L.) Soó and *Rhizoctonia*. *New Phytologist*, **117**, 665–72.

Blakeman, J. P., Mokahel, M. A. & Hadley, G. (1976). Effect of mycorrhizal infection on respiration and activity of some oxidase enzymes of orchid protocorms. *New Phytologist*, **77**, 697–704.

Blumenfeld, H. (1935). *Beiträge zur Physiologie des Wurzelpilzes von Limodorum abortivum (L.) Sw.* Riga, Lettland: Inaugural Dissertation, Philosophische Fakultät der Universität Basel.

Böckel, W. (1972). Ein Ansamungsversuch mit *Cypripedium calceolus*. *Die Orchidee*, **23**, 120–3.

Boller, A., Corrodi, H., Gäumann, E., Hardegger, E., Kern, H. & Winterhalter-Wild, N. (1957). Über induzierte Abwehrstoffe bei Orchideen: I. *Helvetica Chimica Acta*, **40**, 1062–6.

Borris, H. (1969). Samenvermehrung und Anzucht europäischer Erdorchideen. In *Proceedings of the 2nd European Orchid Congress*, pp. 74–8. Paris.

Borris, H. (1970). Vermehrung europäischer Orchideen aus Samen. *Der Erwerbsgärtner*, **24**, 349–51.

Borris, H. & Albrecht, L. (1969). Rationelle Samenvermehrung und Anzucht europäischer Erdorchideen. *Gartenwelt*, **69**, 511–13.

Borris, H., Jeschke, E. M. & Bartsch, G. (1971). Elektronenmikroskopische Untersuchungen zur Ultrastruktur der Orchideen-Mykorrhiza. *Biologische Rundschau*, **9**, 177–80.

Borris, H. & Voigt, T. (1986). Symbiotische und asymbiotische Samenkeimung von *Orchis mascula* – Ein Beitrag zum Problem der Spezifität der Orchideenpilze. *Die Orchidee*, **37**, 222–6.

Borsos, O. (1983). Anatomisch-histochemische Untersuchung der

Knollen der Wildorchideen Ungarns. *Die Orchidee Sonderheft, März 1983*, 61–4.

Breddy, N. C. & Black, W. H. (1954). Orchid mycorrhiza and their application to seedling raising. *Orchid Journal*, **3**, 57–61.

Brieger, F. G. (1976). On the orchid system: general principles and the distinction of subfamilies. In *Proceedings of the 8th World Orchid Conference*, ed. K. Senghas, pp. 488–504. Frankfurt: German Orchid Society.

Brundin, J. A. Z. (1895). Über Wurzelsprosse bei *Listera cordata* L. *Bihang till Kongliga Svenska Vetenskaps-Akademiens Handlinger*, **21**, III.

Brundrett, M. C. & Kendrick, B. (1988). The mycorrhizal status, root anatomy, and phenology of plants in a sugar maple forest. *Canadian Journal of Botany*, **66**, 1153–73.

Burgeff, H. (1909). *Die Wurzelpilze der Orchideen, ihre Kultur und ihre Leben in der Pflanze*. Jena: Gustav Fischer.

Burgeff, H. (1932). *Saprophytismus und Symbiose. Studien an tropischen Orchideen*. Jena: Gustav Fischer.

Burgeff, H. (1934). Pflanzliche Avitaminose und ihr Behebung durch Vitamin-zuführ. *Bericht der Deutschen Botanischen Gesellschaft*, **52**, 384–90.

Burgeff, H. (1936). *Samenkeimung der Orchideen*. Jena: Gustav Fischer.

Burgeff, H. (1943). Problematik der Mycorhiza. *Naturwissenschaften*, **47/48**, 558–67.

Burgeff, H. (1954). *Samenkeimung und Kultur Europäischer Erdorchideen*. Stuttgart: Gustav Fischer.

Burgeff, H. (1959). Mycorrhiza of orchids. In *The orchids: a scientific survey*, ed. C. L. Withner, pp. 361–95. New York: Ronald Press.

Burges, A. (1936). On the significance of mycorrhiza. *New Phytologist*, **35**, 117–31.

Burges, A. (1939). The defensive mechanism in orchid mycorrhiza. *New Phytologist*, **38**, 273–83.

Butcher, D. & Marlow, S. A. (1989). Asymbiotic germination of epiphytic and terrestrial orchids. In *Modern methods in orchid conservation*, ed. H. W. Pritchard, pp. 31–8. Cambridge: Cambridge University Press.

Buttler, K. P. (1986). *Orchideen. Die wildwachsenden Arten und Unterarten Europas, Vorderasiens und Nordafrikas*. Munich: Mosaik Verlag.

Campbell, E. O. (1970a). Morphology of the fungal association in three species of *Corallorhiza* in Michigan. *Michigan Botanist*, **9**, 108–13.

Campbell, E. O. (1970b). The fungal association of *Yoania australis*. *Transactions of the Royal Society of New Zealand. Biological Science*, **12**, 5–12.

Campbell, R. (1977). *Microbial ecology*. Oxford: Blackwell Scientific.

Carlson, M. C. (1935). The germination of the seed and development of the seedling of *Calopogon pulchellus* (Sw.) R.Br. *Transactions of Illinois State Academy*, **28**, 85–6.

Carlson, M. C. (1938). Origin and development of shoots from the tips of roots of *Pogonia ophioglossoides*. *Botanical Gazette*, **100**, 215–25.

Carlson, M. C. (1940). Formation of the seed of *Cypripedium parviflorum*. *Botanical Gazette*, **102**, 295–301.

Case, F. W. (1964). *Orchids of the western Great Lakes region*. Bloomfield Hills, Michigan: Cranbrook Institute of Science.

Case, F. W. (1987). *Orchids of the western Great Lakes region*. Revised edition. Bulletin 48. Bloomfield Hills, Michigan: Cranbrook Institute of Science.

Case, F. W. (1990). Native orchid habitats: the horticulturist's viewpoint. In *North American native terrestrial orchid propagation and production*, ed. C. E. Sawyers, pp. 1–14. Chadds Ford, Pennsylvania: Brandywine Conservancy.

Case, F. W. Jr. (1983). Notes concerning changes in Great Lakes orchid populations since 1964. In *North American terrestrial orchids. Symposium II. Proceedings and lectures*, ed. E. H. Plaxton, pp. 133–42. Ann Arbor, USA: Michigan Orchid Society.

Catling, P. M. (1990). Biology of North American representatives of the subfamily Spiranthoideae. In *North American native terrestrial orchid propagation and production*, ed. C. E. Sawyers, pp. 49–67. Chadds Ford, Pennsylvania: Brandywine Conservancy.

Champagnat, M. (1971). Recherches sur la multiplication végétative de *Neottia nidus-avis* Rich. *Annales des Sciences Naturelles Botanique et Biologie Végetale, 12 Série*, 12, 209–48.

Chia, T. F., Hew, C. S. & Loh, C. S. (1988). Carbon/nitrogen ratio and greening and protocorm formation in orchid callus tissues. *HortScience*, **23**, 599–601.

Chodat, R. & Lendner, A. (1896). Sur les mycorrhizes du *Listera cordata*. *Bulletin de l'Hérbier Boissier*, **4**, 265–72.

Clement, E. (1926). The non-symbiotic and symbiotic germination of orchid seeds. *Orchid Review*, **34**, 165–9.

Clements, M. A. (1982*a*). Developments in the symbiotic germination of Australian terrestrial orchids. In *Proceedings of the 10th World Orchid Conference 1981*, ed. J. Stewart & C. N. Van der Merwe, pp. 269–73. Johannesburg: South African Orchid Council.

Clements, M. A. (1982*b*). The germination of Australian orchid seed. In *Proceedings of the Orchid Symposium, 13th International Botanical Congress, Sidney*, ed. L. Lawler & R. D. Kerr, pp. 5–8. New South Wales: Orchid Society.

Clements, M. A. (1982c). Australian native orchids. In *Orchid biology II*, ed. J. Arditti, pp. 295–303. Ithaca, New York: Cornell University Press.

Clements, M. A. (1988). Orchid mycorrhizal associations. *Lindleyana*, **3**, 73–86.

Clements, M. A., Muir, H. & Cribb, P. J. (1986). A preliminary report on the symbiotic germination of European terrestrial orchids. *Kew Bulletin*, **41**, 437–45.

Clifford, J. B. (1899). The mycorrhiza of *Tipularia unifolia*. *Bulletin of the Torrey Botanical Club*, **26**, 635–8, Plate 372.

Cocucci, A. & Jensen, W. A. (1969). Orchid embryology: megagametophyte of *Epidendrum scutella* following fertilization. *American Journal of Botany*, **56**, 629–40.

Coke, J. L. (1990). Aseptic germination and growth of some terrestrial orchids. In *North American native terrestrial orchid propagation and production*, ed. C. E. Sawyers, pp. 90–1. Chadds Ford, Pennsylvania: Brandywine Conservancy.

Correll, D. S. (1950). *Native orchids of North America north of Mexico*. Massachusetts: Waltham.

Costantin, J. & Dufour, L. (1920). Sur la biologie du *Goodyera repens*. *Revue Générale de Botanique*, **32**, 529–33.

Crackles, E. (1975). The Monkey Orchid in Yorkshire. *The Naturalist*, **932**, 25–6.

Cribb, P. & Bailes, C. (1989). *Hardy orchids: orchids for the garden and frost-free greenhouse*. Portland, Oregon: Timber Press.

Currah, R. S. (1987). *Thanatephorus pennatus* sp. nov. isolated from mycorrhizal roots of *Calypso bulbosa* (Orchidaceae) from Alberta. *Canadian Journal of Botany*, **65**, 1957–60.

Currah, R. S. & Zelmer, C. (1992). A key and notes for the genera of fungi mycorrhizal with orchids and a new species in the genus *Epulorhiza*. *Reports of the Tottori Mycological Institute*, **30**, 43–59.

Currah, R. S., Hambleton, S. & Smreciu, A. (1987b). Ecological and mycorrhizal studies of Alberta orchids. *Restoration and Management Notes*, **5**, 40.

Currah, R. S., Hambleton, S. & Smreciu, A. (1988). Mycorrhizae and mycorrhizal fungi of *Calypso bulbosa*. *American Journal of Botany*, **75**, 739–52.

Currah, R. S., Sigler, L. & Hambleton, S. (1987a). New records and new taxa of fungi from the mycorrhizae of terrestrial orchids of Alberta. *Canadian Journal of Botany*, **65**, 2473–82.

Currah, R. S., Smreciu, E. A. & Hambleton, S. (1990). Mycorrhizae and mycorrhizal fungi of boreal species of *Platanthera* and *Coeloglossum* (Orchidaceae). *Canadian Journal of Botany*, **68**, 1171–81.

Curtis, J. T. (1936). The germination of native orchid seeds. *American Orchid Society Bulletin,* **5,** 42–7.

Curtis, J. T. (1939). The relation of specificity of orchid mycorrhizal fungi to the problem of symbiosis. *American Journal of Botany,* **26,** 390–9.

Curtis, J. T. (1943). Germination and seedling development in five species of *Cypripedium* L. *American Journal of Botany,* **30,** 199–206.

Curtis, J. T. & Spoerl, E. (1948). Studies on the nitrogen nutrition of orchid embryos. II. Comparative utilization of nitrate and ammonium nitrogen. *American Orchid Society Bulletin,* **17,** 111–14.

Dafni, A. & Bernhardt, P. (1990). Pollination of terrestrial orchids of southern Australia and the Mediterranean region. *Evolutionary Biology,* **24,** 193–252.

Dangeard, M. M. & Armand, L. (1898). Observations de biologie cellulaire (Mycorhizes d'*Ophrys aranifera*). *Revue de Mycologie, Toulouse,* **20,** 13–8, Table 182.

Darnell, A. W. (1952). Propagation of terrestrial orchids by seed. *Orchid Journal,* **1,** 242.

Davis, A. (1946). Orchid seed and seed germination. *American Orchid Society Bulletin,* **15,** 218–23.

Dexheimer, J. & Serrigny, J. (1983). Étude ultrastructurale des endomycorhizes d'une orchidée tropicale: *Epidendrum ibaguense* H.B.K. I. Localisation des activités phosphatiques acides *et al.*,calines. *Bulletin de la Societé Botanique de France, Lettres Botanique,* **130,** 187–94.

Diels, L. & Mattick, F. (1958). *Pflanzengeographie,* 5th edn., Berlin: Sammlung Göschen.

Dijk, E. (1988*a*). Mykorrhizen der Orchideen. II. Die Pilze. *Die Orchidee,* **39,** 116–20.

Dijk, E. (1988*b*). Mykorrhizen der Orchideen. III. Physiologische Aspekte bezüglich Kohlenstoff und Stickstoff. *Die Orchidee,* **39,** 196–200.

Dijk, E. (1990). Effects of mycorrhizal fungi on *in vitro* nitrogen response of juvenile orchids. *Agriculture, Ecosystems and Environment,* **29,** 91–7.

Dixon, K. (1991). Seeder/clonal concepts in Western Australian orchids. In *Population ecology of terrestrial orchids,* ed. T. C. E. Wells & J. H. Willems, pp. 111–23. The Hague: SPB Academic Publishing.

Dixon, K. & Sivasithamparam, K. (1991). Research activities: propagation of Australian terrestrial orchids. *Orchadian,* **10,** 107.

Dörr, I. & Kollmann, R. (1969). Fine structure of mycorrhiza in *Neottia nidus-avis* (L.) L.C. Rich. (Orchidaceae). *Planta,* **89,** 372–5.

Dowdell, R. J., Smith, K. A., Cress, R. & Restall, R. F. (1972). Field studies of ethylene in the soil atmosphere – equipment and preliminary results. *Soil Biology and Biochemistry,* **4,** 325–31.

Downie, D. G. (1940). On the germination and growth of *Goodyera repens*.

Transactions of the Botanical Society of Edinburgh, **33**, 36–51.

Downie, D. G. (1941). Notes on the germination of some British orchids. *Transactions of the Botanical Society of Edinburgh*, **33**, 94–103.

Downie, D. G. (1943a). Notes on the germination of *Corallorhiza innata*. *Transactions of the Botanical Society of Edinburgh*, **33**, 380–82.

Downie, D. G. (1943b). Source of the symbiont of *Goodyera repens*. *Transactions of the Botanical Society of Edinburgh*, **33**, 383–90.

Downie, D. G. (1949a). The germination of *Goodyera repens* (L.) R.Br. in fungal extract. *Transactions of the Botanical Society of Edinburgh*, **35**, 120–5.

Downie, D. G. (1949b). The germination of *Listera ovata* (L.) R.Br. *Transactions of the Botanical Society of Edinburgh*, **35**, 126–30.

Downie, D. G. (1957). *Corticium solani* – an orchid endophyte. *Nature*, **179**, 160.

Downie, D. G. (1959a). *Rhizoctonia solani* and orchid seed. *Transactions of the Botanical Society of Edinburgh*, **37**, 279–85.

Downie, D. G. (1959b). The mycorrhiza of *Orchis purpurella*. *Transactions of the Botanical Society of Edinburgh*, **38**, 16–29.

Dressler, R. L. (1965). Notes on the genus *Govenia* in Mexico (Orchidaceae). *Brittonia*, **17**, 266–77.

Dressler, R. L. (1981). *The orchids – natural history and classification*. Cambridge, Massachusetts: Harvard University Press.

Dressler, R. L. (1993). *Phylogeny and classification of the orchid family*. Cambridge: Cambridge University Press.

Duddridge, J. A. (1985). A comparative ultrastructural analysis of the host-fungus interface in mycorrhizal and parasitic associations. In *Developmental biology of higher fungi*, ed. D. Moore, L. A. Casselton, D. A. Wood & J. C. Frankland, pp. 141–73. Cambridge: Cambridge University Press.

Eiberg, H. (1969). Keimung europäischer Erdorchideen. *Die Orchidee*, **20**, 266–70.

Eiberg, H. (1970). *Asymbiotisk frøspiring og kulturforsøg hos nogle europæiske jordorkideer*. MSc thesis. University of Copenhagen: Plant Physiological Laboratory.

Eilhardt, K. H. (1980). Einige Beobachtungen bei der asymbiotischen Aussat von Orchideensamen. *Die Orchidee*, **31**, 183–5, 260.

Ek, M., Ljungquist, P. O. & Stenström, E. (1983). Indole-3–acetic acid production by mycorrhizal fungi determined by gas chromatography–mass spectrometry. *New Phytologist*, **94**, 401–7.

Ernst, R. (1967). Effect of carbohydrate selection on the growth rate of freshly germinated *Phalaenopsis* and *Dendrobium* seed. *American Orchid Society Bulletin*, **36**, 1068–73.

Ernst, R. (1976). Charcoal or glass wool in asymbiotic culture of orchids. In *Proceedings of the 8th World Orchid Conference*, ed. K. Senghas, pp. 379–83. Frankfurt: German Orchid Society.

Ernst, R., Arditti, J. & Healey, P. L. (1970). The nutrition of orchid seedlings. *American Orchid Society Bulletin*, **39**, 559–65, 691–700.

Ernst, R., Arditti, J. & Healey, P. L. (1971). Carbohydrate physiology of orchid seedlings. II. Hydrolysis and effects of oligosaccharides. *American Journal of Botany*, **58**, 827–35.

Fabre, J.-H. (1856). De la germination des Ophrydées. *Annales des Sciences Naturelles IV, Série Botanique*, **5**, 163–86, Plate II.

Farrell, L. (1985). Biological flora of the British Isles. *Journal of Ecology*, **73**, 1041–53.

Fast, G. (1974). Über eine Methode der kombinierten generativen-vegetativen Vermehrung von *Cypripedium calceolus*. *Die Orchidee*, **25**, 125–9.

Fast, G. (1978). Über das Keimverhalten europäischer Erdorchideen bei asymbiotischer Aussat. *Die Orchidee*, **29**, 270–4.

Fast, G. (1980). Vermehrung und Anzucht. In *Orchideenkultur. Botanische Grundlagen, Kulturverfahren, Pflanzenbeschreibungen*, ed. G. Fast, pp. 207–23. Stuttgart: Ulmer.

Fast, G. (1982). European terrestrial orchids (Symbiotic and asymbiotic methods). In *Orchid Biology. II. Reviews and perspectives. Orchid seed germination and seedling culture – a manual*, ed. J. Arditti, pp. 309–26. Ithaca, New York: Cornell University Press.

Fast, G. (1983). Stand und Aussichten bei der Anzucht europäischer Orchideen. *Die Orchidee (Sonderheft März)*, 97–100.

Fast, G. (1985). Zur Ökologie einiger mitteleuropäischer Waldorchideen unter besonderer Berücksichtigung der Bodenverhältnisse in Bayern. *Die Orchidee*, **36**, 148–52.

Fay, M. F. & Muir, H. J. (1990). The role of micropropagation in the conservation of European plants. In *Conservation techniques in botanic gardens*, ed. J. E. Hernández Bermejo, M. Clemente & V. Heywood, pp. 27–32. Königstein: Koeltz.

Filipello Marchisio, V., Berta, G., Fontana, A. & Marzetti Mannina, F. (1985). Endophytes of wild orchids native to Italy: their morphology, caryology, ultrastructure and cytochemical characterization. *New Phytologist*, **100**, 623–41.

Fitter, A. H. & Hay, R. K. M. (1987). *Environmental Physiology of Plants*, 2nd edn. London: Academic Press.

Fonnesbech, M. (1972a). Growth hormones and propagation of *Cymbidium in vitro*. *Physiologia Plantarum*, **27**, 310–16.

Fonnesbech, M. (1972*b*). Organic nutrients in the media for propagation of *Cymbidium in vitro*. *Physiologia Plantarum*, **27**, 360–4.

Frank, A. B. (1891). Über die auf Verdauung von Pilzen abziehende Symbiose der mit endotrophen Mykorrhizen begabten Pflanzen, sowie der Leguminosen und Erlen. *Bericht der Deutschen Botanischen Gesellschaft*, **9**, 244–53.

Fredrikson, M. (1991). An embryological study of *Platanthera bifolia* (Orchidaceae). *Plant Systematics and Evolution*, **174**, 213–20.

Freudenstein, J. V. (1992). *Systematics of* Corallorhiza *and the* Corallorhizinae *(Orchidaceae)*. PhD thesis. Cornell University.

Frosch, W. (1980). Asymbiotische Aussat von *Orchis morio*. *Die Orchidee*, **31**, 123–4.

Frosch, W. (1982). Beseitigung der durch die innere Hülle bedingten Keimhemmung bei europäischen Orchideen. *Die Orchidee*, **33**, 145–6.

Frosch, W. (1983*a*). Asymbiotische Vermehrung von *Ophrys holosericea* mit Blüten nach 22 Monaten. *Die Orchidee*, **34**, 58–61.

Frosch, W. (1983*b*). Asymbiotische Vermehrung von *Orchis morio* mit der ersten Blüte nach 23 Monaten. *Die Orchidee (Sonderheft März)*, 101–4.

Frosch, W. (1985). Asymbiotische Vermehrung von *Cypripedium reginae* mit Blüten drei Jahre nach der Aussat. *Die Orchidee*, **36**, 30–2.

Frosch, W. (1986). Möglichkeiten und Grenzen einer Langzeit-lagerung von Saatgut europäischer Orchideen. *Die Orchidee*, **37**, 239–40.

Fuchs, A. & Ziegenspeck, H. (1922). Aus der Monographie des *Orchis Traunsteineri* Saut. *Botanisches Archiv*, **2**, 238–48.

Fuchs, A. & Ziegenspeck, H. (1924*a*). Aus der Monographie des *Orchis Traunsteineri* Saut. III. Entwicklungsgeschichte einiger deutscher Orchideen. *Botanisches Archiv*, **5**, 120–32.

Fuchs, A. & Ziegenspeck, H. (1924*b*). Aus der Monographie des *Orchis Traunsteineri* Saut. V. Die Pilzverdauung der Orchideen. *Botanisches Archiv*, **6**, 193–206.

Fuchs, A. & Ziegenspeck, H. (1925). Bau und Form der Wurzeln der einheimischen Orchideen in Hinblick auf ihre Aufgaben. *Botanisches Archiv*, **11**, 290–379.

Fuchs, A. & Ziegenspeck, H. (1926*a*). Entwicklungsgeschichte der Axen der einheimischen Orchideen und ihre Physiologie und Biologie. I. *Cypripedium, Helleborine, Limodorum, Cephalanthera. Botanisches Archiv*, **14**, 165–260.

Fuchs, A. & Ziegenspeck, H. (1926*b*). Entwicklungsgeschichte der Axen der einheimischen Orchideen und ihre Physiologie und Biologie. II. *Listera, Neottia, Goodyera. Botanisches Archiv*, **16**, 360–413.

Fuchs, A. & Ziegenspeck, H. (1927*a*). Entwicklungsgeschichte der Axen der einheimischen Orchideen und ihre Physiologie und Biologie. III. *Botanisches Archiv*, **18**, 378–475.

Fuchs, A. & Ziegenspeck, H. (1927*b*). Die Dactylorchisgruppe der Ophrydineen. *Botanisches Archiv*, **19**, 163–274.

Fuchs, A. & Ziegenspeck, H. (1927*c*). Entwicklung, Axen und Blätter einheimischer Orchideen. IV. *Botanisches Archiv*, **20**, 275–422.

Gaddy, L. L. (1983). Studies on the biology of the small whorled pogonia (*Isotria medeoloides*) in South Carolina. *Net Areas Journal*, **3**, 14–37.

Galunder, R. (1984). Persönliche Erfahrungen bei der asymbiotischen Aussat mit unterschiedlichen Agar-Agar-Mengen. *Die Orchidee*, **35**, 224–6.

Galunder, R. (1986). Der Einfluss von Brauereihefe auf die Keimung von mitteleuropäischen Erdorchideen. *Die Orchidee*, **37**, 135–7.

Gamborg, O. L. (1970). The effects of amino acids and ammonium on the growth of plant cells in suspension culture. *Plant Physiology*, **45**, 372–5.

Garay, L. A. (1960). On the origin of the Orchidaceae. *Botanical Museum Leaflets, Harvard University*, **19**, 57–87.

Garay, L. A. (1986). Olim Vanillaceae. *Botanical Museum Leaflets*, **30**, 223–37.

Gäumann, E. & Hohl, H. R. (1960). Weitere Untersuchungen über die chemischen Abwehrreaktionen der Orchideen. *Phytopathologische Zeitschrift*, **38**, 93–104.

Gäumann, E. & Kern, H. (1959). Über chemische Abwehrreaktionen bei Orchideen. *Phytopathologische Zeitschrift*, **36**, 1–26.

Gäumann, E., Müller, E., Nüesch, J. & Rimpau, R. H. (1961). Über die Wurzelpilze von *Loroglossum hircinum* (L.) Rich. *Phytopathologische Zeitschrift*, **41**, 89–96.

Gäumann, E., Nüesch, J. & Rimpau, R. H. (1960). Weitere Untersuchungen über die chemischen Abwehrreaktionen der Orchideen. *Phytopathologische Zeitschrift*, **38**, 274–308.

Gehlert, R. & Kindl, H. (1991). Induced formation of dihydrophenanthrene and bibenzyl synthase upon destruction of orchid mycorrhiza. *Phytochemistry*, **30**, 457–60.

Geitler, L. (1956). Embryologie und mechanistischer Deutung der Embryogenese von *Epipogium aphyllum*. *Österreichische Botanische Zeitschrift*, **103**, 312–35.

Gentry, A. H. & Dodson, C. H. (1987). Diversity and biogeography of neotropical vascular epiphytes. *Annals of the Missouri Botanical Garden*, **74**, 205–33.

George, E. F., Puttock, D. J. M. & George, H. J. (1988). *Plant culture media. 2. Commentary and analysis*. Edington, Westbury: Exegetics Ltd.

Gill, D. E. (1989). Fruiting failure, pollinator inefficiency, and speciation in orchids. In *Speciation and its consequences*, ed. D. Otte & J. A. Endler, pp. 458–81. Sunderland, Massachusetts: Sinauer Associates.

Gillman, H. (1876). *Aplectrum* with coral-like root. *Bulletin of the Torrey Botanical Club*, **6**, 94–5.

Godfery, M. J. (1924). *Spiranthes romanzoffiana*. *Orchid Review*, **32**, 357–8.

Good, R. (1936). On the distribution of the lizard orchid (*Himantoglossum hircinum* Koch). *New Phytologist*, **35**, 142–70.

Greenwood, J. S. (1989). Phytin synthesis and deposition. In *Recent advances in the development and germination of seeds*, ed. R. B. Taylorson, pp. 109–25. New York: Plenum Press.

Gregg, K. B. (1989). Reproductive biology of the orchid *Cleistes divaricata* (L.) Ames var. *bifaria* Fernald growing in a West Virginia meadow. *Castanea*, **54**, 57–78.

Gregg, K. B. (1990). The natural life cycle of *Platanthera*. In *North American native terrestrial orchid propagation and production*, ed. C. E. Sawyers, pp. 25–39. Chadds Ford, Pennsylvania: Brandywine Conservancy.

Gregg, K. B. (1991). Variation in behavior of four populations of the orchid *Cleistes divaricata*, an assessment using transition matrix models. In *Population ecology of terrestrial orchids*, ed. T. C. E. Wells & J. H. Willems, pp. 139–59. The Hague: SPB Academic Publishing.

Griesbach, R. J. (1979). The albino form of *Epipactis helleborine*. *American Orchid Society Bulletin*, **48**, 808–9.

Grime, J. P. (1979). *Plant strategies and vegetation processes*. Chichester: Wiley.

Groom, P. (1895). Contributions to the knowledge of monocotyledonous saprophytes. *Journal of the Linnean Society, Botany*, **31**, 149–215.

Gruenschneder, A. (1973). Protocorm Proliferation von *Dactylorhiza maculata* – eine Möglichkeit zur Massenvermehrung und rationelle Anzucht. *Die Orchidee*, **24**, 249–50.

Guignard, L. (1886). Sur la pollination et ses effets chez les Orchidées. *Annales des Sciences Naturelles, Série Botanique*, **7**, **4**, 202–40.

Gumprecht, R. (1980). Pflegemassnahmen zur Förderung von Erdorchideen am natürlichen Standort. *Jahresbericht des Naturwissenschaftlichen Vereins Wuppertal*, **33**, 164–7.

Haas, N. F. (1973). Erste Ergebnisse zur Samenvermehrung von *Cypripedium reginae*. *Die Orchidee*, **24**, 247–9.

Haas, N. F. (1977a). Asymbiotische Vermehrung europäischer Erdorchideen. I. *Dactylorhiza sambucina* (L.) Soó. *Die Orchidee*, **28**, 27–31.

Haas, N. F. (1977b). Asymbiotische Vermehrung europäischer

Erdorchideen. II. *Nigritella nigra* (L.) Rchb.f. und *Nigritella miniata* (Cr.) Janchen. *Die Orchidee*, **28**, 69–73.

Hadley, G. (1969). Cellulose as a carbon source for orchid mycorrhiza. *New Phytologist*, **68**, 933–9.

Hadley, G. (1970*a*). The interaction of kinetin, auxin and other factors in the development of North temperate orchids. *New Phytologist*, **69**, 549–55.

Hadley, G. (1970*b*). Non-specificity of symbiotic infection in orchid mycorrhiza. *New Phytologist*, **69**, 1015–23.

Hadley, G. (1975). Organization and fine structure of orchid mycorrhiza. In *Endomycorrhizas*, ed. F. E. Sanders, B. Mosse & P. B. Tinker, pp. 335–51. London: Academic Press.

Hadley, G. (1982*a*). Orchid mycorrhiza. In *Orchid Biology. II. Reviews and perspectives*, ed. J. Arditti, pp. 83–118. Ithaca, New York: Cornell University Press.

Hadley, G. (1982*b*). European terrestrial orchids. In *Orchid Biology. II. Reviews and perspectives. Orchid seed germination and seedling culture – a manual*, ed. J. Arditti, pp. 326–9. Ithaca, New York: Cornell University Press.

Hadley, G. (1982*c*). Germination of British orchids. *Orchid Review*, **90**, 84–6.

Hadley, G. (1983). Symbiotic germination of orchid seed. *Orchid Review*, **91**, 44–7.

Hadley, G. (1984). Uptake of ^{14}C glucose by asymbiotic and mycorrhizal orchid protocorms. *New Phytologist*, **96**, 263–73.

Hadley, G. (1990). The role of mycorrhizae in orchid propagation. In *North American native terrestrial orchid propagation and production*, ed. C. E. Sawyers, pp. 15–24. Chadds Ford, Pennsylvania: Brandywine Conservancy.

Hadley, G. & Harvais, G. (1968). The effect of certain growth substances on asymbiotic germination and development of *Orchis purpurella*. *New Phytologist*, **67**, 441–5.

Hadley, G., Johnson, R. P. C. & John, D. A. (1971). Fine structure of the host–fungus interface in orchid mycorrhiza. *Planta*, **100**, 191–9.

Hadley, G. & Ong, S. H. (1978). Nutritional requirements of orchid endophytes. *New Phytologist*, **81**, 561–9.

Hadley, G. & Pegg, G. F. (1989). Host–fungus relationships in orchid mycorrhizal systems. In *Modern methods in orchid conservation*, ed. H. W. Pritchard, pp. 51–71. Cambridge: Cambridge University Press.

Hadley, G. & Perombelon, M. (1963). Production of pectic enzymes by *Rhizoctonia solani* and orchid endophytes. *Nature*, **200**, 1337.

Hadley, G. & Purves, S. (1974). Movement of [14]carbon from host to fungus in orchid mycorrhiza. *New Phytologist*, **73**, 475–82.

Hadley, G. & Williamson, B. (1971). Analysis of the post-infection growth stimulus in orchid mycorrhiza. *New Phytologist*, **70**, 445–55.

Hailes, N. S. J. & Seaton, P. T. (1989). The effects of the composition of the atmosphere on the growth of seedlings of *Cattleya aurantiaca*. In *Modern methods in orchid conservation*, ed. H. W. Pritchard, pp. 73–85. Cambridge: Cambridge University Press.

Hamada, M. (1939). Studien über die Mykorrhiza von *Galeola septentrionalis* Reichb. f. – Ein neuer Fall der Mykorrhiza-Bildung durch intraradicale Rhizomorpha. *Japanese Journal of Botany*, **10**, 151–211.

Hanke, M. & Dollwet, H. H. A. (1976). The production of ethylene by certain soil fungi. *Science of Biology Journal, Stillwater*, **2**, 227–30.

Harbeck, M. (1963). Einige Beobachtungen bei der Aussat verschiedener europäischer Erdorchideen auf sterilem Nährboden. *Die Orchidee*, **14**, 58–65.

Harbeck, M. (1964). Anzucht von *Orchis maculata* vom Samen bis zur Blüte. *Die Orchidee*, **15**, 57–61.

Harbeck, M. (1968). Versuche zur Samenvermehrung einiger *Dactylorhiza*-Arten. *Jahresbericht der Naturwissenschaftlichen Vereins in Wuppertal*, **21/22**, 112–18.

Harley, J. L. (1969). *Biology of mycorrhiza*, 2nd edn. London: Leonard Hill.

Harley, J. L. (1984). The mycorrhizal associations. In *Cellular interactions*, ed. H. F. Linskens & J. Heslop-Harrison, pp. 148–86. Berlin: Springer.

Harrison, C. R. (1977). Ultrastructural and histochemical changes during the germination of *Cattleya aurantiaca* (Orchidaceae). *Botanical Gazette*, **138**, 41–5.

Harrison, C. R. & Arditti, J. (1978). Physiological changes during the germination of *Cattleya aurantiaca* (Orchidaceae). *Botanical Gazette*, **139**, 180–9.

Hartmann, H. T. & Kester, D. E. (1983). *Plant propagation. Principles and practices*, 4th edn. Englewood Cliffs, New Jersey: Prentice-Hall.

Harvais, G. (1972). The development and growth requirements of *Dactylorhiza purpurella* in asymbiotic cultures. *Canadian Journal of Botany*, **50**, 1223–9.

Harvais, G. (1973). Growth requirements and development of *Cypripedium reginae* in axenic culture. *Canadian Journal of Botany*, **51**, 327–32.

Harvais, G. (1974). Notes on the biology of some native orchids of Thunder Bay, their endophytes and symbionts. *Canadian Journal of Botany*, **52**, 451–60.

Harvais, G. (1980). Scientific notes on a *Cypripedium reginae* of Northwestern Ontario, Canada. *American Orchid Society Bulletin*, **49**, 237–44.

Harvais, G. (1982). An improved culture medium for growing the orchid *Cypripedium reginae* axenically. *Canadian Journal of Botany*, **60**, 2547–55.

Harvais, G. & Hadley, G. (1967a). The relation between host and endophyte in orchid mycorrhiza. *New Phytologist*, **66**, 205–15.

Harvais, G. & Hadley, G. (1967b). The development of *Orchis purpurella* in asymbiotic and inoculated cultures. *New Phytologist*, **66**, 217–30.

Harvais, G. & Pekkala, D. (1975). Vitamin production by a fungus symbiotic with orchids. *Canadian Journal of Botany*, **53**, 156–63.

Harvais, G. & Raitsakas, A. (1975). On the physiology of a fungus symbiotic with orchids. *Canadian Journal of Botany*, **53**, 144–55.

Hayes, B. A. (1969). Observations on orchid seed mycorrhizae. *Mycopathologia et Mycologia Applicata*, **38**, 139–44.

Healey, P. L., Michaud, J. D. & Arditti, J. (1980). Morphometry of orchid seeds. III. Native California and related species of *Goodyera, Piperia, Platanthera, Spiranthes*. *American Journal of Botany*, **67**, 508–18.

Henrich, J. E., Stimart, D. P. & Ascher, P. D. (1981). Terrestrial orchid seed germination *in vitro* on a defined medium. *Journal of the American Society for Horticultural Science*, **106**, 193–6.

Henrikson, L. E. (1951). Asymbiotic germination of orchids and some effects of vitamins on *Thunia marshalliana*. *Svensk Botanisk Tidskrift*, **45**, 447–59.

Hesselman, H. (1900). Om mykorrhizabildningor hos arktiska växter. *Bihang till Svenska Vetenskapsakademins Handlinger*, **26**, 1–46.

Hew, C. S. & Khoo, S. I. (1980). Photosynthesis of young orchid seedlings. *New Phytologist*, **86**, 349–57.

Hijner, J. A. & Arditti, J. (1973). Orchid mycorrhiza: vitamin production and requirements by the symbionts. *American Journal of Botany*, **60**, 829–35.

Hildebrand, F. (1863). Die Fruchtbildung der Orchideen, ein Beweis für die doppelte Wirkung des Pollen. *Botanische Zeitung*, **44**, 329–45, Table XII.

Hilhorst, H. W. M. & Karssen, C. M. (1989). The role of light and nitrate in seed germination. In *Recent advances in the development and germination of seeds*, ed. R. B. Taylorson, pp. 191–205. New York: Plenum Press.

Hirt, W. (1906). Semina scobiformia, ihre Verbreitung im Pflanzenreich, Morphologie, Anatomie und biologische Bedeutung. *Mitteilungen aus dem Botanischen Museum der Universität Zürich*, **30**, 1–108.

Hofsten, A. von. (1973). The ultrastructure of mycorrhiza in *Ophrys insectifera*. *Zoon*, Suppl 1, 93–6.

Holländer, S. (1932). *Ernährungsphysiologische Untersuchungen an Wurzelpilzen saprophytisch lebender Orchideen*. Würzberg: Dissertation Julius-Maximilian-Universität.

Holm, T. (1900). *Pogonia ophioglossoides* Mitt.: a morphological and anatomical study. *American Journal of Science, Series 4*, **9**, 13–9.

Holm, T. (1904). The root-structure of North American terrestrial Orchideae. *American Journal of Science*, **18**, 197–212.

Holman, R. T. (1976). Cultivation of *Cypripedium calceolus* and *Cypripedium reginae*. *American Orchid Society Bulletin*, **45**, 415–22.

Homès, J. & Vanséveren-Van Espen, N. (1973). Effets du saccharose et de la lumière sur le développement et la morphologie de protocormes d'orchidées cultivés *in vitro*. *Bulletin de la Société royale de Botanique de Belgique*, **106**, 89–106.

Huber, B. (1921). Zur Biologie der Torfmoororchidee *Liparis loeselii* Rich. *Sitzungsberichte der Akademie der Wissenschaften Mathematisch-Naturwissenschaftliche Classe, Wien, Abteilung 1*, **130**, 307–28, 1 plate.

Hutchings, M. J. (1987a). The population biology of the early spider orchid, *Ophrys sphegodes* Mill. I. A demographic study from 1975 to 1984. *Journal of Ecology*, **75**, 711–27.

Hutchings, M. J. (1987b). The population biology of the early spider orchid, *Ophrys sphegodes* Mill. II. Temporal patterns in behavior. *Journal of Ecology*, **75**, 729–42.

Hutchings, M. J. (1989). Population biology and conservation of *Ophrys sphegodes*. In *Modern methods in orchid conservation*, ed. H. W. Pritchard, pp. 101–15. Cambridge: Cambridge University Press.

Ichihashi, S. (1978). Studies on the media for orchid seed germination. II. The effects of anionic and cationic combinations relevent to seeding populations and culture periods on the growth of *Bletilla striata* seedlings. *Engei Gakkai Zasshi [Journal of the Japanese Society for Horticultural Science]*, **46**, 521–9.

Ichihashi, S. (1979). Studies on the media for orchid seed germination. III. The effect of total ionic concentration, cation/anion ratio, NH_4/NO_3 ratio and minor elements on the growth of *Bletilla striata* seedlings. *Engei Gakkai Zasshi [Journal of the Japanese Society for Horticultural Science]*, **47**, 524–36.

Ichihashi, S. (1990). Effects of light on root formation of *Bletilla striata* seedlings. *Lindleyana*, **5**, 140–3.

Ichihashi, S. & Yamashita, M. (1977). Studies on the media for orchid seed germination. I. The effects of balances inside each cation and anion group for the germination and seedling development of *Bletilla striata* seeds. *Engei Gakkai Zasshi [Journal of the Japanese Society for Horticultural Science]*, **45**, 407–13.

Ilag, C. & Curtis, R. W. (1968). Production of ethylene by fungi. *Science*, **159**, 1357–8.

Inghe, O. & Tamm, C. O. (1988). Survival and flowering of perrenial herbs. V. Patterns of flowering. *Oikos*, **51**, 203–19.

Irmisch, T. (1853). *Beiträge zur Biologie und Morphologie der Orchideen*. Leipzig: Ambrosius Abel.

Jackson, M. B. (1985). Ethylene and the responses of plants to excess water in their environment – a review. In *Ethylene and plant development*, ed. J. A. Roberts & G. A. Tucker, pp. 241–65. London: Butterworth.

Janse, J. M. (1897). Les endophytes radicaux de quelques plantes javanaises. *Annales du Jardin Botanique de Buitenzorg*, **14**, 53–201, Plates 5–15.

Jaretzky, R. & Bereck, E. (1938). Der Schleim in der Knollen von *Orchis purpureus* Huds. und *Platanthera bifolia* (L.) Rchb. *Archiv der Pharmazie*, **276**, 17–27.

Jennings, A. V. & Hanna, H. (1898). *Corallorhiza innata* R. Br., and its mycorrhiza. *Scientific Proceedings of the Royal Dublin Society*, N.S., **9**, 1–11.

Johansen, B. & Rasmussen, H. N. (1992). *Ex situ* conservation of orchids. *Opera Botanica*, **113**, 43–8.

Johow, F. (1885). Die chlorophyllfreien Humusbewohner West Indiens, biologisch-morphologisch dargestellt. *Jahrbücher für wissenschaftliche Botanik*, **16**, 415–49, Plates 16–18.

Johow, F. (1889). Die chlorophyllfreien Humuspflanzen nach ihren biologischen und anatomisch-entwicklungsgeschichtlichen Verhältnissen. *Jahrbücher für wissenschaftliche Botanik*, **20**, 475–525, Plates 19–22.

Jonsson, L. & Nylund, J. E. (1979). *Favolaschia dybowskyana* (Singer)Singer (Aphyllophorales), a new orchid mycorrhizal fungus from tropical Africa. *New Phytologist*, **83**, 121–8.

Jorgensen, E. (1982). Lake Park – revisited after fifty years. *American Orchid Society Bulletin*, **51**, 41–2.

Jørgensen, B. I. (n.d.). Development of *Dactylorhiza majalis* (Orchidaceae) seedlings in symbiotic culture. (Manuscript.)

Kano, K. (1968). Acceleration of the germination of so-called hard-to-germinate orchid seeds. *American Orchid Society Bulletin*, **37**, 690–8.

Karasawa, K. (1966). On the media with banana and honey added for seed germination and subsequent growth of orchids. *Orchid Review*, **74**, 313–18.

Keenan, P. E. (1988). Three-Birds orchid at Golden Pond. *American Orchid Society Bulletin*, **57**, 25–7.

Kermode, A. R. (1990). Regulatory mechanisms involved in the transition from seed development to germination. *Critical Reviews in Plant Science*, **9**, 155–95.

Knudson, L. (1922). Non-symbiotic germination of orchid seeds. *Botanical Gazette*, **73**, 1–25.

Knudson, L. (1924). Further observations on non-symbiotic germination of orchid seeds. *Botanical Gazette*, **77**, 212–19.

Knudson, L. (1925). Physiological studies of the symbiotic germination of orchid seeds. *Botanical Gazette*, **79**, 345–79.

Knudson, L. (1927). Symbiosis and asymbiosis relative to orchids. *New Phytologist*, **26**, 328–36.

Knudson, L. (1929). Physiological investigations on orchid seed germination. *Proceedings of the International Congresses of Plant Science*, **2**, 1183–9.

Knudson, L. (1941). Germination of seed of *Goodyera pubescens*. *American Orchid Society Bulletin*, **9**, 119–201.

Knudson, L. (1943). Nutrient solutions for orchid seed germination. *American Orchid Society Bulletin*, **12**, 77–9.

Knudson, L. (1946). A new nutrient solution for orchid seed germination. *American Orchid Society Bulletin*, **15**, 214–17.

Kober, V. (1972). *Cypripedium calceolus* – Anzucht aus Samen. *Die Orchidee*, **23**, 77–8.

Koch, L. & Schulz, D. (1975). Über Samen und Samenkeimung der *Phalaenopsis* Heideperle. *Die Orchidee*, **26**, 27–30.

Kull, T. (1988). Identification of clones in *Cypripedium calceolus* L. (Orchidaceae). *Proceeding of the Academy of Sciences of the Estonian SSR, Biology*, **37**, 195–9.

Kull, T. & Kull, K. (1991). Preliminary results from a study of populations of *Cypripedium calceolus* in Estonia. In *Population ecology of terrestrial orchids*, ed. T. C. E. Wells & J. H. Willems, pp. 69–76. The Hague: SPB Academic Publishing.

Kumazawa, M. (1958). The sinker of *Platanthera* and *Perularia* – its morphology and development. *Phytomorphology*, **8**, 137–45.

Kurtzweil, H. (1994). The unusual seed structure of *Disa uniflora*, with notes on the dispersal of the species. In *Proceedings of the 14th World Orchid Conference, Glasgow*, ed. A. Pridgeon, pp. 394–6. London: Her Majesty's Stationery Office

Kusano, S. (1911). *Gastrodia elata* and its symbiotic association with *Armillaria mellea*. *Journal of the College of Agriculture, Imperial University of Tokyo*, **4**, 1–66.

Kutschera, L. & Lichtenegger, E. (1982). *Wurzelatlas mitteleuropäischer Grünlandspflanzen I. Monocotyledoneae*, pp. 78–90. Stuttgart: Gustav Fischer.

LaGarde, R. V. (1929). Non-symbiotic germination of orchids. *Annals of the Missouri Botanical Garden*, **16**, 499–514.

Lawler, L. J. (1984). Ethnobotany of the Orchidaceae. In *Orchid Biology. III. Review and perspectives*, ed. J. Arditti, pp. 27–149. Ithaca, New York: Cornell University Press.

Leake, J. R. & Read, D. J. (1990). The effects of phenolic compounds on nitrogen mobilisation by ericoid mycorrhizal systems. *Agriculture, Ecosystem & Environment*, **29**, 225–36.

Leeson, E., Haynes, C. & Wells, T. C. E. (1991). Studies of the phenology and dry matter allocation of *Dactylorhiza fuchsii*. In *Population ecology of terrestrial orchids*, ed. T. C. E. Wells & J. H. Willems, pp. 125–38. The Hague: SPB Academic Publishing.

Lehaie, J. H. de (1910). Observations pour servir a l'etude de la dissémination des orchidées indigènes en Belgique. *Bulletin de la Société Royale de Botanique de Belgique*, **47**, 45–52.

Lesica, P. & Antibus, R. K. (1990). The occurrence of mycorrhizae in vascular epiphytes of two Costa Rican rain forests. *Biotropica*, **22**, 250–8.

Lewis, D. H. (1973). Concepts in fungal nutrition and the origin of biotrophy. *Biological Review*, **48**, 261–78.

Lichtenegger, E. & Kutschera-Mitter, L. (1991). Spatial root types. In *Plant roots and their environment*, ed. B. L. McMichael & H. Persson, pp. 359–65. Amsterdam: Elsevier Science Publichers.

Liddell, R. W. (1944). Germinating native orchid seed. *American Orchid Society Bulletin*, **12**, 344–5.

Liddell, R. W. (1952). Germinating native orchid seed. *American Orchid Society Bulletin*, **21**, 92–3.

Light, M. H. S. (1989). Germination in the *Cypripedium/Paphiopedilum* alliance. *Canadian Orchid Journal*, **5**, 11–19.

Light, M. H. S. (1994). In *Forum – the propagation of temperate terrestrial orchids*, ed. M. Ramsay. In *Proceedings of the 14th World Orchid Conference, Glasgow*, ed. A. Pridgeon, p. 404. London: Her Majesty's Stationery Office.

Light, M. H. S. & MacConaill, M. (1989). Albinism in *Platanthera hyperborea*. *Lindleyana*, **4**, 158–60.

Light, M. H. S. & MacConaill, M. (1990). Characterization of the optimal capsule development stage for embryo culture of *Cypripedium calceolus* var. *pubescens*. In *North American native terrestrial orchid propagation and production*, ed. C. E. Sawyers, p. 92. Chadds Ford, Pennsylvania:

Brandywine Conservancy.

Light, M. H. S. & MacConaill, M. (1991). Patterns of appearance in *Epipactis helleborine*. In *Population ecology of terrestrial orchids*, ed. T.C. E. Wells & J. H. Willems, pp. 77–87. The Hague: SPB Academic Publishing.

Lindén, B. (1980). Aseptic germination of seeds of northern terrestrial orchids. *Annales Botanici Fennici*, **17**, 174–82.

Lindén, B. (1992). Two new methods for pretreatment of seeds of Northern orchids to improve germination in axenic culture. *Annales Botanici Fennici*, **29**, 305–13.

Lindquist, B. (1960). The raising of *Disa uniflora* seedlings in Gothenburg. In *Proceedings of the 3rd World Orchid Conference 1960*, ed. P. M. Synge, pp. 207–11. London: Royal Horticultural Society.

Ling, H., Mowen, P. & Reilly, B. (1990). Propagation of *Cypripedium acaule*: Tissue culture, greenhouse, and garden trials. In *North American native terrestrial orchid propagation and production*, ed. C. E. Sawyers, pp. 93–4. Chadds Ford, Pennsylvania: Brandywine Conservancy.

Lucas, R. L. (1977). The movement of nutrients through fungal mycelium. *Transactions of the British Mycological Society*, **69**, 1–9.

Lucke, E. (1971a). Zur Samenkeimung mediterraner *Ophrys*. *Die Orchidee*, **22**, 62–5.

Lucke, E. (1971b). Zur Ernte unreifer Samenkapseln. *Die Orchidee*, **22**, 146–7.

Lucke, E. (1975). *Orchideenkultur für alle*. Minden: A. Philler Verlag.

Lucke, E. (1976). Erste Ergebnisse zur asymbiotischen Samenkeimung von *Himantoglossum hircinum*. *Die Orchidee*, **27**, 60–1.

Lucke, E. (1977). Naturstoffzusätze bei der asymbiotischen Samenvermehrung der Orchideen *in vitro*. *Die Orchidee*, **28**, 185–91.

Lucke, E. (1978a). Samenkeimung von *Cypripedium reginae*. *Die Orchidee*, **29**, 42.

Lucke, E. (1978b). Fragen zur aseptischen Samenernte bzw. Desinfektion der Orchideensamen. *Die Orchidee*, **29**, 42–3.

Lucke, E. (1981). Samenstruktur und Samenkeimung europäischer Orchideen nach Veyret sowie weitere Untersuchungen. I. *Die Orchidee*, **32**, 182–8.

Lucke, E. (1982). Samenstruktur und Samenkeimung europäischer Orchideen nach Veyret sowie weitere Untersuchungen. II–III. *Die Orchidee*, **33**, 8–16, 108–15.

Lucke, E. (1984). Samenstruktur und Samenkeimung europäischer Orchideen nach Veyret sowie weitere Untersuchungen. IV–V. *Die Orchidee*, **35**, 13–20, 153–8.

Lucke, E. (1985). Zur Lagerung und Vitalitätsprüfung von Orchideensamen. *Die Orchidee*, **36**, 111–12.

Lund, I. D. (1988). The genus *Cremastra* (Orchidaceae), a taxonomic revision. *Nordic Journal of Botany*, **8**, 197–203.

Lynch, J. M. & Harper, S. H. T. (1974). Formation of ethylene by a soil fungus. *Journal of General Microbiology*, **80**, 187–95.

MacDougal, D. T. (1898). The mycorrhizae of *Aplectrum*. *Bulletin of the Torrey Botanical Club*, **25**, 110–12.

MacDougal, D. T. (1899a). Symbiotic saprophytism. *Annals of Botany*, **13**, 1–47, Plates 1–2.

MacDougal, D. T. (1899b). Symbiosis and saprophytism. *Bulletin of the Torrey Botanical Club*, **26**, 511–30, Plates 367–9.

MacDougal, D. T. & Dufrenoy, J. (1944). Mycorrhizal symbiosis in *Aplectrum, Corallorhiza*, and *Pinus*. *Plant Physiology*, **19**, 440–65.

Mackay, R. D. (1972). The measurement of viability. In *Viability of seeds*, ed. E. H. Roberts, pp. 172–208. London: Chapman & Hall.

Magnus, W. (1900). Studien an der endotrophen Mycorrhiza von *Neottia nidus-avis* L. *Jahrbücher für Wissenschaftlische Botanik*, **35**, 205–72.

Magrou, J. (1924). A propos du pouvoir fungicide des tubercules d'Ophrydees. *Annales des Sciences Naturelles (Paris) 10. sér. VI*, **18**, 256–70.

Malmgren, S. (1988). Fröförökning av *Dactylorhiza* i stor skala – en kort manual. *Svensk Botanisk Tidskrift*, **82**, 161–6.

Malmgren, S. (1989a). Asymbiotisk förökning från frö av guckusko, flugblomster, brunkulla och några andra svenska orkidéarter. *Svensk Botanisk Tidskrift*, **83**, 347–54.

Malmgren, S. (1989b). Fröförökning (4). *Orchidéer*, **10**, 154–7, 163.

Malmgren, S. (1993). Asymbiotisk fröförökning i stor skala av *Anacamptis, Ophrys, Orchis* och andra orkideer med runda rotknölar. *Svensk Botanisk Tidskrift*, **87**, 221–34.

Manning, J. C. & Van Staden, J. (1987). The development and mobilisation of seed reserves in some African orchids. *Australian Journal of Botany*, **35**, 343–53.

Marcuse, M. (1902). *Anatomisch-biologischer Beitrag zur Mycorrhizenfrage*. Jena: Inaugural Dissertation.

Mariat, F. (1952). Recherches sur la physiologie des embryons d'orchidée. *Revue Générale de Botanique*, **59**, 324–77.

Masamune, G. (1969). *Illustrated Flora of Nippon*, vol. 8, *Orchidaceae*. Japan: Koyo Shoin.

Masuhara, G. & Katsuya, K. (1989). Effects of mycorrhizal fungi on seed germination and early growth of three Japanese terrestrial orchids. *Scientia Horticulturae*, **37**, 331–7.

Masuhara, G., Kimura, S. & Katsuya, K. (1988). Seasonal changes in the mycorrhizas of *Bletilla striata* (Orchidaceae). *Transactions of the Mycological Society of Japan*, **29**, 25–31.

McAlpine, K. L. (1947). Germination of orchid seeds. *Orchid Review*, **55**, 8–22.

McIntyre, D. K., Veitch, G. J. & Wrigley, J. W. (1974). Australian terrestrial orchids from seed. II. Improvements in techniques and further successes. *American Orchid Society Bulletin*, **43**, 52–3.

McRill, M. & Sagar, G. R. (1973). Earthworms and seeds. *Nature*, **243**, 482.

Mead, J. W. & Bulard, C. (1975). Effects of vitamins and nitrogen sources on asymbiotic germination and development of *Orchis laxiflora* and *Ophrys sphegodes*. *New Phytologist*, *74*, 33–40.

Mead, J. W. & Bulard, C. (1979). Vitamins and nitrogen requirements of *Orchis laxiflora* Lamk. *New Phytologist*, *83*, 129–36.

Mei-sheng, T., Fu-xiong, W., Nan-fen, Q. & An-ci, S. (1985). *In vitro* seed germination and developmental morphology of seedling in *Cymbidium ensifolium*. *Acta Botanica Sinica*, **27**, 455–9, Plate I.

Mitchell, R. B. (1988). A new mycorrhizal fungus for germinating *Dactylorhiza*. *Orchid Review*, **96**, 282–5.

Mitchell, R. B. (1989). Growing hardy orchids from seeds at Kew. *The Plantsman*, **11**, 152–69.

Miyoshi, K. & Mii, M. (1987). Breakage or avoidance of secondary seed dormancy induced by dry storage in *Calanthe discolor*. In *Proceedings of the 12th World Orchid Conference 1987*, ed. K. Saito & R. Tanaka, p. 292. Tokyo, Japan.

Miyoshi, K. & Mii, M. (1988). Ultrasonic treatment for enhancing seed germination of terrestrial orchid, *Calanthe discolor*, in asymbiotic culture. *Scientia Horticulturae*, **35**, 127–30.

Möller, O. (1966). Kulturversuche mit einigen heimischen Orchideen. *Die Orchidee*, **17** (Sonderheft Juni), 136–41.

Möller, O. (1967). Der Vegetationsrhythmus von *Orchis mascula*. *Die Orchidee*, **18**, 67–9.

Möller, O. (1968). Wachstumsrhythmus und vegetative Vermehrung von *Cypripedium calceolus*. *Die Orchidee*, **19**, 222–4.

Möller, O. (1971). Ergänzungen zu Herbert Oesterreichs Beobachtungen über die Nestwurz und den Vegetationsrhythmus der Erdorchideen. *Die Orchidee*, **22**, 207–10.

Möller, O. (1977). Kritische Anmerkungen zu Vermehrung von *Cypripedium* aus Samen. *Die Orchidee*, **28**, 184–5.

Möller, O. (1985*a*). Die Mineralsalze der Standortböden der europäischer Orchideen. *Die Orchidee*, **36**, 118–21.

Möller, O. (1985b). Die Mineralsalze der Böden von *Cypripedium calceolus* und *Orchis purpurea*. *Die Orchidee*, **36**, 124–6.

Möller, O. (1985c). Die Nährstoffaufnahme der Erdorchideen. *Die Orchidee*, **36**, 32–5.

Möller, O. (1986a). Der Einfluss der Nährsalze auf die Keimung und den Wuchs der *Orchis ustulata* in der Natur. *Die Orchidee*, **37**, 133–5.

Möller, O. (1986b). Die Notwendigkeit einer Intensivierung der Erdorchideenkunde. *Die Orchidee*, **37**, 243.

Möller, O. (1987a). Die subterrane Innovation und der Wachstumszyklus einiger Erdorchideen. *Die Orchidee*, **38**, 13–22.

Möller, O. (1987b). Vom Samenkorn bis zur ersten echten Knolle: Das Protocormstadium von *Orchis mascula*. *Die Orchidee*, **38**, 297–302.

Möller, O. (1988). Der Wuchs und die Innovation der *Spiranthes spiralis*. *Die Orchidee*, **39**, 15–9.

Möller, O. (1989). Die Samenkeimung, das Protocormstadium und der Aufwuchs der *Orchis militaris*. *Die Orchidee*, **40**, 29–32.

Möller, O. (1990a). Der Wuchs der Winterblätter der europäischen Erdorchideen. *Die Orchidee*, **41**, 21.

Möller, O. (1990b). Beobachtungen und Bemerkungen über den Wuchs der *Dactylorhiza maculata*. *Die Orchidee*, **41**, 22–6.

Mollison, J. E. (1943). *Goodyera repens* and its endophyte. *Transactions and Proceedings of the Botanical Society of Edinburgh*, **33**, 391–403.

Montfort, C. (1940). Beziehungen zwischen morphologischen und physiologischen Reduktionserscheinungen im Bereicht der Licht-Ernährung bei saprophytischen Orchideen. *Bericht der Deutschen Botanischen Gesellschaft*, **58**, 41–9.

Montfort, C. & Küsters, G. (1940). Saprophytismus und Photosynthese. I. Biochemische und physiologische Studien an Humus-Orchideen. *Botanisches Archiv*, **40**, 571–633.

Moore, R. T. (1987). The genera of *Rhizoctonia*-like fungi: *Ascorhizoctonia*, *Ceratorhiza* gen. nov., *Epulorhiza* gen. nov., *Moniliopsis*, and *Rhizoctonia*. *Mycotaxon*, **29**, 91–9.

Mossberg, B. & Nilsson, S. (1977). *Nordens orkidéer*. Stockholm: Wahlström & Widstrand.

Mossberg, B. & Nilsson, S. (1980). *Syd- och Mellemeuropas orkidéer*. Stockholm: Wahlström & Widstrand.

Mousley, H. (1925). Further notes on *Calypso*. *Torreya*, **25**, 54–9.

Mrkvicka, A. C. (1990a). Neue Beobachtungen zu Samenkeimung und Entwicklung von *Liparis loeselii* (L.) Rich. *Mitteilungsblatt, Arbeitskreis heimische Orchideen Baden-Württenberg*, **22**, 172–80.

Mrkvicka, A. C. (1990b). Über die Kultur und Vermehrung einiger heimischer Orchideenarten in künstlich angelegten Feuchtbiotopen. *Die Orchidee*, **41**, 188–91.

Mrkvicka, A. C. (1991). *Spiranthes aestivalis* (Poir.) Rich. – Beobachtungen zu Keimung, Entwicklung und Ökologie. *Mitteilungsblatt, Arbeitskreis heimische Orchideen Baden-Württenberg*, **23**, 473–86.

Mrkvicka, A. C. (1992). Zur Einfluss von Temperatur, Feuchte und Tageslänge auf Spross- und Blütenentwicklung europäischer Erdorchideen. *Die Orchidee*, **43**, 28–33.

Muick, F. (1977). Zur Vermehrung von *Cypripedium* aus Samen. *Die Orchidee*, **28**, 116–18.

Muick, F. (1978). Propagation of *Cypripedium* species from seeds. *American Orchid Society Bulletin*, **47**, 307–8.

Muir, H. J. (1987). Symbiotic micropropagation of *Orchis laxiflora*. *Orchid Review*, **95**, 27–9.

Muir, H. J. (1989). Germination and mycorrhizal fungus compatibility in European orchids. In *Modern methods in orchid conservation*, ed. H. W. Pritchard, pp. 39–56. Cambridge: Cambridge University Press.

Myers, P. J. & Ascher, P. D. (1982). Culture of North American orchids from seed. *HortScience*, **17**, 550.

Nagashima, T. (1989). Embryogenesis, seed formation and immature seed germination *in vitro* in *Ponerorchis graminifolia* Reichb.f. *Engei Gakkai Zasshi (Journal of the Japanese Society for Horticultural Science)*, **58**, 187–94.

Nagl, W. (1972). Evidence of DNA amplification in the orchid *Cymbidium in vitro*. *Cytobios*, **5**, 145–54.

Nakamura, S. I. (1962). Zur Samenkeimung einer chlorophyllfreien Erdorchidee *Galeola septentrionalis* Reichb. f. *Zeitschrift für Botanik*, **50**, 487–97.

Nakamura, S. I. (1964). Einige Experimente zur Samenkeimung einer chlorophyllfreien Erdorchidee *Galeola septentrionalis* Reichb. f. *Memoirs of the College of Agriculture, Kyoto, University Botany 4*, **86**, 1–48.

Nakamura, S. I. (1976). Atmospheric conditions required for the growth of *Galeola septentrionalis* seedlings. *Botanical Magazine Tokyo*, **89**, 211–18.

Nakamura, S. J. (1982). Nutritional conditions required for the non-symbiotic culture of an achlorophyllous orchid *Galeola septentrionalis*. *New Phytologist*, **90**, 701–15.

Nakamura, S. J., Uchida, T. & Hamada, M. (1975). Atmospheric conditions controlling the seed germination of an achlorophyllous orchid, *Galeola septentrionalis*. *Botanical Magazine Tokyo*, **88**, 103–9.

Neger, F. W. (1913). *Biologie der Pflanzen auf experimentelle Grundlage.* Stuttgart: Ferdinand Enke.

Nieuwdorp, P. J. (1972). Some observations with light and electron microscope on the endotrophic mycorrhiza of orchids. *Acta Botanica Neerlandica,* **21,** 128–44.

Niewieczerzalowna, B. (1933). Recherches morphologiques sur la mycorrhize des orchidées indigenes. *Compte Rendu des Séances de la Société scientique de Varsovie,* **25,** 85–115.

Nishikawa, T. & Ui, T. (1976). *Rhizoctonia*'s isolated from wild orchids in Hokkaido. *Transactions of the Mycological Society of Japan,* **17,** 77–84.

Nishimura, G. (1982). Japanese orchids. In *Orchid Biology II. Reviews and perspectives. Orchid seed germination and seedling culture – a manual,* ed. J. Arditti, pp. 331–46. Ithaca, New York: Cornell University Press.

Nishimura, G. (1991). Comparative morphology of cotyledonous orchid seedlings. *Lindleyana,* **6,** 140–6.

Nobecourt, P. (1923). Sur la production d'anticorps par les tubercules des Ophrydées. *Compte Rendu Hebdomadaire des Séances de l'Académie Sciences, Paris,* **17,** 1055–7.

Noggle, G. R. & Wynd, F. L. (1943). Effects of vitamins on germination and growth of orchids. *Botanical Gazette,* **104,** 455–9.

Norstog, K. J. (1973). New synthetic medium for the culture of premature barley embryos. *In Vitro,* **8,** 307–8.

Ogura, Y. (1953). Anatomy and morphology of the subterranean organs in some Orchidaceae. *Journal of the Faculty of Science, University of Tokyo. Sect. III,* **6,** 135–57.

Ohwi, J. (1965). *Flora of Japan.* Washington, DC: Smithsonian Institute.

Oliva, A. P. & Arditti, J. (1984). Seed germination of North American orchids. II. Native California and related species of *Aplectrum, Cypripedium,* and *Spiranthes. Botanical Gazette,* **145,** 495–501.

Olver, S. (1981). Growing *Cypripedium reginae* in controlled environment chambers. *American Orchid Society Bulletin,* **50,** 1091–2.

Pais, M. S. & Barroso, J. (1983). Localization of polyphenoloxidases during the establishment of *Ophrys lutea* endomycorrhizas. *New Phytologist,* **95,** 219–22.

Pais, M. S. S. & Barroso, J. (1990). A review on *Ophrys lutea* mycorrhizas. Host/endophyte interaction. *Agriculture, Ecosystem & Environment,* **29,** 345–8.

Pate, J. S. & Dixon, K. W. (1982). *Tuberous, cormous and bulbous plants. Biology of an adaptive strategy in Western Australia.* Nedlands, Australia: University of Western Australia Press.

Pauw, M. A. de & Remphrey, W. R. (1993). *In vitro* germination of three *Cypripedium* species in relation to time of seed collection, media, and cold treatment. *Canadian Journal of Botany*, **71**, 879–85.

Peklo, J. (1906). Zur Lebensgeschichte von *Neottia nidus-avis* L. *Flora*, **96**, 260–75.

Penningsfeld, F. (1990*a*). Anzucht und Ernährung heimischer Orchideen. *Die Orchidee*, **41**, 185–8.

Penningsfeld, F. (1990*b*). Anzucht und Ernährung heimischer Orchideen. II. *Die Orchidee*, **41**, 221–5.

Perombelon, M. & Hadley, G. (1965). Production of pectic enzymes by pathogenic and symbiotic *Rhizoctonia* strains. *New Phytologist*, **64**, 144–51.

Peterson, R. L. & Currah, R. S. (1990). Synthesis of mycorrhizae between protocorms of *Goodyera repens* (Orchidaceae) and *Ceratobasisium cereale*. *Canadian Journal of Botany*, **68**, 1117–25.

Pfeffer, W. (1897). *Pflanzenfysiologie 1, Stoffwechsel*. Leipzig: Wilhelm Engelmann.

Pinfield, N. J. & Stobart, A. K. (1972). Hormonal regulation of germination and early seedling development in *Acer pseudoplatanus*. *Planta*, **104**, 134–45.

Poddubnaya-Arnoldi, V. A. & Zinger, N. V. (1961). Application of histochemical technique to the study of embryonic processes in some orchids. *Recent Advances in Botany (Toronto)*, **8**, 711–14.

Ponchet, M., Beck, D. & Poupet, A. (1985). Multiplication végétative *in vitro*, par prolifération de bourgeons de deux espèces de *Serapias*: *S. olbia* Verguin et *S. pseudocordigera* Moric. (Orchidaceae). *Bulletin de la Société Botanique de France, Lettres Botanique*, **132**, 289–300.

Pottinger, M., Tasker, S. & Klein, B. (1988). Notes from Kew. *Orchid Review*, **96**, 36–45.

Prillieux, E. (1856). De la structure anatomique et du mode de végétation de *Neottia nidus-avis*. *Annales des Sciences Naturelles IV, Série Botanique*, **5**, 267–82, Plates 17–18.

Prillieux, E. & Rivière, A. (1856). Observations sur la germination et le développement d'une orchidée. *Annales des Sciences Naturelles IV, Série Botanique*, **5**, 119–36, Plates 5–7.

Pritchard, H. W. (1984). Liquid nitrogen preservation of terrestrial and epiphytic orchid seed. *Cryo-Letters*, **5**, 295–300.

Pritchard, H. W. (1985*a*). Determination of orchid seed viability using fluorescein diacetate. *Plant, Cell and Environment*, **8**, 727–30.

Pritchard, H. W. (1985*b*). Growth and storage of orchid seeds. In *Proceedings of the 11th World Orchid Conference 1984, Miami, Florida*, ed. K. W. Tan, pp. 290–3. Singapore: International Press Company.

Purves, S. & Hadley, G. (1975). Movement of carbon compounds between the partners in orchid mycorrhiza. In *Endomycorrhizas*, ed. F. E.

Sanders, B. Mosse & P. B. Tinker, pp. 175–94. London: Academic Press.

Purves, S. & Hadley, G. (1976). The physiology of symbiosis in *Goodyera repens. New Phytologist*, **77**, 689–96.

Raghavan, V. & Torrey, J. G. (1964). Inorganic nitrogen nutrition of the seedlings of the orchid *Cattleya. American Journal of Botany*, **51**, 264–74.

Ramin, I. von (1973). Erfahrungen beim Umpflanzen europäischer Orchideen. *Die Orchidee*, **24**, 121.

Ramin, I. von (1976). Erfahrungen in Kultur und Vermehrung von Erdorchideen. In *Proceedings of the 8th World Orchid Conference*, ed. K. Senghas, pp. 364–66. Frankfurt: German Orchid Society.

Ramin, I. von (1983). Aussaten von Orchideen auf verschiedenen Agar-Nährböden. *Die Orchidee Sonderheft März 1983*, 109–11.

Ramsay, R. R., Sivasithamparam, K. & Dixon, K. W. (1987). Anastomosis groups among *Rhizoctonia*-like endophytic fungi in Southwestern Australian *Pterostylis* species (Orchidaceae). *Lindleyana*, **2**, 161–6.

Ramsbottom, J. (1923). Orchid mycorrhiza. *Transactions of the British Mycological Society*, **8**, 28–61.

Ramsbottom, J. (1929). Orchid mycorrhiza. *Proceedings of the International Congress Plant Science*, **2**, 1676–87.

Rasmussen, H. N. (1986). The vegetative architecture of orchids. *Lindleyana*, **1**, 42–50.

Rasmussen, H. N. (1990). Cell differentiation and mycorrhizal infection in *Dactylorhiza majalis* (Rchb. f.) Hunt & Summerh. (Orchidaceae) during germination *in vitro. New Phytologist*, **116**, 137–47.

Rasmussen, H. N. (1992*a*). Germination and growth of mycorrhizal seedlings of *Tipularia discolor* (Orchidaceae) on woody debris. *American Journal of Botany*, Suppl., **79**, 68.

Rasmussen, H. N. (1992*b*). Seed dormancy patterns in *Epipactis palustris* (Orchidaceae): requirements for germination and establishment of mycorrhiza. *Physiologia Plantarum*, **86**, 161–7.

Rasmussen, H. N. (1994*a*). The roles of fungi in orchid life history. In *Proceedings of the 14th World Orchid Conference, Glasgow*, ed. A. Pridgeon, pp. 130–7 London: Her Majesty's Stationery Office.

Rasmussen, H. N. (1994*b*). 'Easy' and 'difficult' species of terrestrial orchids. In *Proceedings of the 14th World Orchid Conference, Glasgow*, ed. A. Pridgeon, p. 404. London: Her Majesty's Stationery Office.

Rasmussen, H. N., Andersen, T. F. & Johansen, B. (1990*a*). Temperature sensitivity of *in vitro* germination and seedling development of *Dactylorhiza majalis* (Orchidaceae) with and without a mycorrhizal fungus. *Plant, Cell and Environment*, **13**, 171–7.

Rasmussen, H. N., Andersen, T. F. & Johansen, B. (1990*b*). Light stimulation and darkness requirement for the symbiotic germination of *Dactylorhiza majalis in vitro*. *Physiologia Plantarum*, **79**, 226–30.

Rasmussen, H. N., Johansen, B. & Andersen, T. F. (1989). Density-dependent interactions between seedlings of *Dactylorhiza majalis* (Orchidaceae) in symbiotic *in vitro* culture. *Physiologia Plantarum*, **77**, 473–8.

Rasmussen, H. N., Johansen, B. & Andersen, T. F. (1991). Symbiotic *in vitro* culture of immature embryos and seeds from *Listera ovata*. *Lindleyana*, **6**, 134–9.

Rasmussen, H. N. & Rasmussen, F. N. (1991). Climatic and seasonal regulation of seed plant establishment in *Dactylorhiza majalis* inferred from symbiotic experiments *in vitro*. *Lindleyana*, **6**, 221–7.

Rasmussen, H. N. & Whigham, D. (1993). Seed ecology of dust seeds *in situ*: a new study technique and its application in terrestrial orchids. *American Journal of Botany*, **80**, 1374–8.

Rauh, W., Barthlott, W. & Ehler, N. (1975). Morphologie und Funktion der Testa staubförmiger Flugsamen. *Botanische Jahrbücher*, **96**, 353–74.

Reed, H. S. & MacDougal, D. T. (1945). Growth processes of *Corallorhiza*. *Growth*, **9**, 235–58.

Reinke, J. (1873). Zur Kenntniss des Rhizoms von *Corallorhiza* und *Epipogon*. *Flora*, **56**, 145–52, 161–7, 177–84, 209–24.

Reissek, S. (1847). Über Endophyten der Pflanzenzelle. *Naturwissenschaften*, **1**, 31–46, Table 2.

Renner, O. (1938). Über blasse, saprophytische *Cephalanthera alba* und *Epipactis latifolia*. *Flora*, **32**, 225–36.

Reyburn, A. N. (1978). The effects of pH on the expression of a darkness-requiring dormancy in seeds of *Cypripedium reginae* Walt. *American Orchid Society Bulletin*, **47**, 798–802.

Richardson, J. A. (1956). The role of soil fungi in the spread of orchid species in clay pits. *Proceedings of the University of Durham Philosophical Society*, **12**, 183–9.

Richardson, K. A., Peterson, R. L. & Currah, R. S. (1992). Seed reserves and early symbiotic protocorm development of *Platanthera hyperborea* (Orchidaceae). *Canadian Journal of Botany*, **70**, 291–300.

Ridley, H. N. (1930). *The dispersal of plants throughout the world*. Ashford, Kent: Reeve & Co.

Riess, S. & Scrugli, A. (1987). Associazioni micorriziche nelle orchidee spontanee della Sardegna. *Micologia Italiana*, **16**, 21–8.

Riether, W. (1990). Keimverhalten terrestrischer Orchideen gemässigter Klimate. *Die Orchidee*, **41**, 100–9.

Riley, C. T. (1983). Hardy orchids – horticultural seed germination and commercial potential. In *North American Terrestrial Orchids. Symposium II. Proceedings and Lectures*, ed. E. H. Plaxton, pp. 9–12. Ann Arbor, USA: Michigan Orchid Society.

Roberts, E. H. (1973). Loss of seed viability: chromosomal and genetical aspects. *Seed Science and Technology*, **1**, 515–27.

Rose, F. (1948). *Orchis purpurea. Journal of Ecology*, **36**, 366–77.

Rücker, W. (1974). Einfluss von Cytokininen auf Wachstum und Differenzierung *in vitro* kultivierter Protokorme von *Cymbidium*. *Zeitschrift für Pflanzenphysiologie*, **72**, 338–51.

Rudnicki, R. (1969). Studies on abscisic acid in apple seeds. *Planta*, **86**, 63–8.

Ruiz-Herrera, J. (1992). *Fungal cell wall: structure, synthesis, and assembly*. Boca Raton, Florida: CRC Press.

Saksena, H. K. & Vaartaja, O. (1961). Taxonomy, morphology and pathogenecity of *Rhizoctonia* species from forest nurseries. *Canadian Journal of Botany*, **39**, 627–47.

Salisbury, F. B. & Ross, C. W. (1985). *Plant physiology*, 3rd edn. Belmont, California: Wadsworth.

Salmia, A. (1986). Chlorophyll-free form of *Epipactis helleborine* (Orchidaceae) in SE Finland. *Annales Botanici Fennici*, **23**, 49–57.

Salmia, A. (1988). Endomycorrhizal fungus in chlorophyll-free and green forms of the terrestrial orchid *Epipactis helleborine*. *Karstenia*, **28**, 3–18.

Salmia, A. (1989a). Features of endomycorrhizal infection of chlorophyll-free and green forms or *Epipactis helleborine* (Orchidaceae). *Annales Botanici Fennici*, **26**, 15–26.

Salmia, A. (1989b). General morphology and anatomy of cholorophyll-free and green forms of *Epipactis helleborine* (Orchidaceae). *Annales Botanici Fennici*, **26**, 95–105.

Sawa, K., Taneda, M. & Fujimori, R. (1979). Studies on germination of orchids native to Japan. I. Germination of orchids native to Shikoku Island. In *Proceedings of the Japanese Society of Horticulture Scientific Conference*, pp. 278–9.

Schwabe, M. (1953). Eine Beobachtung an *Orchis maculata* L. *Die Orchidee*, **4**, 57–9.

Seaton, P. T. & Hailes, N. S. J. (1989). Effect of temperature and moisture content on the viability of *Cattleya aurantiaca* seed. In *Modern methods in orchid conservation*, ed. H. W. Pritchard, pp. 17–29. Cambridge: Cambridge University Press.

Seeman, G. (1953). Über eine neue Wässerungsmethode für schwerkeimende Orchideensamen. *Die Orchidee*, **4**, 56.

Serrigny, J. & Dexheimer, J. (1986). Endomycorhize d'une orchidée tropicale: *Epidendrum ibaguense*. Étude comparative des activités phosphatasiques acides entre le champignon symbiote associé et isolé. In *Physiological and genetical aspects of mycorrhizae*, ed. V. Gianinazzi-Pearson & S. Gianinazzi, pp. 271–5. Paris: INRA.

Sharman, B. C. (1939). The development of the sinker of *Orchis mascula*. *Journal of the Linnean Society, Botany*, **52**, 145–58.

Sheviak, C. J. (1974). An introduction to the ecology of the Illinois Orchidaceae. *Illinois State Museum Science Papers*, **14**, 1–89.

Sheviak, C. J. (1983). United States terrestrial orchids. Patterns and problems. In *North American Terrestrial Orchids. Symposium II. Proceedings and Lectures*, ed. E. H. Plaxton, pp. 49–60. Ann Arbor, USA: Michigan Orchid Society.

Shun-xing, G. & Jin-tang, X. (1990). Studies on the changes of cell ultrastructure in the course of seed germination of *Bletilla striata* under fungus infection conditions. *Acta Botanica Sinica*, **32**, 594–8.

Smith, K. A. & Dowdell, R. J. (1974). Field studies of the soil atmosphere. I. Relationship between ethylene, oxygene, soil moisture content and temperature. *Journal of Soil Science*, **25**, 217–30.

Smith, K. A. & Russell, R. S. (1969). Occurrence of ethylene, and its significance, in anaerobic soil. *Nature*, **222**, 769–71.

Smith, S. E. (1966). Physiology and ecology of orchid mycorrhizal fungi with reference to seedling nutrition. *New Phytologist*, **65**, 488–99.

Smith, S. E. (1967). Carbohydrate translocation in orchid mycorrhizas. *New Phytologist*, **66**, 371–8.

Smith, S. E. (1973). Asymbiotic germination of orchid seeds on carbohydrates of fungal origin. *New Phytologist*, **72**, 497–9.

Smith, S. E. (1974). Mycorrhizal fungi. *CRC Critical Reviews in Microbiology*, **3**, 275–313.

Smith, S. E., Long, C. M. & Smith, F. A. (1990). Infection of roots with a dimorphic hypodermis: possible effects on solute uptake. *Agriculture, Ecosystem & Environment*, **29**, 403–7.

Smreciu, E. A. & Currah, R. S. (1989a). Symbiotic germination of seeds of terrestrial orchids of North America and Europa. *Lindleyana*, **1**, 6–15.

Smreciu, E. A. & Currah, R. S. (1989b). *A guide to the native orchids of Alberta*. Edmonton, Alberta, Canada: University of Alberta Devonian Botanic Garden.

Snow, A. A. & Whigham, D. F. (1989). Costs of flower and fruit production in *Tipularia discolor* (Orchidaceae). *Ecology*, **70**, 1286–93.

Sondheimer, E., Tzou, D. S. & Galson, E. C. (1968). Abscisic acid levels and seed dormancy. *Plant Physiology*, **43**, 1443–7.

Stahl, E. (1900). Der Sinn der Mycorrhizenbildung. *Jahrbücher für wissenschaftliche Botanik*, **34**, 539–668.

Stahl, H. (1989). Zur Entwicklung einer Population von *Ophrys apifera* im Gebiet von Stuttgart. *Mitteillungsblatt Arbeitskreis heimische Orchideen Baden-Württenberg*, **21**, 1015–39.

St-Arnaud, M., Lauzer, D. & Barabé, D. (1992). In vitro germination and early growth of seedlings of *Cypripedium acaule* (Orchidaceae). *Lindleyana*, **7**, 22–7.

Stephan, G. (1988). Ergebnisse der asymbiotischen Samenvermehrung von *Spiranthes spiralis* (L.C. Rich) und einige darüber hinausgehende Betrachtungen. *Die Orchidee*, **39**, 19–25.

Stephen, R. C. & Fung, K. K. (1971a). Nitrogen requirements of the fungal endophytes of *Arundina chinensis*. *Canadian Journal of Botany*, **49**, 407–10.

Stephen, R. C. & Fung, K. K. (1971b). Vitamin requirements of the fungal endophytes of *Arundina chinensis*. *Canadian Journal of Botany*, **49**, 411–15.

Stern, W. L., Cheadle, V. I. & Thorsch, J. (1993). Apostasiads, systematic anatomy, and the origins of Orchidaceae. *Botanical Journal of the Linnean Society*, **111**, 411–55.

Stevenson, J. C. (1972). Evolutionary strategies and ecology of *Goodyera* and *Spiranthes* species (Orchidaceae). PhD thesis, Chapel Hill, North Carolina: University of North Carolina.

Stewart, J. & Mitchell, R. (1991). Test tube orchids. *The Garden*, **116**, 32–6.

Stiles, W. (1960). The composition of the atmosphere (oxygen content of air, water, soil, intercellular spaces, diffusion, carbon dioxide and oxygen tension). In *Encyclopedia of Plant Physiology XII/2*, ed. Ruhland, pp. 114–48. Berlin: Springer Verlag.

Stoessl, A. & Arditti, J. (1984). Orchid phytoalexins. In *Orchid biology. Reviews and perspectives III*, ed. J. Arditti, pp. 153–75. Ithaca, New York: Cornell University Press.

Stojanow, N. (1916). Über die vegetative Fortpflanzung der Ophrydineen. *Flora, N.S.*, **9**, 1–39, Tables I–II.

Stokes, M. J. (1974). The in vitro propagation of *Dactylorhiza fuchsii* (Druce)Vermeul. *Orchid Review*, **82**, 62–5.

Stoutamire, W. P. (1963). Terrestrial orchid seedlings. *Australian Plants*, **2**, 119–22.

Stoutamire, W. P. (1964). Seeds and seedlings of native orchids. *Michigan Botanist*, **3**, 107–19.

Stoutamire, W. P. (1974). Terrestrial orchid seedlings. In *The orchids. Scientific studies*, ed. C. L. Withner, pp. 101–28. New York: Wiley.

Stoutamire, W. P. (1983). Early growth in North American terrestrial seedlings. In *North American Terrestrial Orchids. Symposium II*.

Proceedings and lectures, ed. E. H. Plaxton, pp. 14–24. Ann Arbor, USA: Michigan Orchid Society.

Stoutamire, W. P. (1990). Eastern American *Cypripedium* species and the biology of *Cypripedium candidum*. In *North American native terrestrial orchid propagation and production*, ed. C. E. Sawyers, pp. 40–8. Chadds Ford, Pennsylvania: Brandywine Conservancy.

Stoutamire, W. P. (1991). Annual growth cycle of *Cypripedium candidum* Mühl. root system in an Ohio prairie. *Lindleyana*, **6**, 235–40.

Stoutamire, W. P. (1992). Seed germination of *Cypripedium acaule* and *Cypripedium reginae* (Orchidaceae) in axenic culture. *Ohio Journal Science*, **92**(2), 16.

Strauss, M. S. & Reisinger, D. M. (1976). Effects of naphthaleneacetic acid on seed germination. *American Orchid Society Bulletin*, **45**, 722–3.

Strullu, D.-G. & Gourret, J.-P. (1974). Ultrastructure et evolution du champignon symbiotique des racines de *Dactylorchis maculata* (L.)Vermeul. *Journal de Microscopie*, **20**, 285–94.

Stuckey, I. H. (1967). Environmental factors and the growth of native orchids. *American Journal of Botany*, **54**, 232–41.

Summerhayes, V. S. (1951). *Wild orchids of Britain*. London: Collins.

Sundermann, H. (1961). Standorte europäischer Orchideen. I. Gliederung in Standorttypen. *Die Orchidee*, **12**, 131–7.

Sundermann, H. (1962a). Standorte europäischer Orchideen. II. Die Halbtrockenrasen (Mesobrometen) des Kaiserstuhlgebietes. *Die Orchidee*, **13**, 5–9.

Sundermann, H. (1962b). Standorte europäischer Orchideen. III. Sumpfwiesen und Moore. *Die Orchidee*, **13**, 87–92.

Sundermann, H. (1962c). Standorte europäischer Orchideen. IV. Mittelmeergebiet. *Die Orchidee*, **13**, 125–32.

Sundermann, H. (1962d). Standorte europäischer Orchideen. V. Orchideenwälder. *Die Orchidee*, **13**, 205–11.

Swamy, B. G. L. (1949). Embryological studies in the Orchidaceae. II. Embryogeny. *The American Midland Naturalist*, **41**, 202–32.

Tamm, C. O. (1948). Observations on reproduction and survival of some perennial herbs. *Botaniska Notiser*, [1948], 305–21.

Tamm, C. O. (1972). Survival and flowering of perrenial herbs. II. The behavior of some orchids on permanent plots. *Oikos*, **23**, 23–8.

Tamm, C. O. (1991). Behaviour of some orchid populations in a changing environment: observations on permanent plots, 1943–1990. In *Population ecology of terrestrial orchids*, ed. T. C. E. Wells & J. H. Willems, pp. 1–13. The Hague: SPB Academic Publishing.

Tashima, Y., Terashita, T., Umata, H. & Matsumoto, M. (1978). *In vitro* development from seed to flower in *Gastrodia verrucosa* under fungal symbiosis. *Transactions of the Mycological Society of Japan*, **19**, 449–53.

Taylorson, R. B. & Hendricks, S. B. (1977). Dormancy in seeds. *Annual Review of Plant Physiology*, **28**, 331–54.

Terashita, T. (1985). Fungi inhabiting wild orchids in Japan (III). A symbiotic experiment with *Armillaria mellea* and *Galeola septentrionalis*. *Transactions of the Mycological Society of Japan*, **26**, 47–53.

Terashita, T. & Chuman, S. (1987). Fungi inhabiting wild orchids in Japan (III). *Armillaria tabescens*, a new symbiont of *Galeola septentrionalis*. *Transactions of the Mycological Society of Japan*, **28**, 145–54.

Thomale, H. (1957). *Die Orchideen. Einführung in die Kultur und Vermehrung tropischen und einheimischen Orchideen*, 2nd edn, Stuttgart: Eugen Ulmer.

Thomas, M. B. (1893). The genus *Corallorhiza. Botanical Gazette*, **18**, 166–70, Tables 16–17.

Thompson, P. A. (1974). Growing orchids from seed. *Journal of the Royal Horticultural Society*, **99**, 117–22.

Tokunaga, Y. & Nakagawa, T. (1974). Mycorrhiza of orchids in Japan. *Transactions of the Mycological Society of Japan*, **15**, 121–33.

Tomita, M. & Tsutsui, K. (1988). The effect of the concentration of powdered oats in the medium on the growth of symbiotic seedlings of *Spiranthes sinensis* Ames and *Liparis nervosa* Lindl. *Journal of the Faculty of Agriculture, Hokkaido University (Japan)*, **63**, 354–62.

Treub, M. (1879). *Notes sur l'embryogènie de quelques orchidées*. Amsterdam: Johannes Müller.

Tsutsui, K. & Tomita, M. (1986). Symbiotic germination of *Spiranthes sinensis* associated with some orchid endophytes. *Journal of the Faculty of Agriculture, Hokkaido University (Japan)*, **86**, 440–52.

Tsutsui, K. & Tomita, M. (1988). Differences in the symbiotic capacity among isolates of mycorrhizal fungi on some terrestrial orchids. *Journal of the Faculty of Agriculture, Hokkaido University (Japan)* **63**, 345–53.

Tsutsui, K. & Tomita, M. (1989). Effect of plant density on the growth of seedlings of *Spiranthes sinensis* Ames and *Liparis nervosa* Lindl. in symbiotic culture. *Engei Gakkai Zasshi (Journal of the Japanese Society for Horticultural Science)*, **57**, 668–73.

Tsutsui, K. & Tomita, M. (1990). Suitability of several carbohydrates as the carbon sources for symbiotic seedling growth of two orchid species. *Lindleyana*, **5**, 134–9.

Ueda, H. & Torikata, H. (1969a). Organogenesis in the meristem tissue cultures of cymbidiums. II. Effects of growth substances on the organogenesis in dark culture. *Engei Gakkai Zasshi (Journal of the*

Japanese Society for Horticultural Science), **38**, 188–93.

Ueda, H. & Torikata, H. (1969*b*). Organogenesis in the meristem cultures of cymbidiums. III. Histological studies on the shoot formation at the rhizome-tips of *Cymbidium goeringii* Reichb.f. cultured *in vitro*. *Engei Gakkai Zasshi* [*Journal of the Japanese Society for Horticultural Science*], **38**, 262–6.

Ueda, H. & Torikata, H. (1970). Organogenesis in the meristem cultures of cymbidiums. V. Anatomical and histochemical studies on phagocytosis in the mycorrhizome of *Cymbidium goeringii* Reichb.f. (*C. virescens* Lindl.). *Engei Gakkai Zasshi* [*Journal of the Japanese Society for Horticultural Science*], **39**, 256–60.

Ueda, H. & Torikata, H. (1972). Effects of light and culture medium on adventitious root formation by *Cymbidiums* in aseptic culture. *American Orchid Society Bulletin*, **41**, 322–7.

Urech, J., Fechtig, B., Nüesch, J. & Vischer, E. (1963). Hircinol, eine antifungisch wirksame Substanz aus Knollen von *Loroglossum hircinum* (L.)Rich. *Helvetica Chimica Acta*, **46**, 2758–66.

Vacin, F. & Went, F. W. (1949). Some pH changes in nutrient solutions. *Botanical Gazette*, **110**, 605–13.

Van der Kinderen, G. (1987). Abscisic acid in terrestrial orchid seeds: a possible impact on their germination. *Lindleyana*, **2**, 84–7.

Van Staden, J., Webb, D. P. & Wareing, P. F. (1972). The effect of stratification on endogenous cytokinin levels in seeds of *Acer saccharum*. *Planta*, **104**, 110–14.

Vanséveren-Van Espen, N. (1973). Effets du saccharose sur le contenu en chlorophylles de protocormes de *Cymbidium* Sw. (Orchidaceae) cultivés *in vitro*. *Bulletin de le Société royale de Botanique de Belgique*, **106**, 107–15.

Van Waes, J. (1984). *In vitro* studie van de kiemingsfysiologie van Westeuropese orchideeën. Thesis. Rijkuniversiteit Gent.

Van Waes, J. (1987). Effect of activated charcoal on *in vitro* propagation of western European orchids. *Acta Horticulturae*, **212**, 131–8.

Van Waes, J. & Debergh, P. C. (1986*a*). Adaptation of the tetrazolium method for testing the seed viability and scanning electron microscopy of some Western European orchids. *Physiologia Plantarum*, **66**, 435–42.

Van Waes, J. & Debergh, P. C. (1986*b*). *In vitro* germination of some Western European orchids. *Physiologia Plantarum*, **67**, 253–61.

Vermeulen, P. (1947). *Studies on Dactylorchids*. Utrecht: Schotanus & Jens.

Vermeulen, P. (1966). The system of Orchidales. *Acta Botanica Neerlandica*, **15**, 224–53.

Veyret, Y. (1969). La structure des semences des orchidaceae et leur aptitude à la germination *in vitro* en cultures pures. *Musée d'Histoire Naturelle de Paris, Travaux du Laboratoire La Jaysinia*, **3**, 89–98, Plates III–IV.

Veyret, Y. (1974). Development of the embryo and the young seedling stages of orchids. In *The orchids: scientific studies*, ed. C. L. Withner, pp. 223–65. New York: Wiley.

Villiers, T. A. (1975). Genetic maintenance of seeds in imbibed storage. In *Crop genetic resources for today and tomorrow*, ed. O. H. Frankel & J. G. Hawkes, pp. 297–315. Cambridge: Cambridge University Press.

Villiers, T. A. & Edgcumbe, D. J. (1975). On the cause of seed deterioration in dry storage. *Seed Science and Technology*, **3**, 761–74.

Vöth, W. (1966). Eine Beobachtung an Orchideen und Pilzen. *Die Orchidee*, **17**, 307–10.

Vöth, W. (1971). Knollenentwicklung und Vegetationsrhythmus von *Dactylorhiza romana* und *sambucina*. *Die Orchidee*, **22**, 254–6.

Vöth, W. (1976). Aussat und Kultur von *Serapias parviflora* und *S. orientalis*. In *Proceedings of the 8th World Orchid Conference*, ed. K. Senghas, pp. 351–8. Frankfurt: German Orchid Society.

Vöth, W. (1980a). Beschreibung der Gattungen, Arten und Hybriden mit Kulturhinweisen. In *Orchideenkultur. Botanische Grundlagen, Kulturverfahren, Pflanzenbeschreibungen*, ed. G. Fast, pp. 310–450. Stuttgart: Ulmer.

Vöth, W. (1980b). Naturgegebenes Verhalten von *Gymnadenia conopsea* und *Listera ovata*. *Jahresbericht des Naturwissenschaftlischen Vereins in Wuppertal*, **33**, 136–45.

Wahrlich, W. K. (1886). Beiträge zur Kenntnis der Orchideenwurzelpilze. *Botanische Zeitung*, **44**, 481–97.

Waite, S. (1989). Predicting population trends in *Ophrys sphegodes* Mill. In *Modern methods in orchid conservation*, ed. H. W. Pritchard, pp. 117–26. Cambridge: Cambridge University Press.

Warcup, J. H. (1971). Specificity of mycorrhizal association in some Australian terrestrial orchids. *New Phytologist*, **70**, 41–6.

Warcup, J. H. (1973). Symbiotic germination of some Australian terrestrial orchids. *New Phytologist*, **72**, 387–92.

Warcup, J. H. (1975). Factors affecting symbiotic germination of orchid seed. In *Endomycorrhizas*, ed. F. E. Sanders, B. Mosse & P. B. Tinker, pp. 87–104. London: Academic Press.

Warcup, J. H. (1981). The mycorrhizal relationships of Australian orchids. *New Phytologist*, **87**, 371–81.

Warcup, J. H. (1982). Orchid mycorrhizal fungi. In *Proceedings of the Orchid Symposium, 13th International Botanical Congress, Sidney*, ed. L.

Lawler & R. D. Kerr, pp. 57–63. New South Wales: Orchid Society.

Warcup, J. H. (1985). *Rhizanthella gardneri* (Orchidaceae), its *Rhizoctonia* endophyte and close association with *Melaleuca uncinata* (Myrtaceae) in Western Australia. *New Phytologist,* **99**, 273–80.

Warcup, J. H. (1991). The *Rhizoctonia* endophytes of *Rhizanthella* (Orchidaceae). *Mycological Research,* **95**, 656–9.

Warcup, J. H. & Talbot, P. H. B. (1966). Perfect states of some *Rhizoctonia*'s. *Transactions of the British Mycological Society,* **49**, 427–35.

Warcup, J. H. & Talbot, P. H. B. (1967). Perfect states of *Rhizoctonia*'s associated with orchids. *New Phytologist,* **66**, 631–41.

Warcup, J. H. & Talbot, P. H. B. (1971). Perfect states of *Rhizoctonia*'s associated with orchids. II. *New Phytologist,* **70**, 35–40.

Warcup, J. H. & Talbot, P. H. B. (1980). Perfect states of *Rhizoctonia*'s associated with orchids. III. *New Phytologist,* **86**, 267–72.

Ward, E. W. B., Unwin, C. H. & Stoessl, A. (1975). Post-infectional inhibitors from plants, part 15. Antifungal activity of the phytoalexin orchinol and related phenanthrenes and stilbenes. *Canadian Journal of Botany,* **53**, 964–71.

Warming, E. (1874). Om rødderne hos *Neottia nidus-avis* Lin. *Videnskabelige Meddelelser fra den Naturhistoriske Forening i Kjøbenhavn,* 1874(**1–2**), 26–32.

Webb, D. P. & Wareing, P. F. (1972). Seed dormancy in *Acer*: Endogenous germination inhibitors and dormancy in *Acer pseudoplatanus* L. *Planta,* **104**, 115–25.

Weber, F. (1920). Notiz zur Kohlensäureassimilation von *Neottia*. *Bericht der Deutschen Botanischen Gesellschaft,* **38**, 233–42.

Weber, H. C. (1981). Orchideen auf dem Weg zum Parasitismus? Über die Möglichkeit einer phylogenetischen Umkonstruktion der Infektionsorgane von *Corallorhiza trifida* Chat. (Orchidaceae) zu Kontaktorganen parasitischer Blütenpflanzen. *Bericht der Deutschen Botanischen Gesellschaft,* **94**, 275–86.

Wei-Jing, Z. & Bi-Fung, L. (1980). The biological relationship of *Gastrodia elata* and *Armillaria mellea*. *Acta Botanica Sinica,* **22**, 57–62.

Weinert, M. (1990). Keimungsfördernde Faktoren bei schwerkeimenden europäischen Orchideen. 1. Bodenpilze und Agarbedeckung. *Die Orchidee,* **41**, 127–33.

Wells, T. C. E. (1967). Changes in a population of *Spiranthes spiralis* (L.) Chevall. at Knocking Hoe National Nature Reserve, Bedfordshire, 1962–65. *Journal of Ecology,* **55**, 83–99.

Wells, T. C. E. (1981). Population ecology of terrestrial orchids. In *The biological aspects of rare plant conservation,* ed. H. Synge, pp. 281–95.

Chichester: Wiley.

Wells, T. C. E. & Cox, R. (1989). Predicting the probability of the bee orchid (*Ophrys apifera*) flowering or remaining vegetative from the size and number of leaves. In *Modern methods in orchid conservation*, ed. H. W. Pritchard, pp. 127–39. Cambridge: Cambridge University Press.

Wells, T. C. E. & Cox, R. (1991). Demographic and biological studies of *Ophrys apifera*: some results from a 10-year study. In *Population ecology of terrestrial orchids*, ed. T. C. E. Wells & J. H. Willems, pp. 47–61. The Hague: SPB Academic Publishing.

Wells, T. C. E. & Kretz, R. (1987). Asymbiotische Anzucht von *Spiranthes spiralis* (L.)Cheval. vom Samen bis zur Blüte in fünf Jahren. *Die Orchidee*, **38**, 245–7.

Werckmeister, P. (1971). Light induction of geotropism, and the control of proliferation and growth of *Cymbidium* in tissue culture. *Botanical Gazette*, **132**, 346–50.

Wherry, E. T. (1918). The reaction of the soils supporting the growth of certain native orchids. *Journal of the Washington Academy of Sciences*, **8**, 589–98.

Wherry, E. T. (1927). The soil reaction of some saprophytic orchids. *Journal of the Washington Academy of Sciences*, **17**, 35–8.

Whigham, D. F. (1984). Biomass and nutrient allocation of *Tipularia discolor* (Orchidaceae). *Oikos*, **42**, 303–13.

Whigham, D. F. (1990). The effect of experimental defoliation on the growth and reproduction of a woodland orchid, *Tipularia discolor*. *Canadian Journal of Botany*, **68**, 1812–16.

Whigham, D. F. & O'Neill, J. (1988). The importance of predation and small scale disturbance to two woodland herb species. In *Plant form and vegetation structure: adaptation, plasticity, and relation to herbivory*, ed. W. J. A. Werger, P. J. M. Van der Aart, H. J. During & J. T. A. Verhoeven, pp. 243–52. The Hague: SPB Academic Publishing.

Whigham, D. F. & O'Neill, J. (1991). The dynamics of flowering and fruit production in two eastern North American terrestrial orchids, *Tipularia discolor* and *Liparis lilifolia*. In *Population ecology of terrestrial orchids*, ed. T. C. E. Wells & J. H. Willems, pp. 89–101. The Hague, SPB Academic Publishing.

White, J. H. (1907). On polystely in roots of Orchidaceae. *University of Toronto Studies, Biological Series*, **6**, 1–20, Plates I–II, diagram 1–27.

Whitlow, C. E. (1983). *Cypripedium* culture. In *North American terrestrial orchids. Symposium II. Proceedings and lectures*, ed. E. H. Plaxton, pp. 25–31. Ann Arbor, USA: Michigan Orchid Society.

Wilkinson, K. G., Dixon, K. W. & Sivasithamparam, K. (1989). Interaction of soil bacteria, mycorrhizal fungi, and orchid seed in relation to germination of Australian orchids. *New Phytologist*, 112, 429–35.

Willems, J. H. (1982). Establishment and development of a population of *Orchis simia* Lamk. in the Netherlands, 1972 to 1981. *New Phytologist*, 91, 757–65.

Willems, J. H. (1989). Population dynamics of *Spiranthes spiralis* in South Limburg, The Netherlands. *Mémoires de la Societé Royale de Botanique de Belgique*, 11, 115–21.

Willems, J. H. & Bik, L. (1991). Population biology of *Orchis simia* in the Netherlands, 1972–1990. In *Population ecology of terrestrial orchids*, ed. T. C. E. Wells & J. H. Willems, pp. 33–45. The Hague: SPB Academic Publishing.

Williams, J. G. & Williams, A. E. (1983). *Field guide to orchids of North America*. New York: Universe Books.

Williams, P. G. (1985). Orchidaceous *rhizoctonias* in pot cultures of vesicular-arbuscular mycorrhizal fungi. *Canadian Journal of Botany*, 63, 1329–33.

Williamson, B. (1970). Induced DNA synthesis in orchid mycorrhiza. *Planta*, 92, 347–54.

Williamson, B. (1973). Acid phosphatase and esterase activity in orchid mycorrhiza. *Planta*, 112, 149–58.

Williamson, B. & Hadley, G. (1969). DNA content of nuclei in orchid protocorms symbiotically infected with *Rhizoctonia*. *Nature*, 222, 582–3.

Williamson, B. & Hadley, G. (1970). Penetration and infection of orchid protocorms by *Thanatephorus cucumeris* and other *Rhizoctonia* isolates. *Phytopathology*, 60, 1092–6.

Wirth, M. & Withner, C. L. (1959). Embryology and development in the Orchidaceae. In *The orchids. A scientific survey*, ed. C. L. Withner, pp. 155–88. New York: Ronald Press.

Withner, C. L. (1953). Germination of *Cyps*. *Orchid Journal*, 2, 473–7.

Withner, C. L. (1959). Orchid Physiology. In *The orchids. A scientific survey*, ed. C. L. Withner, pp. 315–60. New York: Ronald Press.

Wolff, H. (1927). Zur Physiologie des Wurzelpilzes von *Neottia nidus-avis* Rich. und einigen grünen Orchideen. *Jahrbücher für wissenschaftliche Botanik*, 66, 1–34.

Wolff, H. (1933). Zur Assimilation atmosphärischen Stickstoffs durch die Wurzelpilze von *Coralliorhiza innata* R.Br., sowie der Epiphyten *Cattleya bowringiana* und *Laelia anceps* Ldl. *Jahrbücher für wissenschaftliche Botanik*, 77, 657–84.

Wooster, K. R. (1935). The Lizard orchid. *Gardener's Chronicle*, 98, 244.

Xu, J. T. & Mu, C. (1990). The relation between growth of *Gastrodia elata* protocorms and fungi. *Acta Botanica Sinica*, **32**, 26–31, Plates I–II.

Yannetti, R. A. (1990). *Arethusa bulbosa*: Germination and culture. In *North American native terrestial orchid propagation and production*, ed. C. E. Sawyers, pp. 97–8. Chadds Ford, Pennsylvania: Brandywine Conservancy.

Yates, R. C. & Curtis, J. T. (1949). The effect of sucrose and other factors on the shoot–root ratio of orchid seedlings. *American Journal of Botany*, **36**, 390–6.

Yeung, E. C. & Law, S. K. (1992). Embryology of *Calypso bulbosa*. II. Embryo development. *Canadian Journal of Botany*, **70**, 461–8.

Zettler, L. W., Fairey, J. E. & McInnis, T. M. J. R. (1990). The status and seed germination of *Platanthera integrilabia* (Correll)Luer, an endangered terrestrial orchid. *Association of Southeastern Biologists' Bulletin*, **37**, 86.

Zettler, L. W. & McInnis, T. M. Jr. (1992). Propagation of *Platanthera integrilabia* (Correll)Luer, an endangered terrestrial orchid, through symbiotic seed germination. *Lindleyana*, **7**, 154–61.

Zettler, L. W. & McInnis, T. M. JR. (1994). The effect of white light on the symbiotic seed germination of an endangered orchid, *Platanthera integrilabia*. *Association of Southeastern Biologists' Bulletin*, **41**, 129.

Ziegenspeck, H. (1935). Was bedingt die Schwimmfähigkeit der Samen der einheimischen Orchideen und der Sporen von *Lycopodium*. *Botanisches Archiv*, **37**, 373–82.

Ziegenspeck, H. (1936). Orchidaceae. In *Lebensgeschichte der Blütenpflanzen Mitteleuropas*, *I,4*, ed. O. von Kirschner, E. Loew, & C. Schröter. Stuttgart: Eugen Ulmer.

Zimmerman, J. K. & Whigham, D. F. (1992). Ecological functions of carbohydrates stored in corms of *Tipularia discolor* (Orchidaceae). *Functional Ecology*, **6**, 575–81.

Index

433